T5-CVD-945

CHICAGO PUBLIC LIBRARY
HAROLD WASHINGTON LIBRARY CENTER
R0006990261

Tensors and group theory
for the physical properties of crystals

W. A. WOOSTER

Tensors and group theory for the physical properties of crystals

CLARENDON PRESS · OXFORD
1973

Oxford University Press, Ely House, London W. 1

GLASGOW NEW YORK TORONTO MELBOURNE WELLINGTON
CAPE TOWN IBADAN NAIROBI DAR ES SALAAM LUSAKA ADDIS ABABA
DELHI BOMBAY CALCUTTA MADRAS KARACHI LAHORE DACCA
KUALA LUMPUR SINGAPORE HONG KONG TOKYO

PRINTED IN GREAT BRITAIN BY
BELL & BAIN LTD. GLASGOW

Preface

THIS book owes its origin to a request which the author received in 1968 to conduct a Summer School in the École Polytechnique, Montreal, Canada. The original intention of the organizers of this School was that lectures and practical demonstrations should be given on tensors applicable to the physical properties of crystals. The first part of this book gives the theoretical part of this course. What was not anticipated was that the students would ask for lectures on group theory. The course was accordingly arranged to include this subject as well as the study of tensors.

Past experience has shown the author that students learn little from a lecture given continuously for an hour. They learn much more from a lecture during the course of which they do exercises on the subject of the lecture. This plan was followed throughout in the course, and the book has been written to correspond with it.

The author has always had difficulty in keeping himself awake while reading books on group theory. He has been anxious to avoid this happening to readers of this book. The principal difficulty seems to be the abstract symbolism usually employed in books on group theory. It is difficult to keep clearly in mind what is implied by each of the symbols. A different presentation has been attempted in this book. The methods of using group theory to deal with a number of problems that depend on crystal symmetry are developed, starting with the knowledge which crystallographers usually have. Group theory enlarges the applications of crystal symmetry and is not just an exercise in mathematics. Whereas 'classical' crystallography may be called 'static', the group-theoretical treatment of the same symmetry may be called 'dynamic', since it brings vibrations of atoms and molecules into the range of what may be studied. No attempt at comprehensiveness has been made in this book. For example, the treatment is limited to the thirty-two point groups and 230 space groups. Magnetic, black-and-white, and coloured groups are not discussed. The aim of this book is to help the student to understand how to study the subject, rather than to serve as a reference work containing all the conclusions.

The author is greatly indebted to Dr Norman Henry of the

Department of Mineralogy and Petrology, Cambridge, for helpful criticism of the second part of the book.

W. A. WOOSTER

Cambridge, July 1972

Contents

Part I

Tensors and the physical properties of crystals

1

Symmetry and the physical properties of crystals

1.1 Axes of symmetry

THE elements of symmetry found in crystals include axes, planes, centres, and the identity operator. An axis may have two-fold, three-fold, four-fold, or six-fold symmetry, but not five-fold or any fold higher than six. This limitation is a consequence of the assemblage of the atoms in a crystal on a three-dimensional lattice. When crystals having axes of n-fold symmetry are rotated through an angle $2\pi/n$ about these axes, they come into congruent positions. Thus, if normals, passing through a common origin, are drawn to the faces of a crystal, the rotation of $2\pi/n$ brings the normals into the same positions as those they previously occupied. For example, a four-sided pyramid standing on a square base, having its apex vertically above the centre of its base, has a vertical four-fold axis (Fig. 1.1). If this pyramid is rotated through 90° it comes into a congruent position. In the same way a regular pyramid standing on an equilateral triangle can be rotated through 120° and is then in a position congruent with its original position (Fig. 1.2).

There may be several axes of symmetry present in a crystal. For example, a triangular prism, Fig. 1.3, has one vertical three-fold axis and three two-fold axes perpendicular to the first axis. The cube has the largest number of axes of symmetry. There are four three-fold, or triad axes, directed along the body-diagonals running between opposite corners (Fig. 1.4). There are three four-fold, or tetrad axes, passing normally through the centre point of each cube face and also six two-fold, or diad axes. The axes of crystal symmetry are denoted 1, 2, 3, 4, and 6 respectively in the *International tables for X-ray crystallography* (reference 9). These are represented

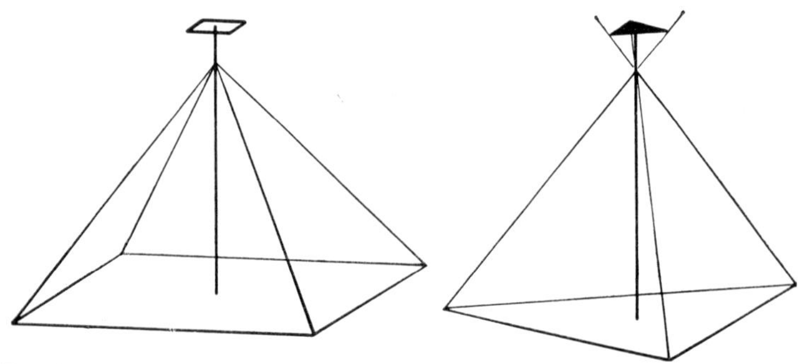

FIG. 1.1. Diagram showing a four-fold axis passing vertically downwards through a regular pyramid standing on a square base.

FIG. 1.2. Diagram showing a three-fold axis passing vertically downwards through a regular pyramid standing on an equilateral triangular base.

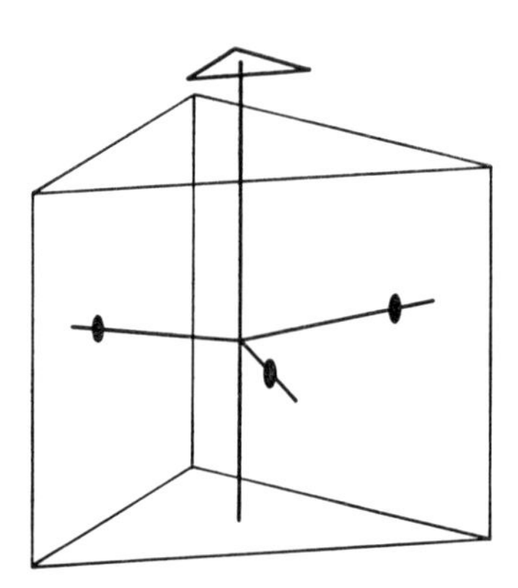

FIG. 1.3. Diagram showing one three-fold and three two-fold axes of a triangular prism having an equilateral triangle as base.

pictorially as shown below.

Q.1.1. How many axes of symmetry are present in a rectangular parallelepiped? Of which kind are they?

Q.1.2. There are six two-fold or diad axes in a cube. Between which points on the cube do they run?

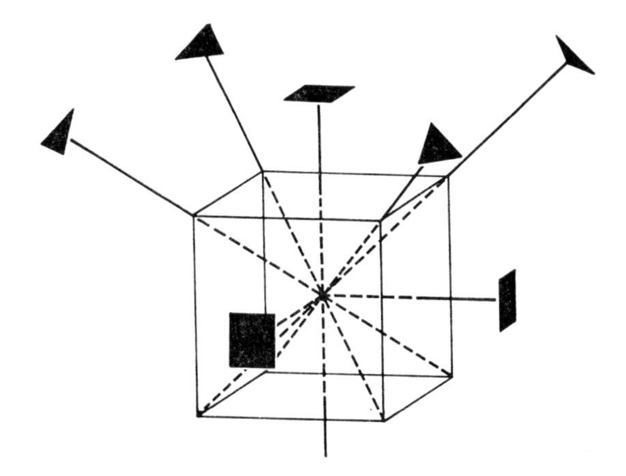

FIG. 1.4. Diagram showing the four three-fold and three four-fold axes of a cube.

1.2 Planes of symmetry

A plane of symmetry is present when a distribution of faces or atoms is repeated in the same way as it would be optically by a mirror whose reflecting surface coincides with the plane of symmetry. There may be only one plane of symmetry present in the external form of a crystal. Figure 1.5 represents a gypsum crystal, $CaSO_4 . 2H_2O$, which has one plane of symmetry, namely 1 2 3. Figure 1.6 represents a triangular prism based on an equilateral triangle. One plane of symmetry is cross-hatched; it bisects the angle between two

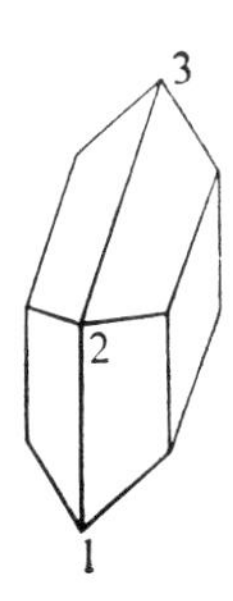

FIG. 1.5. Diagram representing a gypsum crystal, showing a plane of symmetry defined by the points 1, 2, 3.

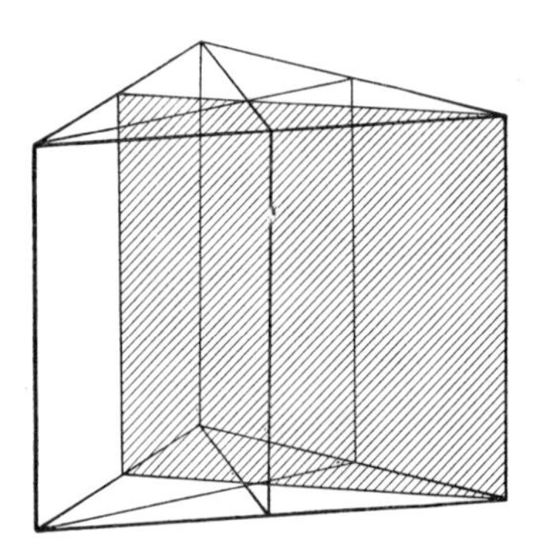

FIG. 1.6. Diagram showing three planes of symmetry in a prism standing on an equilateral triangular base.

vertical faces. The other two planes of symmetry are also marked in Fig. 1.6. A cube has three planes of symmetry passing through its centre, which are parallel to the faces of the cube.

Q.1.3. What other plane of symmetry is present in Fig. 1.6 in addition to those marked on the diagram?

Q.1.4. What other planes of symmetry than those mentioned above has a cube?

1.3 Centres of symmetry

The third type of symmetry is that which is associated with a centre of symmetry. If every face on a crystal has a similar face parallel to it on the other side of the crystal, then a centre of symmetry is said to be present. In a crystal structure possessing a centre of symmetry at a given point, there are pairs of atoms on either side of and at equal distances from the centre, O (Fig. 1.7), lying on straight lines 11′, 22′, 33′ passing through the centre. In such a crystal, every atom has one opposite to it at an equal distance from the centre.

1.4 The identity operator

Whether or not the above symmetry elements are present in a crystal, it is still possible to say that all face normals superpose on

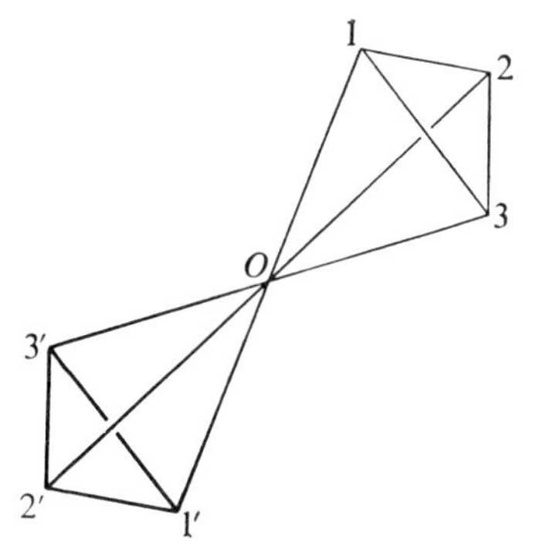

FIG. 1.7. Diagram showing points 1, 2, 3 related by a centre of symmetry to points 1′, 2′, 3′ respectively.

themselves. This may seem an unnecessary statement because it is self-evident. However, the operation of superposition, without prior operation of an axis, plane, or centre, is called the identity operation and is represented by the number 1.

1.5 Reflection–rotation and inversion–rotation axes

Planes of symmetry may be combined with axes of symmetry to give what are known as reflection–rotation axes and are denoted S_n where n varies from 1 to 6. In Figs. 1.8–1.12 the axes S_1, S_2, S_3, S_4, S_6 are shown. In Fig. 1.8, the starting location is marked 1 and the effect of rotation through 2π followed by reflection across the plane is to give location 2. This is, of course, not different from the operations of a plane of symmetry. It is denoted S_1. In Fig. 1.9 the location 1 is rotated about the two-fold axis through π and then reflected across the plane to the location marked 2. The location 2 is also clearly derived from location 1 by the operation of a centre of symmetry at the point where the diad axis cuts the plane of symmetry. This combined symmetry operation is denoted S_2. In Fig. 1.10 is shown the combination of triad axis and plane of symmetry. Location 1 is rotated about the triad axis to location 5 and then reflected across the plane to location 2. This location is then rotated in the same sense to 6 and reflected to give location 3. Continuation of this process develops locations 4, 5, and 6, and the symmetry is denoted S_3. In Fig. 1.11 the location 1 is rotated through 90° about the vertical axis and reflected in the horizontal plane to give location

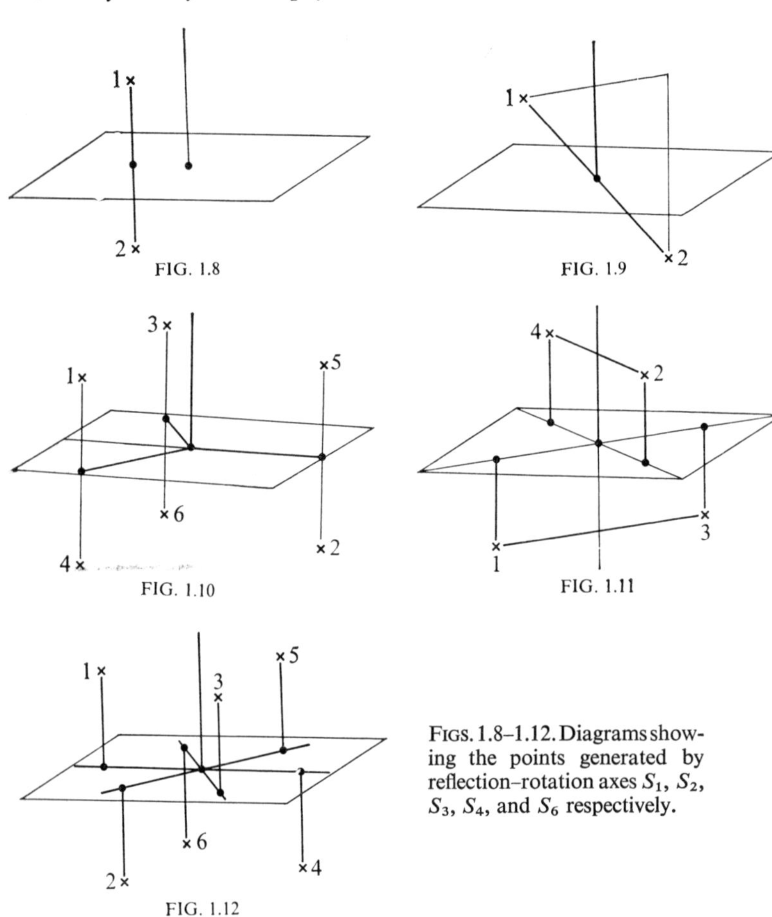

FIGS. 1.8–1.12. Diagrams showing the points generated by reflection–rotation axes S_1, S_2, S_3, S_4, and S_6 respectively.

2. A repetition of the process leads to locations 3 and 4. This symmetry operation is denoted S_4. In Fig. 1.12 the hexad axis is combined with the horizontal plane to give the points 1–6 corresponding to S_6.

Axes of symmetry may also be combined with a centre of symmetry lying on the axis to give what are known as inversion–rotation axes $\bar{1}$–$\bar{6}$. These are illustrated in the same way as Figs. 1.13–1.17. Taking Fig. 1.15 as an example, we see that the location 1 is rotated through 120° to location 5 and then inverted through the centre of symmetry

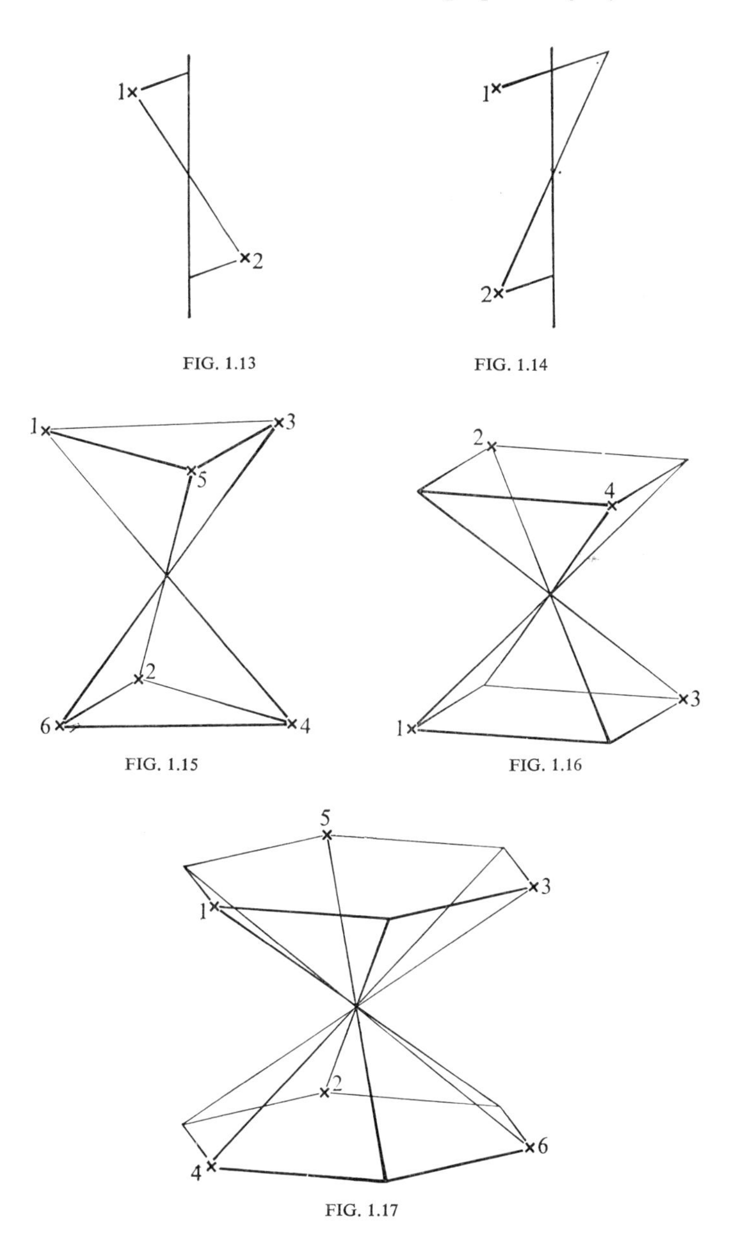

FIG. 1.13 FIG. 1.14

FIG. 1.15 FIG. 1.16

FIG. 1.17

FIGS. 1.13–1.17. Diagrams showing the points generated by inversion-rotation axes $\bar{1}$, $\bar{2}$, $\bar{3}$, $\bar{4}$, and $\bar{6}$ respectively.

to point 2. Location 2 is rotated to 6 and inverted to 3, and so on. Comparing Figs. 1.8–1.12 with Figs. 1.13–1.17, we see that

$$S_1 = \bar{2}, \quad S_2 = \bar{1}, \quad S_3 = \bar{6}, \quad S_4 = \bar{4}, \quad S_6 = \bar{3}.$$

The space groups given in the *International tables for X-ray crystallography* are expressed in terms of the inversion–rotation axes, but the character tables of group theory usually employ the reflection–rotation axes (4, p. 23; 5, p. 14; 7, p. 56; 8, p. 32; 11, p. 44; 14, p. 103).†

1.6 Representation of directions by means of stereograms

In all descriptions of directions in space the method of projection called 'stereographic' is very useful. All the directions considered are taken to radiate from the centre of a sphere, and the poles on the surface of the sphere where they cut the sphere are joined to the south pole. In Fig. 1.18, the centre of the sphere is denoted C, the

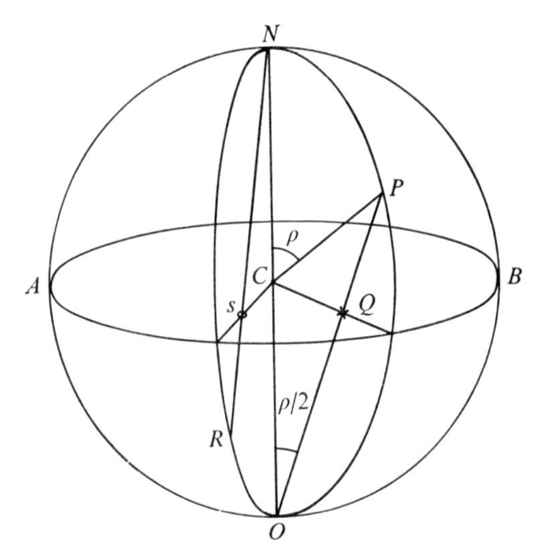

FIG. 1.18. Diagram showing the projection sphere $ANBO$, the plane of the stereographic projection $ASCQB$, a pole P projected from O as cross Q, and a pole R projected from N as open circle S.

† In these references the first number gives the textbook listed under 'References' (see page 337).

direction to be represented is *CP*, the south pole is *O*. Where the line *OP* cuts the equatorial plane *AB*, namely the cross *Q*, is the stereographic projection representing the direction *CP*. If a direction such as *CR* is towards the lower half of the projection sphere it is usual to represent it in the projection by an open circle *S*, obtained by joining *R* to *N*, the north pole of the projection sphere.

There are two important properties of the stereographic projection. The first of these is that all circles drawn on the surface of the projection sphere project as circles, or parts of circles, on the plane of the projection. The second property is that the angle between two arcs on the projection sphere is equal to the angle between the corresponding projections of these arcs on the equatorial plane. This is the property known as 'the preservation of angular truth'. The stereographic projections corresponding to Figs 1.1 and 1.2 are shown in Figs 1.19 and 1.20 respectively. If the normal to a

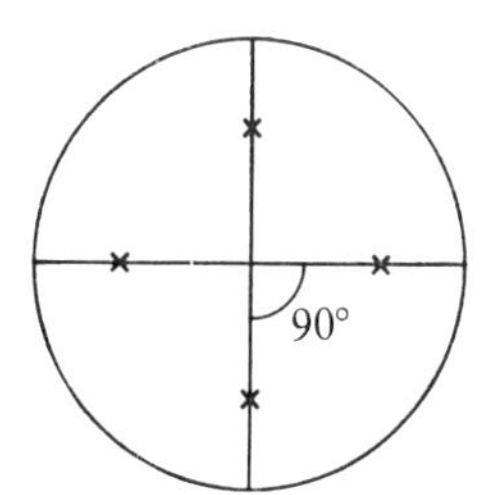

FIG. 1.19. Stereographic projection of pyramidal faces shown in Fig. 1.1.

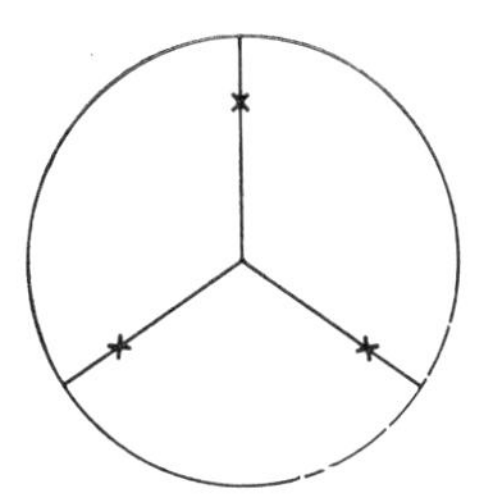

FIG. 1.20. Stereographic projection of pyramidal faces shown in Fig. 1.2.

plane makes an angle ρ with the vertical direction (taken normal to the paper) then the distance of the cross in the stereographic projection from the centre of the projection is equal to $r \tan \rho/2$, where r is the radius of the projection sphere (and of the stereogram). The truth of this statement can be seen from Fig. 1.18. The stereogram of a cubic crystal, Fig. 1.21, consists of lines passing through

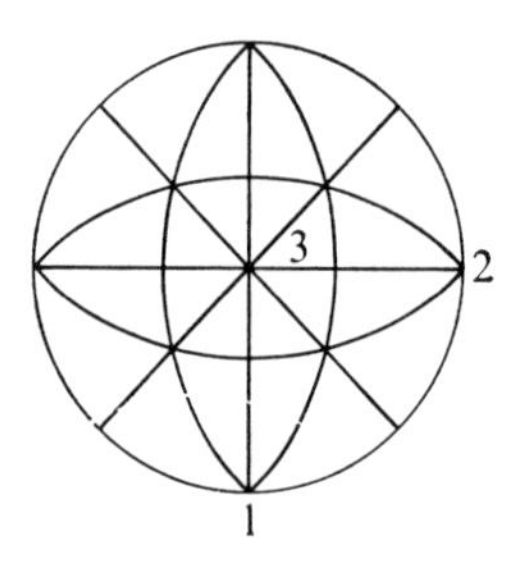

FIG. 1.21. Cubic stereogram representing the planes bisecting the angles between pairs of the axes 1–2, 2–3, 3–1.

the centre, making angles of 45° and 90° with one another, and circles that have centres at points such as 1 and 2 and radii equal to $r\sqrt{2}$. These circles represent planes making an angle of 45° with the plane of the stereogram. It is convenient to use this set of lines and arcs of circles in nearly all problems involving cubic crystals (7, p. 34; 10, p. 66; 11, p. 34; 14, p. 20).

1.7 Point groups

Some symmetry elements can exist together in the same crystal. Altogether there are 32 such possible combinations (see Appendix 2). A systematic survey of these combinations, which are called point groups, will not be given here, but a few examples will be described.

Point group 222 (D_2)

Three mutually perpendicular diad axes form a mutually compatible group of symmetry elements. These two-fold axes are denoted by filled ovals on the stereogram of Fig. 1.22. The lowest oval is called C_{2x}, the one on the right C_{2y}, and the one at the centre of the stereogram C_{2z}. We start with a direction represented by the cross 1 and operate with the diad axis C_{2x}. This produces the small circle 2 representing a direction pointing downwards from the paper at

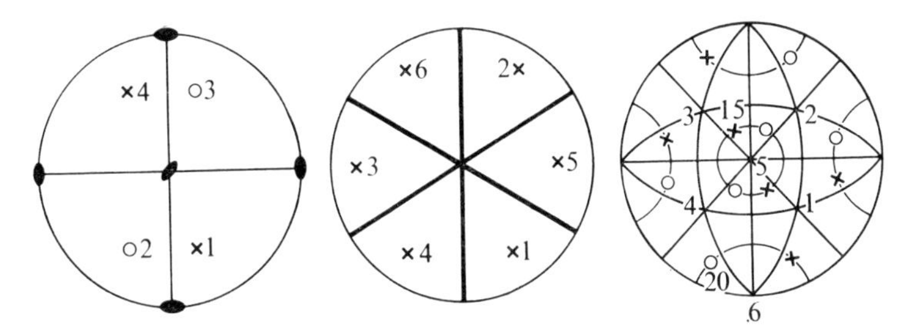

FIGS. 1.22–1.24. Stereograms representing the equivalent directions in the point groups 222 (D_2), 3*m* (C_{3v}), and 23 (*T*) respectively.

the same angle as the direction 1 points upwards. If we operate on 1 with C_{2y} we obtain 3, and with C_{2z} we obtain 4. If now we operate on 2 with C_{2y} we obtain 4, and if we operate on 3 with C_{2x} we obtain 4. Thus the three axes C_{2x}, C_{2y}, C_{2z} are mutually compatible. A point on the line having the direction represented by 1 may be given the coordinates x, y, z referred to the axes C_{2x}, C_{2y}, C_{2z} respectively. Then the points 2, 3, 4 must have the coordinates $x\bar{y}\bar{z}$, $\bar{x}y\bar{z}$, and $\bar{x}\bar{y}z$ respectively.

Point group 3*m* (C_{3v})

A vertical triad axis to which three planes of symmetry are parallel is a mutually compatible assembly if the planes make angles of 120° with one another. In Fig. 1.23, the stereogram shows the distribution of directions 1, 2, 3, 4, 5, 6 that corresponds to this combination of symmetry elements. An operation on 1 with the triad axis produces cross 2 and, after a second operation, cross 3. Operation on cross 1 with the vertical plane of symmetry σ_{v_1}, produces cross 4. From 4 the triad axis produces crosses 5 and 6. Then it can be seen that 2 is reflected to 5 by the plane σ_{v_2} which is at 120° to σ_{v_1}. Similarly, 3 is reflected to 6 by the plane σ_{v_3} which is at 120° to σ_{v_2}. Thus this combination of symmetry elements is mutually compatible.

Point group 23 (*T*)

This is a cubic point group consisting of four triad and three diad axes. The triad axes are denoted C_{3a}, C_{3b}, C_{3c}, and C_{3d} and are marked 1, 3, 2, 4 in the stereogram of Fig. 1.24. The four triad axes are mutually compatible and generate the three diad axes, as may be seen from the distribution of crosses and small circles in Fig. 1.24.

The axes 1 and 3 are considered as sloping upwards from the centre of the projection sphere, while the other two triad axes are considered to be sloping downwards at the same angle with respect to the plane of the paper. The three diad axes are necessarily present when there are four triad axes. The triad axis 1 rotates the pole 3 to the pole 4 and then to pole 2. The axis C_{2z}, pole 5, is taken to C_{2x}, pole 6, by the triad axis C_{3a}, pole 1. Hence the line joining poles 3 and 5 is rotated by C_{3a}, pole 1, to the arc joining poles 4 and 6. Pole 15, which is close to the line joining poles 3 and 5, therefore goes to pole 20, which is near the arc joining poles 4 and 6, under the action of the triad axis C_{3a}, pole 1. Similarly it may be shown that, starting from any one of the crosses or circles and operating with the given symmetry elements, we can generate all the other crosses and circles.

The thirty-two point groups

A complete set of stereograms for all the point groups is given in Appendix 1. This information may be found in many books as well as in Volume 1 of the *International tables for X-ray crystallography*. The Hermann–Mauguin notation for the point groups will in general be used here, though the Schoenflies notation will be referred to, since it occurs in so many books on group theory. The stereograms representing the distribution of equivalent points show the symmetry clearly and from them the x, y, z coordinates of these points can be obtained. The Hermann–Mauguin symbols denoting the point groups consist of a digit giving the symmetry of the principal axis 2, 3, 4, or 6 and an m or a 2 to indicate each plane or diad axis that exists along with the principal axis. A plane perpendicular to the principal axis is distinguished from one parallel to that axis by the use of a sloping stroke placed before the letter m. Thus $4mm$ applies to a tetrad axis and two sets of planes of symmetry that intersect in that axis; $4/m$ refers to a tetrad axis having a plane of symmetry perpendicular to it. The $\bar{3}$, $\bar{4}$, and $\bar{6}$ axes are inversion axes of symmetry (see §1.5). The Schoenflies symbols consist of a capital letter followed by one or two subscripts. The capital letters are C, standing for 'cyclic' and applying to those point groups having only one principal axis; D, standing for 'diagonal', and used for point groups having one or more diad axes normal to the principal axis; S, standing for 'sphenoidal' and applying to the alternating or reflection–rotation axes described in §1.5; T, standing for tetrahedral, and O, standing for octahedral. T and O occur only in the cubic system. The subscripts following the capital letter are numbers and in some cases letters as well. The numbers 1, 2, 3, 4, and 6 following *C*, *D*, and *S* give the order of the principal axis. The letters i, s, h, v, d indicate respectively a centre of symmetry, a single plane of symmetry, a plane of symmetry normal to the principal symmetry axis (h = horizontal), a plane of symmetry parallel to the principal axis (v = vertical), and a plane of symmetry bisecting the angle between two of the diad axes that are normal to the principal axis (1, p. 40; 3, p. 24; 6, p. 42; 7, p. 56; 8, p. 33; 9, p. 25; 11, p. 56; 14, p. 106).

1.8 Transformation of axes

One of the fundamental postulates concerning the physical properties of crystals is known as Neumann's Principle. It states that the physical properties measured in a given way, relative to a coordinate system that is fixed in space, are of the same sign and magnitude when the crystal is rotated, reflected, or inverted into a new orientation corresponding to one of its symmetry elements. For example, if a crystal has a two-fold axis and we measure its thermal conductivity in any given direction in space, and then we turn it through 180° about its two-fold axis and measure the thermal conductivity in the same direction as before, the result will be the same as in the first measurement. This general principle applies to all physical properties and to all kinds and combinations of elements of symmetry.

It is therefore necessary to consider how the rotation, reflection, or inversion symmetries can be treated in an analytical manner applicable to all cases. In Cartesian coordinates it is easy to express these symmetry operations by a set of nine numbers. For example, if we have a rotation of 90° about an axis which we call X_3, then we can denote a set of mutually orthogonal axes X_1, X_2, X_3 before rotation and the same set X'_1, X'_2, X'_3 after rotation (Fig. 1.25). The nine numbers we use are the direction cosines of the axes. The relation between the original and the rotated set of axes can, in this case, be given as follows.

	X_1	X_2	X_3
X'_1	0	1	0
X'_2	−1	0	0
X'_3	0	0	1

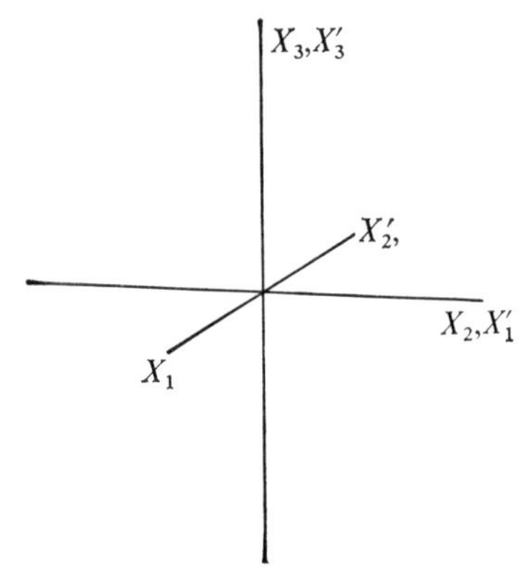

FIG. 1.25. Diagram representing the change of directions of the axes X_1, X_2, X_3 to X_1', X_2', X_3' on rotation through 90° about the axis X_3.

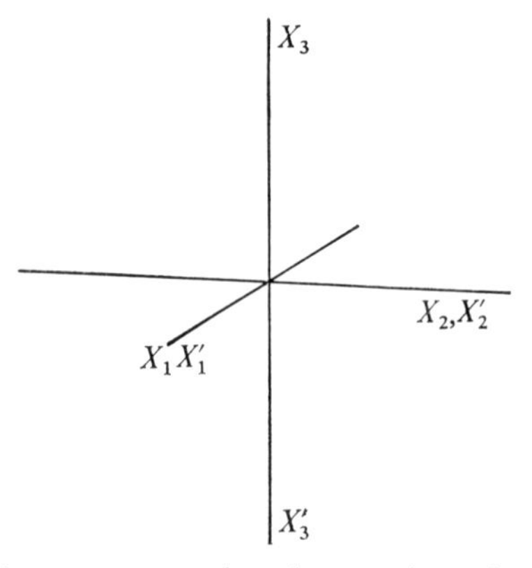

FIG. 1.26. Diagram representing the transformation of the axial system due to a plane of symmetry containing X_1 and X_2.

This matrix means that X_1' is perpendicular to X_1 (cos 90°), it coincides with X_2 (cos 0°), and is perpendicular to X_3 (cos 90°). Similar meanings are given to X_2' and X_3' by the numbers.

Planes of symmetry can be represented in a similar manner by sets of nine numbers. Thus a plane of symmetry perpendicular to the X_3 axis leaves X_1 and X_2 unchanged, but makes X_3' correspond to $-X_3$ (Fig. 1.26). Thus the matrix is

	X_1	X_2	X_3
X_1'	1	0	0
X_2'	0	1	0
X_3'	0	0	-1

A centre of symmetry placed at the origin inverts all three axes, i.e. X_1 is turned into $-X_1$, X_2 into $-X_2$, and X_3 into $-X_3$, so that the matrix representing this inversion is

	X_1	X_2	X_3
X_1'	-1	0	0
X_2'	0	-1	0
X_3'	0	0	-1

Q.1.5. What matrices apply to rotations of 180° and of 60° about the X_3 axis?

Q.1.6. What is the matrix for a plane of symmetry bisecting internally the angle between planes X_1OX_3 and X_2OX_3?

1.9. Transformation of vectorial properties

Any property, such as an electric moment, which can be represented by a straight line of a given length drawn in a given direction is known as a vectorial property. This vector may be denoted $\mathbf{p}$ and relative to Cartesian axes it has components p_1, p_2, p_3. We assume that the axes are changed so that, in the general case, the relation between the old and new orientations is given by the nine direction cosines

	X_1	X_2	X_3
X_1'	c_{11}	c_{12}	c_{13}
X_2'	c_{21}	c_{22}	c_{23}
X_3'	c_{31}	c_{32}	c_{33}

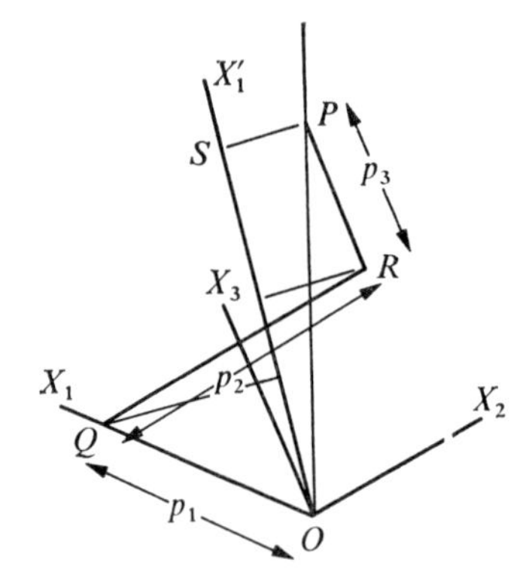

FIG. 1.27. *OP* represents a vector having components *OQ*, *QR*, *RP* on axes X_1, X_2, X_3 respectively. *OS* is the direction of the transformed axis X_1'. The projection of *OP* on to *OS* equals the sum of the projections of *OQ*, *QR*, and *RP* on to *OS*.

In Fig. 1.27 the lines *OS* and *OP* lie in the plane of the paper. *OS* is in the direction of X_1' and *OP* in the direction of **p**. The three lines *OQ*, *QR*, *RP* represent respectively the directions of the axes X_1, X_2, X_3, and do not lie in the plane of the paper. Their lengths correspond to the components p_1, p_2, p_3 respectively. From Fig. 1.27, it can be seen that the sum of the projections of p_1, p_2, and p_3 on to *OS* is equal to the projection of **p** on to *OS*, i.e. to p_1'. Thus we have

$$p_1' = c_{11}p_1 + c_{12}p_2 + c_{13}p_3.$$

Similarly, it may be shown that

$$p_2' = c_{21}p_1 + c_{22}p_2 + c_{23}p_3,$$

$$p_3' = c_{31}p_1 + c_{32}p_2 + c_{33}p_3.$$

This is written in the abbreviated form

$$p_i' = c_{ik}p_k, \tag{1.1}$$

and it is understood that when a letter is repeated, as it is on the right-hand side of equation (1.1), it must be given, in turn, all three values 1, 2, and 3 and the resulting quantities summed. The converse transformation can also be carried out. Since the point *P* can be reached from *O* by travelling distances p_1', p_2', p_3' along the axes

X_1', X_2', X_3' respectively, the sum of the projections of these components on to X_1 is equal to p_1. Thus we have

$$p_1 = c_{11}p_1' + c_{21}p_2' + c_{31}p_3'$$

or in general

$$p_i = c_{ki}p_k'. \tag{1.2}$$

Equations (1.1) and (1.2) are very similar; as the dashes change sides so the order of the suffixes i and k of the direction cosines is reversed (1, p. 1; 9, p. 15; 13, p. 9; 15, p. 1).

1.10. Relation of vectorial properties to symmetry

Centre of symmetry

A crystal showing a vectorial property cannot possess a centre of symmetry. This is almost self-evident, since a polarity, N–S, or plus–minus, is incompatible with a centre of symmetry. This would demand at opposite sides of the crystal N–N poles or plus–plus charges, which is physically impossible.

In an analytical way this can be expressed as follows. The centre of symmetry requires a set of direction cosines,

$$\begin{matrix} -1 & 0 & 0 \\ 0 & -1 & 0 \\ 0 & 0 & -1 \end{matrix}$$

and hence, applying equation (1.1), we get

$$p_1' = -p_1; \quad p_2' = -p_2; \quad p_3' = -p_3.$$

The operation of symmetry demands, however, that, according to Neumann's Principle,

$$p_1' = p_1; p_2' = p_2; p_3' = p_3.$$

These statements about p_i' are compatible only if all the three components are zero, i.e. no vectorial property such as a pyroelectric moment can exist in a centrosymmetrical crystal.

Axes of symmetry

For a diad axis parallel to X_3 the direction-cosine matrix is

$$\begin{matrix} -1 & 0 & 0 \\ 0 & -1 & 0 \\ 0 & 0 & 1 \end{matrix}$$

Then we have, applying Neumann's Principle,

$$p_1' = c_{11}p_1 = -p_1 = 0,$$
$$p_2' = c_{22}p_2 = -p_2 = 0,$$
$$p_3' = c_{33}p_3 = \quad p_3 \neq 0.$$

The last equation is based on the assumption that the vector is not of zero magnitude.

Thus a vector property must be parallel to the diad axis.

For a triad axis parallel to X_3 the transformation matrix is

$$\begin{matrix} -\frac{1}{2} & \frac{\sqrt{3}}{2} & 0 \\ -\frac{\sqrt{3}}{2} & -\frac{1}{2} & 0 \\ 0 & 0 & 1 \end{matrix}$$

The transformed components are therefore

$$p_1' = c_{11}p_1 + c_{12}p_2 = -\frac{1}{2}p_1 + \frac{\sqrt{3}}{2}p_2 = p_1$$

$$\text{or } p_1 = p_2/\sqrt{3}, \tag{1.3}$$

$$p_2' = c_{21}p_1 + c_{22}p_2 = -\frac{\sqrt{3}}{2}p_1 - \frac{1}{2}p_2 = p_2$$

$$\text{or } -p_1 = p_2\sqrt{3}. \tag{1.4}$$

Equations (1.3) and (1.4) can be true only if $p_1 = p_2 = 0$,

$$p_3' = c_{33}p_3 = p_3 \neq 0.$$

Also in this case the vector must be parallel to the triad axis.

Q.1.7. Which non-zero components of a vector property are, according to Neumann's Principle, possible in a crystal having a hexad axis parallel to X_3?

Plane of symmetry

For a plane of symmetry normal to X_3 the transformation matrix is

$$\begin{matrix} 1 & 0 & 0 \\ 0 & 1 & 0 \\ 0 & 0 & -1 \end{matrix} \quad .$$

The transformed components are thus

$$p'_1 = c_{11}p_1 = p_1 \neq 0,$$

$$p'_2 = c_{22}p_2 = p_2 \neq 0,$$

$$p'_3 = c_{33}p_3 = -p_3 = 0.$$

Thus the vector must be normal to X_3, i.e. lie anywhere in the plane containing X_1 and X_2.

Q.1.8. Which non-zero components of a vector property are possible in a crystal having a plane of symmetry parallel to X_3 and bisecting the angle between axes X_1 and X_2?

Answers

Q.1.1. Three two-fold axes are present in a rectangular parallelepiped. They pass through the centres of opposite pairs of faces.

Q.1.2. The diad axes of a cube join the mid-points of parallel edges which lie in a plane passing through the body-centre of the cube.

Q.1.3. A plane parallel to the top and bottom triangular faces and mid-way between them.

Q.1.4. A cube has six other planes of symmetry. Each of these planes contains a face diagonal on each of two opposite faces of the cube.

Q.1.5.

$$\begin{matrix} -1 & 0 & 0 \\ 0 & -1 & 0 \\ 0 & 0 & 1 \end{matrix}$$

is the matrix for a diad axis parallel to X_3.

If we assume an anticlockwise rotation (looking downwards from the positive end of X_3 towards O, the centre), the matrix for a hexad axis is

$$\begin{matrix} \frac{1}{2} & \frac{\sqrt{3}}{2} & 0 \\ -\frac{\sqrt{3}}{2} & \frac{1}{2} & 0 \\ 0 & 0 & 1 \end{matrix}$$

Q.1.6.

$$\begin{matrix} 0 & 1 & 0 \\ 1 & 0 & 0 \\ 0 & 0 & 1 \end{matrix}$$

is the matrix for a plane of symmetry bisecting internally the angle between OX_1 and OX_2.

Q.1.7. (a) From the matrix for a hexad axis we obtain the transformation of the components of a vector as follows

$$p_1' = +\frac{1}{2}p_1+\frac{\sqrt{3}}{2}p_2 = p_1,$$

$$p_2' = -\frac{\sqrt{3}}{2}p_1+\frac{1}{2}p_2 = p_2,$$

$$p_3' = p_3.$$

From the first equation we obtain

$$p_1 = p_2\sqrt{3}$$

and from the second equation

$$-p_1\sqrt{3} = p_2.$$

This is impossible unless

$$p_1 = p_2 = 0.$$

Thus only p_3 is non-zero.

Q.1.8. A plane of symmetry containing X_3 and bisecting the angle between X_1 and X_2 reflects X_1 into X_2, X_2 into X_1, and leaves X_3 unchanged. The transformation matrix is therefore as given in the answer to Q.1.6. Hence,

$$p_1' = p_2 = p_1,$$

$$p_2' = p_1 = p_2,$$

$$p_3' = p_3.$$

Thus the vector property must have components p_1 and p_2 equal to one another and p_3 not zero. The vector can therefore lie in any direction in the plane of symmetry.

2

Second-order tensors

2.1. Second-order tensors

THERE are a number of physical properties that depend on the relation between two vector quantities. Thus thermal conduction expresses the relation between heat flow and temperature gradient. Heat flow in a given direction in a crystal is measured by the amount of heat crossing unit area normal to that direction in unit time, and is a vector quantity. Similarly, temperature gradient is measured by change in temperature per unit length in a given direction and is also a vector quantity. If components of the vector representing heat flow are h_1, h_2, h_3, and components of the vector representing temperature gradient are $\mathrm{d}\theta/dx_1$, $d\theta/\mathrm{d}x_2$, $\mathrm{d}\theta/\mathrm{d}x_3$, then it is experimentally verifiable that

$$\left.\begin{aligned} h_1 &= k_{11}\frac{\mathrm{d}\theta}{\mathrm{d}x_1}+k_{12}\frac{\mathrm{d}\theta}{\mathrm{d}x_2}+k_{13}\frac{\mathrm{d}\theta}{\mathrm{d}x_3} \\ h_2 &= k_{21}\frac{\mathrm{d}\theta}{\mathrm{d}x_1}+k_{22}\frac{\mathrm{d}\theta}{\mathrm{d}x_2}+k_{23}\frac{\mathrm{d}\theta}{\mathrm{d}x_3} \\ h_3 &= k_{31}\frac{\mathrm{d}\theta}{\mathrm{d}x_1}+k_{32}\frac{\mathrm{d}\theta}{\mathrm{d}x_2}+k_{33}\frac{\mathrm{d}\theta}{\mathrm{d}x_3} \end{aligned}\right\}, \tag{2.1}$$

The ks are coefficients of thermal conductivity, which are constants for a given material in single-crystal form. These equations may be expressed more concisely by writing

$$h_i = k_{in}\frac{\mathrm{d}\theta}{\mathrm{d}x_n}. \tag{2.2}$$

Later we shall consider a number of other examples of properties relating two vector quantities. Collectively, the coefficients such as k_{in} are known as second-order tensors, because they have the common property that, when the axes of reference are changed, the tensors transform according to the following rule (see Appendix 3). Using the conductivity k_{in} as an example and assuming that the axes are transformed according to the matrix of direction cosines

	X_1	X_2	X_3
X'_1	c_{11}	c_{12}	c_{13}
X'_2	c_{21}	c_{22}	c_{23}
X'_3	c_{31}	c_{32}	c_{33},

we have, according to this rule, for the transformed (dashed) ks,

$$k'_{pq} = c_{pn}c_{qi}k_{ni}, \tag{2.3}$$

or, alternatively,

$$k_{ni} = c_{pn}c_{qi}k'_{pq}. \tag{2.4}$$

Q.2.1. What is the full expression for k_{12} in terms of k'_{pq}?

2.2. Representation surfaces

The conductivity of most crystals varies with the direction of measurement. As with all such variations, it is useful to represent the different values of the conductivity by some solid figure. We can obtain the form of this figure by finding the conductivity in a particular direction and then changing this direction to include all possible directions.

An arbitrary direction is denoted X'_3 and it has direction cosines c_{31}, c_{32}, c_{33}. We suppose that a plate is cut with its large surfaces normal to X'_3 (and, therefore, parallel to X'_1 and X'_2) and that a temperature gradient $\mathrm{d}\theta/\mathrm{d}x'_3$ is maintained between these two faces. The component of the heat flowing from one face to the other is h'_3 and the conductivity k'_{33} is the factor in the equation

$$h'_3 = k'_{33}\frac{\mathrm{d}\theta}{dx'_3}.$$

Since the temperature is the same all over each major face,

$$\frac{\mathrm{d}\theta}{\mathrm{d}x'_1} = \frac{\mathrm{d}\theta}{\mathrm{d}x'_2} = 0.$$

The rule for the transformation of second-order tensors given by equation (2.3) requires that

$$k'_{33} = c_{31}c_{31}k_{11}+c_{31}c_{32}k_{12}+c_{31}c_{33}k_{13}+$$
$$+c_{32}c_{31}k_{21}+c_{32}c_{32}k_{22}+c_{32}c_{33}k_{23}+$$
$$+c_{33}c_{31}k_{31}+c_{33}c_{32}k_{32}+c_{33}c_{33}k_{33}.$$

Now it can be shown experimentally that

$$k_{mn} = k_{nm}$$

whatever the values of m and n. If it were not so, heat would, in general, flow out from a point along spiral curves, whereas in fact it flows out along straight lines. Hence

$$k'_{33} = c_{31}^2k_{11}+c_{32}^2k_{22}+c_{33}^2k_{33}+2c_{31}c_{32}k_{12}+$$
$$+2c_{32}c_{33}k_{23}+2c_{31}c_{33}k_{13}. \qquad (2.5)$$

Equation (2.5) is the equation for a second-order quadric, i.e. a tri-axial ellipsoid, or a tri-axial hyperboloid, or the lobed figures with positive and negative parts.

When the axes of the ellipsoid are chosen as the axes X_1, X_2, X_3, the equation (2.5) simplifies to

$$k'_{33} = c_{31}^2k_{11}+c_{32}^2k_{22}+c_{33}^2k_{33}, \qquad (2.6)$$

and the values of k_{mn} are called the principal coefficients of thermal conductivity. Though this choice of axes makes k_{12}, k_{23}, and k_{31} each equal to zero, it has not, in general, reduced the number of quantities that must be found.

Q.2.2. What quantities other than k_{11}, k_{22}, k_{33} need to be defined to specify the conductivities?

The three principal conductivities are usually written k_1, k_2, k_3 for the sake of simplicity in writing. Though the ks must always be positive for thermal conductivity, the corresponding coefficients for some second-order tensor properties, e.g. thermal expansion, may be negative. In this case, the ellipsoid becomes a hyperboloid having positive and negative surfaces.

Q.2.3. A plate of quartz (hexagonal) is cut parallel to a natural rhombohedral face $(10\bar{1}1)$, and the thermal conductivity is measured by determination of the heat flow through the plate and the temperature difference between the opposite sides. Given that the principal conductivities are $k_{11} = 15{\cdot}8$, $k_{33} = 29{\cdot}1 \times 10^{-3}$ cal deg^{-1} cm^{-1} s^{-1} and the inclination of the normal to the plate to the z-axis is 52°, find the thermal conductivity of the plate.

Q.2.4. The principal thermal conductivities of Rochelle salt (orthorhombic) are respectively $k_{11} = 1{\cdot}10$, $k_{22} = 1{\cdot}46$, $k_{33} = 1{\cdot}34 \times 10^{-3}$ cal deg^{-1} cm^{-1} s^{-1}. A plate of crystal is cut so that the normal to the plate is inclined to the x, y, and z axes at angles of 30°, 70°, and 68·6° respectively. Calculate the coefficient of conductivity for this plate.

So far we have considered the flow of heat through a flat plate, and the temperatures of its major surfaces were taken to have fixed values. In this case, the vector representing the temperature gradient is normal to the major surfaces of the plate, but nothing is known about the direction of the vector representing the heat flow. In general this is not normal to the major surfaces of the plate. We shall choose our X_1, X_2, X_3 axes to coincide with the principal axes of the conductivity ellipsoid, so that the components of the heat flow are given by the equations

$$h_1 = k_1 \frac{\mathrm{d}\theta}{\mathrm{d}x_1};\; h_2 = k_2 \frac{\mathrm{d}\theta}{\mathrm{d}x_2};\; h_3 = k_3 \frac{\mathrm{d}\theta}{\mathrm{d}x_3}.$$

If the normal to the major surfaces is represented by a vector $\mathbf{p}$ of length equal to the temperature gradient then

$$h_1 = k_1 p_1; \quad h_2 = k_2 p_2; \quad h_3 = k_3 p_3.$$

The angle ϕ between the vectors $\mathbf{h}$ and $\mathbf{p}$ is given by

$$\begin{aligned}\cos\phi &= \mathbf{h}.\mathbf{p}/|h||p| \\ &= \frac{h_1p_1+h_2p_2+h_3p_3}{\surd\{(h_1^2+h_2^2+h_3^2)(p_1^2+p_2^2+p_3^2)\}} \\ &= \frac{k_1p_1^2+k_2p_2^2+k_3p_3^2}{\surd\{(k_1^2p_1^2+k_2^2p_2^2+k_3^2p_3^2)(p_1^2+p_2^2+p_3^2)\}}. \qquad (2.7)\end{aligned}$$

The condition discussed above is represented by Fig. 2.1, in which the vector representing the temperature gradient, $\mathbf{d}\boldsymbol{\theta}/\mathbf{dx}$, is shown

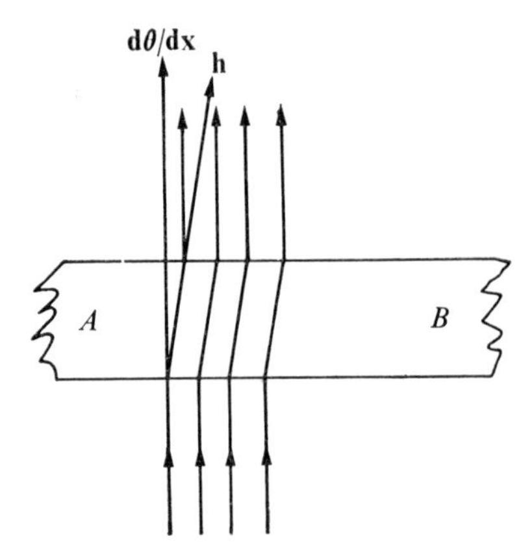

FIG. 2.1. Diagram showing the relation between the vectors of direction of heat flow and of maximum temperature gradient in a thin flat plate.

normal to the faces of the plate *AB*, and the vector representing the heat flow, **h**, is shown inclined to it at an angle ϕ.

When $p_2 = p_3 = 0, p_1 \neq 0,$

$$\cos\phi = \frac{k_1 p_1^2}{\sqrt{(k_1^2 p_1^2 p_1^2)}} = 1,$$

and the angle ϕ becomes zero. The same is true when the major faces are perpendicular to X_2 or X_3. Thus the angle ϕ reaches a maximum for directions lying between the directions of the principal axes.

2.3. Resistance

Instead of using a plate with large surfaces and small thickness, we can use a long bar with relatively small area of cross-section. In this case, care must be taken to ensure that heat is not lost from the long sides of the bar. In this case, the vector **h**, is defined by the length of the bar and the temperature gradient is left to look after itself. This is illustrated by Fig. 2.2. Since the components of **h** are

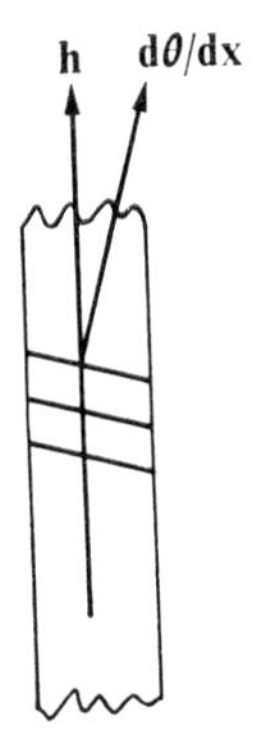

FIG. 2.2. Diagram showing the relation between the vectors of direction of heat flow and of maximum temperature gradient in a thin rod.

defined by the orientation of the length of the bar relative to the principal axes, it is not possible to use the equations having the components of $\mathbf{d\theta/dx}$ on the right-hand side, since these are unknown. We therefore write

$$\frac{d\theta}{dx_m} = r_{mn}h_n, \tag{2.8}$$

where the r_{mn} are called the coefficients of resistance. The connection between the rs and ks follows from equations (2.1) and (2.8), and is expressed in general by the equation

$$\left.\begin{aligned} r_{mn} &= \delta k_{mn}/\Delta_k \\ \text{or} \qquad & \\ k_{mn} &= \delta r_{mn}/\Delta r \end{aligned}\right\}, \tag{2.9}$$

where δ refers to minors and Δ to the whole determinant of k or r. The equations (2.8) can be illustrated by writing down the value of, for example r_{23}, as follows

$$r_{23} = \frac{\begin{vmatrix} k_{31} & k_{32} \\ k_{11} & k_{12} \end{vmatrix}}{\begin{vmatrix} k_{11} & k_{12} & k_{13} \\ k_{21} & k_{22} & k_{23} \\ k_{31} & k_{32} & k_{33} \end{vmatrix}}$$

When the axes X_1, X_2, X_3 are chosen parallel to the principal axes of the ellipsoid, we have

$$r_{11} = \frac{1}{k_{11}}; \quad r_{22} = \frac{1}{k_{22}}; \quad r_{33} = \frac{1}{k_{33}}.$$

The formulation in terms of resistance is particularly appropriate for electrical conduction. Wires or rods are commonly used in the study of electrical properties, and these determine the direction of current flow but not the direction of the potential gradient.

Q.2.5. A long rod is cut from the same crystal as described in Q.2.4 in a direction perpendicular to the plane of the plate. Determine the angle between the normal to the isothermal surfaces in the rod and the axis of the rod.

Q.2.6. A cylindrical rod of quartz is made with its axis normal to the plane $(10\bar{1}1)$ and one end is maintained at a higher temperature than the other. Assuming that there is no heat loss from the cylindrical surface, find the angle between the normal to the isothermal surface and the axis of the rod. ($k_{11} = 15{\cdot}8$, $k_{33} = 29{\cdot}1 \times 10^{-3}$ cal deg^{-1} cm^{-1} s^{-1}.)

Q.2.7. A plate of bismuth is cut from a single crystal and the normal to the plate is inclined at 30° to the trigonal axis of the crystal. The principal resistivities are $\rho_1 = 109$, $\rho_3 = 138 \times 10^{-6}$ Ω cm. Calculate the current flowing through unit area (1 cm^2) when 1 V is applied across a large plate 1 cm thick.

Q.2.8. A rod is made from the same bismuth crystal having a diameter of 1 cm and a direction inclined at 30° to the trigonal axis of the crystal. A potential drop of 1 V cm^{-1} is maintained along the rod. The potential difference between points diametrically opposite on a plane perpendicular to the axis of the rod is measured. Determine the maximum value of this measured potential difference and the direction of the corresponding diameter.

2.4. Conductivity in relation to symmetry

Monoclinic system

The two-fold axis of this system will be assumed parallel to the axis X_3. In this case the change of orientation of the axes caused by the two-fold axis is given by the scheme

	X_1	X_2	X_3
X_1'	-1	0	0
X_2'	0	-1	0
X_3'	0	0	1.

In general we have

$$k'_{nm} = c_{np}c_{mq}k_{pq}.$$

Now, only c_{11}, c_{22}, c_{33} are non-zero and *only terms containing two or no* c_{33}s give equality on the two sides of the equation. Thus,

$$k'_{13} = c_{11}c_{33}k_{13} = -k_{13},$$

But since by Neumann's Principle, k'_{13} must equal k_{13} this constant is zero. From this it follows that the coefficients in this symmetry system are

$$\begin{matrix} k_{11} & k_{12} & 0 \\ k_{12} & k_{22} & 0 \\ 0 & 0 & k_{33}. \end{matrix}$$

Orthorhombic system

If there are three mutually perpendicular axes coinciding with X_1, X_2, and X_3 respectively, then the rule which applied to X_3 in the monoclinic system must also apply to X_1 and X_2 in this system. Thus the non-zero coefficients must contain two 1s, or two 2s, or two 3s. The scheme of coefficients therefore becomes

$$\begin{matrix} k_{11} & 0 & 0 \\ 0 & k_{22} & 0 \\ 0 & 0 & k_{33}. \end{matrix}$$

Tetragonal system

The transformation corresponding to a tetrad axis parallel to X_3 is

	X_1	X_2	X_3			
X_1'	0	1	0	c_{11}	c_{12}	c_{13}
X_2'	-1	0	0	c_{21}	c_{22}	c_{23}
X_3'	0	0	1	c_{31}	c_{32}	c_{33}.

From this we obtain

$$k'_{11} = c_{12}c_{12}k_{22} = k_{22} = k_{11},$$

Q.2.9. What is the value of k'_{12}?

$$k'_{13} = c_{12}c_{33}k_{23} = k_{23} = k_{13},$$

$$k'_{33} = c_{33}c_{33}k_{33} = k_{33},$$

and all other products of two cs are zero. If we continue with the rotation about the tetrad axis we arrive at the same conditions that apply to a two-fold axis. This requires that

$$k_{13} = k_{23} = 0.$$

Finally, the scheme of ks for this system is

$$\begin{matrix} k_{11} & 0 & 0 \\ 0 & k_{11} & 0 \\ 0 & 0 & k_{33}. \end{matrix}$$

It can be shown that the same applies to the rhombohedral and hexagonal systems (13, p. 195; 15, p. 63; 16, p. 44).

Q.2.10. In the cubic system the axes X_1, X_2, X_3 are symmetrically equivalent. What effects does this have on the ks?

Answers

Q.2.1.

$$\begin{aligned} k_{12} = {} & c_{11}c_{12}k'_{11}+c_{11}c_{22}k'_{12}+c_{11}c_{32}k'_{13}+ \\ & + c_{21}c_{12}k'_{21}+c_{21}c_{22}k'_{22}+c_{21}c_{32}k'_{23}+ \\ & + c_{31}c_{12}k'_{31}+c_{31}c_{22}k'_{32}+c_{31}c_{32}k_{33}. \end{aligned}$$

Q.2.2. Three angles relating the principal axes of the conductivity ellipsoid to the crystallographic axes.

Q.2.3.

$$\begin{aligned} k'_{33} &= k_{11}c_{31}^2+k_{33}c_{33}^2 \\ &= 15{\cdot}8\times\sin^2 52^\circ+29{\cdot}1\cos^2 52^\circ \\ &= 15{\cdot}8\times 0{\cdot}621+29{\cdot}1\times 0{\cdot}378 \\ &= 9{\cdot}8+11{\cdot}0 = 20{\cdot}8\times 10^{-3}\ \text{cal deg}^{-1}\,\text{cm}^{-1}\,\text{s}^{-1}. \end{aligned}$$

Q.2.4.
$$
\begin{aligned}
k'_{33} &= 1{\cdot}10\cos^2 30^\circ + 1{\cdot}46\cos^2 70^\circ + 1{\cdot}34\cos^2 68{\cdot}6^\circ \\
&= 1{\cdot}10\times 0{\cdot}750 + 1{\cdot}46\times 0{\cdot}117 + 1{\cdot}34\times 0{\cdot}133 \\
&= 0{\cdot}825 + 0{\cdot}171 + 0{\cdot}178 = 1{\cdot}174\times 10^{-3}\ \text{cal deg}^{-1}\text{cm}^{-1}\text{s}^{-1}.
\end{aligned}
$$

Q.2.5. The components of heat flow are proportional to cos 30°, cos 70°, and cos 68·6° respectively, i.e. 0·866, 0·342, and 0·365. The principal coefficients of resistance are the reciprocals of the corresponding conductivity coefficients, i.e. $r_{11} = 0{\cdot}91$, $r_{22} = 0{\cdot}685$, $r_{33} = 0{\cdot}746$. Hence

$$
\begin{aligned}
\left(\frac{\mathrm{d}\theta}{\mathrm{d}x}\right)_1 &= 0{\cdot}91\times 0{\cdot}866 = 0{\cdot}788, \\
\left(\frac{\mathrm{d}\theta}{\mathrm{d}x}\right)_2 &= 0{\cdot}685\times 0{\cdot}342 = 0{\cdot}234, \\
\left(\frac{\mathrm{d}\theta}{\mathrm{d}x}\right)_3 &= 0{\cdot}746\times 0{\cdot}365 = 0{\cdot}272.
\end{aligned}
$$

The direction cosines of the normal to the isothermal surfaces are given by

$$
\begin{aligned}
&0{\cdot}788/\sqrt{(0{\cdot}788^2 + 0{\cdot}234^2 + 0{\cdot}272^2)} = 0{\cdot}788/0{\cdot}866 = 0{\cdot}910, \\
&0{\cdot}234/0{\cdot}866 = 0{\cdot}270, \\
&0{\cdot}272/0{\cdot}866 = 0{\cdot}314.
\end{aligned}
$$

The angle ϕ between the normal to the isothermal surface and the axis of the rod is therefore given by

$$
\begin{aligned}
\cos\phi &= 0{\cdot}866\times 0{\cdot}910 + 0{\cdot}342\times 0{\cdot}270 + 0{\cdot}365\times 0{\cdot}314 \\
&= 0{\cdot}788 + 0{\cdot}092 + 0{\cdot}115 = 0{\cdot}995. \\
\phi &= 5{\cdot}7^\circ.
\end{aligned}
$$

Q.2.6. The components of the heat flow are defined by the direction of the axis of the cylinder and are respectively proportional to sin 52°, 0, cos 52°. Now since $r_{11} = 1/k_{11}$ and $r_{33} = 1/k_{33}$,

$$
\left(\frac{\mathrm{d}\theta}{\mathrm{d}x}\right)_1 = r_{11}\sin 52^\circ = \frac{0{\cdot}788}{15{\cdot}8} = 0{\cdot}0499,
$$

and

$$
\left(\frac{\mathrm{d}\theta}{\mathrm{d}x}\right)_3 = r_{33}\cos 52^\circ = \frac{0{\cdot}616}{29{\cdot}1} = 0{\cdot}0212.
$$

If ϕ is the inclination to the z axis of the normal to the isothermal surface, then

$$\cot \phi = \frac{0{\cdot}0212}{0{\cdot}0499} = 0{\cdot}425,$$

$$\phi = 67^\circ.$$

The axis of the rod is inclined at 52° to the z axis so that the normal to the isothermal surface is inclined at $67^\circ - 52^\circ = 15^\circ$ to the axis of the rod.

Q.2.7. The principal conductivities, derived from the principal resistivities, are

$$\sigma_1 = \frac{1}{\rho_1} = 0{\cdot}917 \times 10^4\ \Omega^{-1}\ \text{cm}^{-1},$$

$$\sigma_3 = \frac{1}{\rho_3} = 0{\cdot}725 \times 10^4\ \Omega^{-1}\ \text{cm}^{-1}.$$

The conductivity σ'_{33} for the plate is therefore given by

$$\sigma'_{33} = (0{\cdot}917 \sin^2 30^\circ + 0{\cdot}725 \cos^2 30^\circ) \times 10^4$$

$$= (0{\cdot}23 + 0{\cdot}54) \times 10^4 = 0{\cdot}77 \times 10^4.$$

The current is therefore $0{\cdot}77 \times 10^4$ A.

Q.2.8. The components of current flow j_1 and j_3 are defined by the length of the rod and are proportional to sin 30° and cos 30° respectively. The components of potential gradient E_1 and E_3 are therefore given by

$$E_1 = \rho_1 \sin 30^\circ = 109 \times 0{\cdot}5 \times 10^{-6} = 54{\cdot}5 \times 10^{-6}$$

and

$$E_3 = \rho_3 \cos 30^\circ = 138 \times 0{\cdot}866 \times 10^{-6} = 120 \times 10^{-6}.$$

The inclination of the electric vector to the trigonal axis ϕ is therefore given by

$$\tan \phi = \frac{E_1}{E_3} = \frac{54{\cdot}5}{120} = 0{\cdot}454,$$

$$\phi = 24\tfrac{1}{2}^\circ.$$

$$30^\circ - 24\tfrac{1}{2}^\circ = 5\tfrac{1}{2}^\circ, \qquad \tan 5\tfrac{1}{2}^\circ = 0{\cdot}096.$$

Thus the maximum difference in potential is observed between the ends of a diameter lying in a plane containing the trigonal axis and the axis of the rod. Further, since the rod is 1 cm in diameter, the potential difference between points at opposite ends of the diameter Fig. 2.3, must correspond to equipotential surfaces 0·096 cm apart, i.e. to 0.096V.

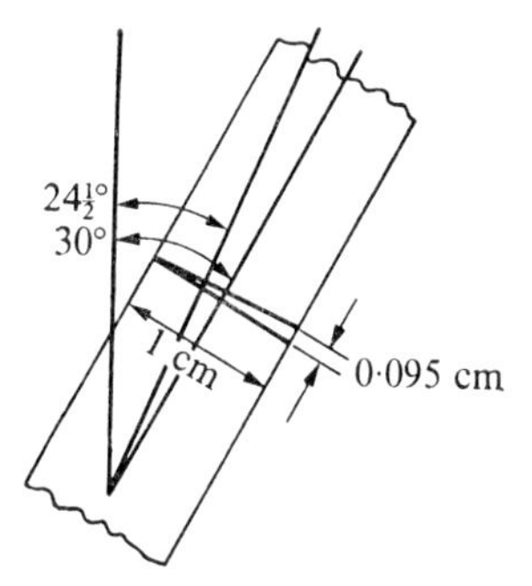

FIG. 2.3. Diagram showing the relation between the direction of current flow and the direction of the maximum potential gradient in the bismuth rod of Q. 2.8.

Q.2.9. $k'_{12} = c_{12}c_{21}k_{21} = -k_{21} = k_{12}$,
and, since $k_{21} = k_{12}, k_{21} = k_{12} = 0$.

Q.2.10. The equivalence of X_1, X_2, X_3 so far as conductivity is concerned requires that

$$k_{11} = k_{22} = k_{33}.$$

3

Thermal expansion and glide-twinning

3.1. Thermal expansion

WHEN crystals are heated they may expand in all directions, or contract in all directions, or expand in some directions and contract in others. We are here concerned with the relation between two vectors, which are (a) the change in length for a change in temperature of 1°C of a given line in the crystal and (b) the vector representing that line in length and direction. These two vectors will be denoted **p** and **r** respectively (Fig. 3.1). Before considering the relation

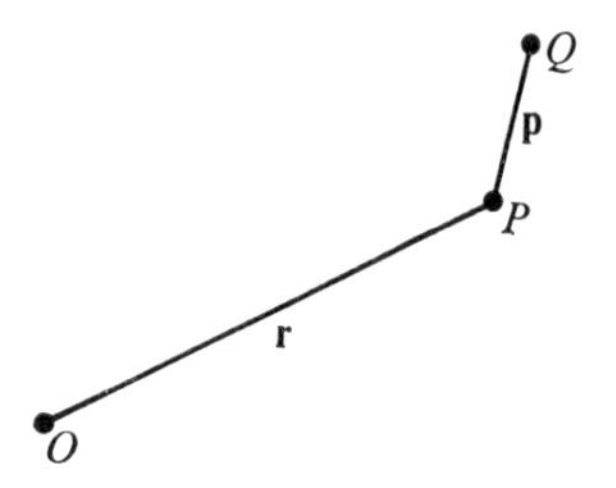

FIG. 3.1. The line marked **r** represents a certain vector drawn from the origin O to any point P in the crystal. The line marked **p** represents the displacement of P to Q caused by the thermal expansion.

between these two vectors during the expansion or contraction of the crystal, we must consider the nature of what is called homogeneous deformation. This is a term which applies both to thermal expansion and to the glide-twinning of calcite.

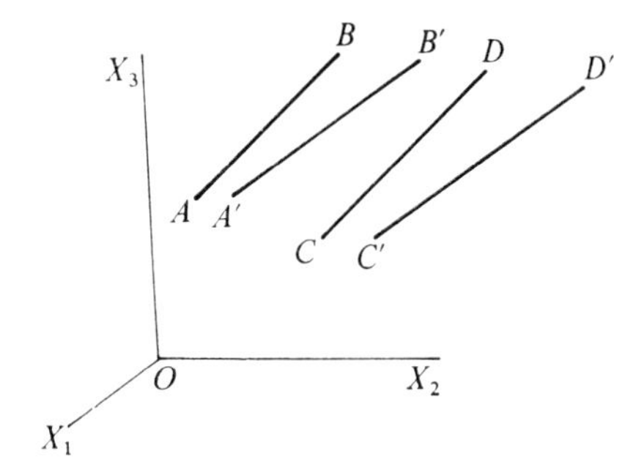

FIG. 3.2. Diagram showing two parallel lines of equal length *AB*, *CD* before homogeneous deformation, and the corresponding parallel lines *A′B′*, *C′D′* after the homogeneous deformation.

Suppose two parallel lines *AB*, *CD* (Fig. 3.2) are each of unit length, and that the coordinates of their end points are

$$A,\ u_1,\ u_2,\ u_3;\ \ B,\ v_1,\ v_2,\ v_3;\ \ C,\ w_1,\ w_2,\ w_3;\ \ D,\ z_1,\ z_2,\ z_3.$$

The direction cosines of these lines of unit length are the same, being parallel, and are denoted m_1, m_2, m_3. Then,

$$m_1 = v_1 - u_1 = z_1 - w_1;\ m_2 = v_2 - u_2 = z_2 - w_2;\ \text{etc.}$$

If it is assumed that the displacement of each point in the crystal is proportional to each of its coordinates, we may write

$$\delta u_1 = \alpha_{11}u_1 + \alpha_{12}u_2 + \alpha_{13}u_3,$$

$$\delta v_1 = \alpha_{11}v_1 + \alpha_{12}v_2 + \alpha_{13}v_3,$$

where the αs are constants.

The direction cosines of the line *A′B′* after the deformation are denoted m_1', m_2', m_3'. We have

$$\begin{aligned} m_1' &= (v_1 + \delta v_1) - (u_1 + \delta u_1) \\ &= v_1 - u_1 + \delta v_1 - \delta u_1 \\ &= v_1 - u_1 + \alpha_{11}(v_1 - u_1) + \alpha_{12}(v_2 - u_2) + \alpha_{13}(v_3 - u_3) \\ &= m_1 + \alpha_{11}m_1 + \alpha_{12}m_2 + \alpha_{13}m_3. \end{aligned}$$

Similar expressions may be written for m_2' and m_3' using α_{2i} and α_{3i}.

The direction cosines of the line $C'D'$ after the deformation are denoted m_1'', m_2'', m_3''. Following the same reasoning as for m_1', we have

$$m_1'' = m_1+\alpha_{11}m_1+\alpha_{12}m_2+\alpha_{13}m_3 = m_1',$$

and similarly for m_2'' and m_3''. Thus after the deformation the lines $A'B'$ and $C'D'$ are parallel to one another.

We have not assumed any particular unit of length, and hence the argument holds whatever the actual length of the lines AB, CD. Thus straight lines remain straight and parallel lines remain parallel under homogeneous deformation, which is defined by the general equation

$$\delta u_i = \alpha_{ik}u_k. \tag{3.1}$$

This equation implies that each component of the vector **p** is proportional to each component of the vector **r**. The coefficients of expansion α_{ik} form a second-order tensor having the same transformation properties as the conductivities considered in Chapter 2. The representation surface is an ellipsoid or a hyperboloid. It is quite possible with the αs to obtain negative values, and in this case there are certain directions of zero expansion.

A simple example of this is afforded by the crystal of calcite, $CaCO_3$, which is rhombohedral with a positive α_{33} but negative α_{11}. The coefficient of expansion in any arbitrary direction is denoted α_{33}' and, as in this system of symmetry $\alpha_{22} = \alpha_{11}$, we have, following equation (2.6),

$$\begin{aligned}\alpha_{33}' &= c_{31}c_{31}\alpha_{11}+c_{32}c_{32}\alpha_{22}+c_{33}c_{33}\alpha_{33}\\ &= \alpha_{11}(c_{31}^2+c_{32}^2)+\alpha_{33}c_{33}^2.\end{aligned}$$

Further,

$$c_{31}^2+c_{32}^2+c_{33}^2 = 1.$$

Hence

$$\begin{aligned}\alpha_{33}' &= \alpha_{11}(1-c_{33}^2)+\alpha_{33}c_{33}^2\\ &= \alpha_{11}+(\alpha_{33}-\alpha_{11})c_{33}^2\end{aligned}$$

If $\alpha_{33}' = 0$, then

$$c_{33}^2 = \frac{\alpha_{11}}{\alpha_{11}-\alpha_{33}}.$$

The angle that this value of c_{33} corresponds to is 64° 43′. There is a cone of such directions of zero expansion, having the trigonal axis of the crystal as its axis. The representation surface for the αs is therefore a figure of revolution about X_3, which in section has the form shown in Fig. 3.3 (though Fig. 3.3 is not drawn to scale).

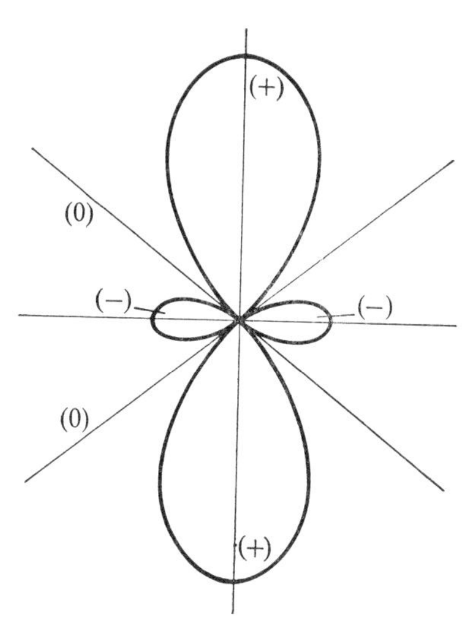

FIG. 3.3. Diagram representing the positive thermal expansion along directions neighbouring the axis X_3 and the contraction along directions neighbouring the axis X_1.

We may, using equation (2.9), represent the expansion by the reciprocal coefficients defined by the equation

$$\beta_{ik} = \delta\alpha_{ik}/\Delta_{\alpha}.$$

The representation surface corresponding to the β_{ik} consists of positive and negative hyperboloidal surfaces, a section of which is shown in Fig. 3.4.

3.2. Change of angle between faces of a crystal due to thermal expansion

In general, the angles between faces of a crystal that are not perpendicular to the principal axes change with change of tempera-

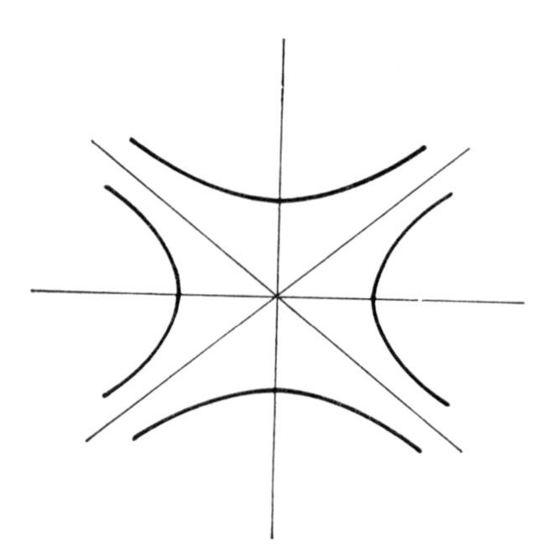

FIG. 3.4. Diagram representing the reciprocal values to those given in Fig. 3.3.

ture. The vectors of unit length normal to the two faces considered are denoted **r** and **s**. The angle between them, ϕ, is given by the equation

$$\cos\phi = r_1 s_1 + r_2 s_2 + r_3 s_3.$$

We shall assume that the axes of reference are chosen parallel to the principle axes of expansion. If the vectors representing the expansion of **r** and **s** due to a change of temperature t are **p** and **q** respectively, we have, as in Chapter 2,

$$p_1 = \alpha_{11} r_1 t; \quad p_2 = \alpha_{22} r_2 t; \quad p_3 = \alpha_{33} r_3 t,$$

and

$$q_1 = \alpha_{11} s_1 t; \quad q_2 = \alpha_{22} s_2 t; \quad q_3 = \alpha_{33} s_3 t.$$

If the vectors **r** and **s** change into $\mathbf{r}^t$ and $\mathbf{s}^t$ as a result of the heating through $t°$, then

$$r_1^t = r_1 + \alpha_{11} r_1 t; \quad r_2^t = r_2 + \alpha_{22} r_2 t; \quad r_3^t = r_3 + \alpha_{33} r_3 t,$$

and

$$s_1^t = s_1 + \alpha_{11} s_1 t; \quad s_2^t = s_2 + \alpha_{22} s_2 t; \quad s_3^t = s_3 + \alpha_{33} s_3 t.$$

If the angle between the normals to the faces after expansion is ϕ^t, then

$$\cos\phi^t = r_1^t s_1^t + r_2^t s_2^t + r_3^t s_3^t.$$

The αs are all small quantities. (The unit in which expansion coefficients are expressed is 10^{-6}.) We may therefore write

$$r_1^t s_1^t = r_1 s_1(1+2\alpha_{11}t), \text{ etc.}$$

With this simplification we have

$$\cos\phi^t = \cos\phi + 2(r_1 s_1 \alpha_{11} + r_2 s_2 \alpha_{22} + r_3 s_3 \alpha_{33})t. \qquad (3.2)$$

From equation (3.2), the difference between the angles ϕ^t and ϕ can be determined.

A method involving this change of angle was used by Mitscherlich about 150 years ago, in combination with a study of the volume expansion, to establish that calcite has a negative coefficient α_{11}.

Q.3.1. The coefficients of thermal expansion of calcite $CaCO_3$ (rhombohedral) are $\alpha_{11} = -5{\cdot}56\times10^{-6}$, $\alpha_{33} = 24{\cdot}91\times10^{-6}\,°C^{-1}$. Calculate the change in angle between two rhombohedral faces that have the direction cosines 0·7022, 0, 0·7120 and −0·3511, 0·6080, 0·7120 respectively, when the change of temperature is 100 °C.

3.3. Determination of the principal expansion coefficients from measurements in non-principal directions

Tetragonal crystal

The most important method of measuring coefficients of expansion is now the one involving the reflection of X-rays at large Bragg angles. If the spacing of the lattice planes d changes by δd as a result of a change of temperature t, it will cause a change of Bragg angle $\delta\theta$. Bragg's law states that

$$\lambda = 2d\sin\theta.$$

On differentiating, we obtain

$$0 = d\cos\theta\,\delta\theta + \delta d\sin\theta,$$

$$\frac{\delta d}{d} = -\cot\theta\,\delta\theta.$$

Thus for a given fractional change in d, $\delta\theta$ is large when $\cot\theta$ is small, i.e. when θ is near 90°. Most of the experimental work is done when θ lies between 80° and 85°. We cannot often use a plane having convenient indices such as $h00$, or $0k0$, or $00l$ to measure α_{11} and

α_{33} directly. It is sometimes possible to find planes of the types $hk0$, $0kl$, or $h0l$ which give high θ values, but often it is necessary to use planes having general indices hkl.

First we shall consider a tetragonal crystal in which two $h0l$ planes, denoted $h_1 0 l_1$ and $h_2 0 l_2$ respectively, give reflections at appropriate angles. The normals to these planes are vectors which are best expressed in terms of the reciprocal lattice (see Fig. 3.5). If a, c are

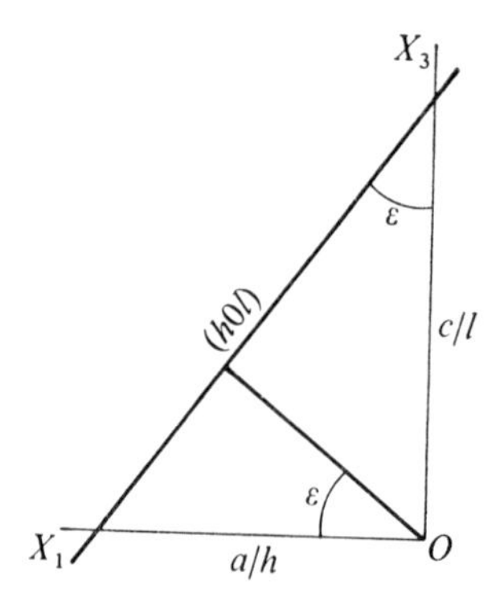

FIG. 3.5. Diagram showing the relation between the inclination of the normal to plane ($h0l$) to the X_1 axis and the indices h and l.

the sides of the Bravais unit cell and a^*, c^* the sides of the reciprocal unit cell, we shall take

$$a^* = \frac{1}{a} \quad \text{and} \quad c^* = \frac{1}{c}. \tag{3.3}$$

The vector normal to the plane $h0l$ is

$$h\mathbf{a}^* + l\mathbf{c}^*,$$

and its Cartesian components are ha^*, 0, lc^*. This is treated as axis X_3'.

If c_{31} and c_{33} are the direction cosines of the normal to the plane $h0l$, then

$$c_{31} = \frac{ha^*}{\sqrt{(h^2a^{*2}+l^2c^{*2})}}; \quad c_{33} = \frac{lc^*}{\sqrt{(h^2a^{*2}+l^2c^{*2})}}.$$

Alternatively we may write (see Fig. 3.5)

$$\frac{la}{hc} = \frac{lc^*}{ha^*} = \tan \varepsilon; \quad c_{31} = \cos \varepsilon; \quad c_{33} = \sin \varepsilon.$$

We have seen, equation (2.6), that for a second-order tensor

$$\alpha'_{33} = c_{31}^2\alpha_{11} + c_{32}^2\alpha_{22} + c_{33}^2\alpha_{33}.$$

In this case $c_{32} = 0$, so we have

$$\alpha'_{33} = c_{31}^2\alpha_{11} + c_{33}^2\alpha_{33}. \tag{3.4}$$

Two values of α'_{33} are measured for $h_1 0 l_1$ and $h_2 0 l_2$, and from the two equations (3.4) we obtain α_{11} and α_{33}.

Q.3.2. Crystals of di-p-xylylene, $C_{16}H_{16}$, are tetragonal and have cell dimensions $a = 7{\cdot}79$ Å, $c = 9{\cdot}30$ Å at 291 K. The coefficients of expansion were determined for the 802 and 109 reflections and found to be 63·4 and $47{\cdot}3 \times 10^{-6}$ respectively. Determine the principal coefficients of thermal expansion.

Orthorhombic crystal

The corresponding problem for an orthorhombic crystal can be solved in a similar manner. For any plane *hkl* the direction cosines of its normal are given

$$c_{31} = \frac{ha^*}{\sqrt{(h^2a^{*2} + k^2b^{*2} + l^2c^{*2})}}, \text{ etc.,}$$

and $\alpha'_{33} = c_{31}^2\alpha_{11} + c_{32}^2\alpha_{22} + c_{33}^2\alpha_{33}.$

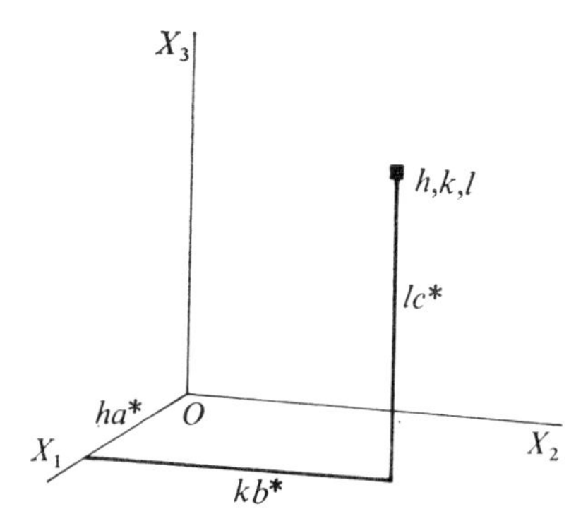

FIG. 3.6. Diagram showing the position in the reciprocal lattice of an orthorhombic crystal of a point having indices *h*, *k*, *l*.

Observations of the expansion normal to three planes are made, and we obtain three simultaneous equations with α_{11}, α_{22}, α_{33} as the unknowns. The solution gives the required expansion coefficients.

Q.3.3. A crystal of sulphur (orthorhombic) has well-developed pairs of parallel faces of indices $\{111\}$, $\{011\}$, and $\{001\}$. The table below gives the angles, and the squares of the cosines of these angles, between the normal to each of these faces and the Cartesian axes. The values of α_{hkl} are the coefficients of thermal expansion in directions normal to these faces.

hkl	PX_1	$\cos^2 PX_1$	PX_2	$\cos^2 PX_2$	PX_3	$\cos^2 PX_3$	α_{hkl}
111	42·9°	0·5365	52·9°	0·3636	71·6°	0·0999	$66{\cdot}3\times10^{-6}$
011	90·0°	0·0000	27·8°	0·7832	62·2°	0·2168	$65{\cdot}3\times10^{-6}$
001	90·0°	0·0000	90·0°	0·0000	0·0°	1·0000	$20{\cdot}0\times10^{-6}$

Determine the principal coefficients of expansion.

Monoclinic crystal

One of the principal axes of every second-order tensor property must coincide with the two-fold axis of a monoclinic crystal. We shall take X_2 as the diad axis. There are four quantities to be determined—the three principal expansion coefficients and the inclination to the axis X_3 of the two αs that lie perpendicular to X_2. It is usually possible to make three measurements on planes of the type $h0l$ and one on planes of the type $0k0$. This provides enough information to determine all four constants.

In Fig. 3.7 are represented the crystallographic c and a axes, the principal axes X_1, X_3, and the direction X_3' normal to the planes for which the expansion coefficient has been measured. The angle ψ is defined as the acute angle between the direction of the greatest coefficient of expansion in the ac plane, and the c axis. It is given a positive sign when the direction of the greatest coefficient lies between $+a$ and $+c$, and a negative sign when it lies (as in Fig. 3.7) on the negative side of the a axis. The angle between X_3' and X_3 is denoted ϕ and this we require to find. The angle between X_3' and c is known from the indices hkl and is denoted ξ.

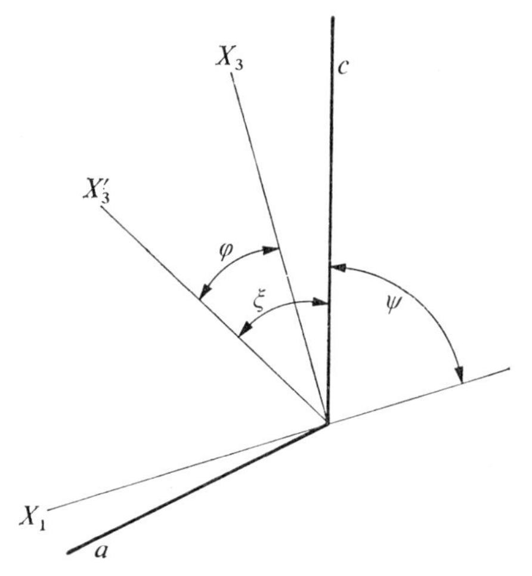

FIG. 3.7. Diagram showing the directions in the (010) plane of a monoclinic crystal of the axes X_1, X_3, and X_3'.

The expansion coefficient for the direction X_3' is called α_{33}', and we have seen that

$$\alpha_{33}' = c_{31}^2\alpha_{11}+c_{33}^2\alpha_{33}$$
$$= (\sin^2\phi)\alpha_{11}+(\cos^2\phi)\alpha_{33}.$$

In this equation there are three unknowns, namely α_{11}, α_{33}, and ϕ, and to find these quantities we must have at least three measurements such as α_{33}'. We may proceed as follows.

$$\alpha_{33}' = \tfrac{1}{2}(1-\cos 2\phi)\alpha_{11}+\tfrac{1}{2}(1+\cos 2\phi)\alpha_{33}$$
$$= \tfrac{1}{2}(\alpha_{11}+\alpha_{33})-\tfrac{1}{2}(\alpha_{11}-\alpha_{33})\cos 2\phi.$$

From Fig. 3.7 it can be seen that

$$\phi = \xi+\psi-(\pi/2),$$
$$2\phi = 2\xi+2\psi-\pi,$$
$$\cos 2\phi = -(\cos 2\xi\cos 2\psi-\sin 2\xi\sin 2\psi).$$

Now put

$$\tfrac{1}{2}(\alpha_{11}+\alpha_{33}) = A,$$
$$\tfrac{1}{2}(\alpha_{11}-\alpha_{33})\cos 2\psi = B,$$
$$-\tfrac{1}{2}(\alpha_{11}-\alpha_{33})\sin 2\psi = C.$$

Then

$$\alpha'_{33} = A + B\cos 2\xi + C\sin 2\xi.$$

Three sets of measurements of α'_{33} and ξ enable the constants A, B, and C to be found. Then we have

$$\frac{C}{B} = -\tan 2\psi,$$

$$\alpha_{11} = A + \frac{B}{\cos 2\psi},$$

$$\alpha_{33} = A - \frac{B}{\cos 2\psi}.$$

This gives all the magnitudes required, provided α_{22} has been measured by observations on the reflections of the type $0k0$ (13, p. 106; 15, p. 15; 16, p. 54).

3.4. Mechanical homogeneous deformation

The same kind of changes in length or shape brought about by thermal expansion can also be produced by mechanical deformation. When a crystal is subjected to mechanical forces it may be stretched, compressed, or twisted, and Fig. 3.1 could show the change in the end point of a vector $\mathbf{r}$ due to the displacement by the vector $\mathbf{p}$. Certain types of changes in length or shape are examples of homogeneous deformation. One of these is shown by calcite.

Glide-twinning in calcite

A particular application of the theory is to the phenomenon shown by a crystal of calcite. A cleavage rhomb looks as in Fig. 3.8(a). If a knife is pressed down at right angles to an edge running up to the tip of the crystal (from which the trigonal axis emerges), a glide process occurs, which results in the formation of a piece of the crystal in another orientation, as shown in Fig. 3.8(b). Each atomic plane in this mechanically twinned portion of the crystal has glided in the direction AO through the same distance relative to the neighbouring plane beneath it. After the gliding is over, the face PQR has the orientation of a mirror image, in a plane parallel to plane BCD, of its

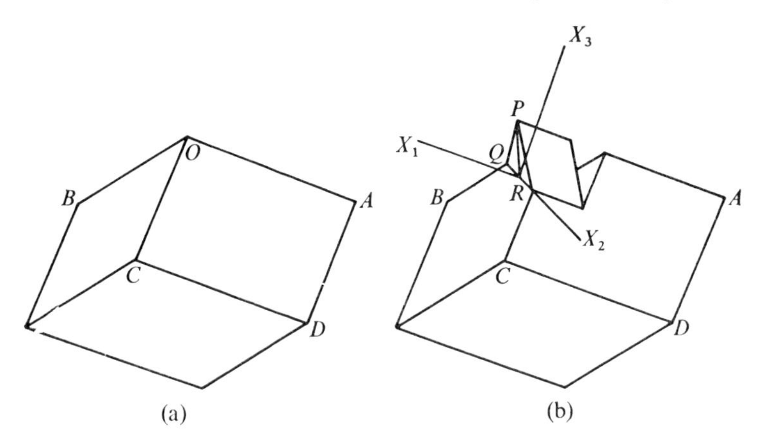

FIG. 3.8. (a) Diagram of a rhombohedron of calcite.
(b) Diagram showing the effect of the glide process in producing a twinned portion of the crystal.

original orientation, i.e. the normals to the planes PQR and OBC are equally inclined at 19° 9′ but on opposite sides of a plane parallel to BCD.

To treat this phenomenon by the same method that we have used for thermal expansion, we require to work out the principal axes of the ellipsoid representing this deformation. In the first place, we may take our axes of reference as follows (Fig. 3.8).

X_3 normal to plane BCD,

X_1 in the direction DC,

and

X_2 perpendicular to X_1 and X_3.

The components of displacement are denoted u_i, the components of the original vectors are denoted x_k, and the coefficients relating these components are denoted r_{ik}. The general expression for homogeneous deformation is

$$u_i = r_{ik}x_k.$$

In Fig. 3.9 a section through *O* by a plane perpendicular to *BC* is drawn. The line *SP* is, as stated above, inclined at the same angle to

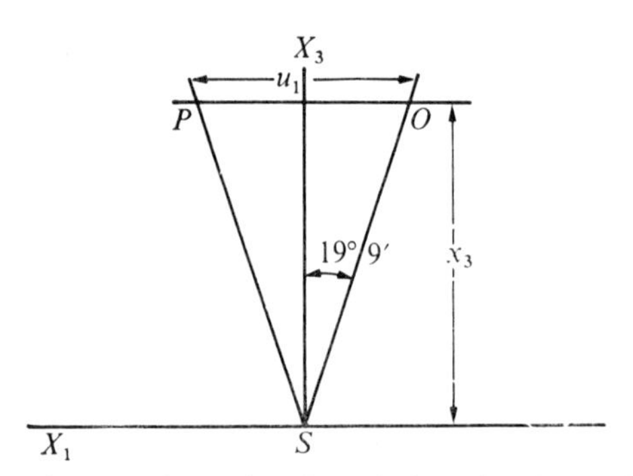

FIG. 3.9. Diagram of a section through the twinned crystal of calcite.

the vertical line through *S* as is the line *SO* but on the other side of it. In terms of the symbols defined above,

$$u_1 = r_{11}x_1 + r_{12}x_2 + r_{13}x_3.$$

Since the amount of gliding in a plane parallel to the plane *BCD* is the same all over the plane, the displacement of a point given by u_1 must be independent of x_1 and x_2. This can only be true if

$$r_{11} = r_{12} = 0.$$

Thus we have

$$u_1 = r_{13}x_3,$$

and

$$r_{13} = u_1/x_3 = 2\tan 19°\,9' = 0{\cdot}69475.$$

The component u_2 remains unchanged for every point in the crystal, since the gliding takes place in a direction normal to x_2. Hence we may write

$$r_{21} = r_{22} = r_{23} = 0.$$

Similarly, u_3 is invariant throughout the crystal during gliding, and hence

$$r_{31} = r_{32} = r_{33} = 0.$$

Thus the gliding is specified by only one coefficient, namely r_{13}.

3.5. The change of direction of any line in a crystal of calcite subjected to the glide–twinning process

If any direction is defined by direction cosines referred to the axes used in § 3.4, then the coordinates of a point on the line at unit distance from the origin are equal to the magnitudes of the direction cosines of that line. The line may be denoted the X_3' axis, and for this axis, we have, from § 1.3,

$$x_i = c_{3i}.$$

During the glide process x_2 and x_3 remain unchanged but x_1 becomes x_1', where

$$x_1' = x_1 + 0{\cdot}69475x_3.$$

The new direction cosines are therefore

$$c_{31}' = (x_1 + 0{\cdot}69475x_3)/D,$$

where D is the length of the line after gliding, i.e.

$$D^2 = (x_1 + 0{\cdot}69475x_3)^2 + x_2^2 + x_3^2.$$

We have also

$$c_{32}' = x_2/D,$$

$$c_{33}' = x_3/D.$$

These direction cosines refer to a set of axes related to the glide process, but by the usual transformation they can be referred to a set of Cartesian axes having X_1 as a diad axis and X_3 as the trigonal axis, (15, p. 39).

Q.3.4. Using the axes of reference specified by Fig. 3.8 and given that the direction cosines of the normals to two of the rhombohedral (cleavage) faces relative to these axes are (a) 0·94463, 0, 0·32814 and (b) 0, 0·60807, 0·79388 respectively,

(1) write down the direction cosines of a line OM perpendicular to axis X_2, lying in the plane of face (a), and also the direction cosines of a line ON perpendicular to X_2, lying in the plane of face (b);

(2) determine the change of orientation of lines OM, ON caused by the glide process, expressed in terms of the new direction cosines;

(3) find the new direction cosines of the rhombohedral faces (a), (b) after the gliding.

Answers

Q.3.1.

$$\cos\phi^t = \cos\phi + 2\{(0{\cdot}7022\times -0{\cdot}3511\times -5{\cdot}56) + (0{\cdot}7120\times 0{\cdot}7120\times \times 24{\cdot}91)\}\times 10^{-6}\times 100$$

$$= \cos\phi + 0{\cdot}0028.$$

$$\cos\phi = (0{\cdot}7022\times -0{\cdot}3511) + (0{\cdot}7120\times 0{\cdot}7120)$$

$$= 0{\cdot}2605,$$

$$\phi = 74^\circ\, 54'.$$

$$\cos\phi^t = 0{\cdot}2633,$$

$$\phi^t = 74^\circ\, 44'.$$

The change of angle is a decrease of 10′.

Q.3.2. The cell dimensions give ($a^* = \lambda/\alpha$ etc, $\lambda = 1{\cdot}54$ Å)

$$a^* = 0{\cdot}1977, \qquad c^* = 0{\cdot}1657.$$

The direction cosines of the normal to a plane $h0l$ are given by c_{31} and c_{33}, where

$$c_{31} = \frac{ha^*}{\sqrt{(h^2a^{*2} + l^2c^{*2})}}, \qquad c_{33} = \frac{lc^*}{\sqrt{(h^2a^{*2} + l^2c^2)^*}}.$$

The coefficient of expansion α_{h0l} in a direction normal to the plane $h0l$ is given by

$$\alpha_{h0l} = \alpha_{11}c_{31}^2 + \alpha_{33}c_{33}^2.$$

For the plane 802,

$$c_{31} = \frac{8\times 0{\cdot}1977}{\sqrt{(64\times 0{\cdot}1977^2 + 4\times 0{\cdot}1657^2)}} = \frac{1{\cdot}5816}{1{\cdot}6160} = 0{\cdot}9787,$$

$$c_{31}^2 = 0{\cdot}9579;$$

$$c_{33} = \frac{2\times 0{\cdot}1657}{1{\cdot}6160} = 0{\cdot}2051,$$

$$c_{33}^2 = 0{\cdot}0421.$$

For the plane 109,

$$c_{31} = \frac{0{\cdot}1977}{\sqrt{(0{\cdot}1977^2 + 81\times 0{\cdot}1657^2)}} = 0{\cdot}1314,$$

$$c_{31}^2 = 0{\cdot}0173;$$

$$c_{33} = \frac{9\times 0{\cdot}1657}{1{\cdot}5043} = 0{\cdot}9914,$$

$$c_{33}^2 = 0{\cdot}9828.$$

Inserting these values into the equation for α_{h0l}, we have (omitting the factor 10^{-6} until the end)

$$63{\cdot}4 = \alpha_{11}\,0{\cdot}9579+\alpha_{33}\,0{\cdot}0421,$$

$$47{\cdot}3 = \alpha_{11}\,0{\cdot}0173+\alpha_{33}\,0{\cdot}9828.$$

From this we obtain

$$\alpha_{11} = \frac{63{\cdot}4\times 0{\cdot}9828-47{\cdot}3\times 0{\cdot}0421}{0{\cdot}9579\times 0{\cdot}9828-0{\cdot}0173\times 0{\cdot}0421} = \frac{60{\cdot}3182}{0{\cdot}9407} = 64{\cdot}1\times 10^{-6}\,{}^\circ\mathrm{C}^{-1},$$

$$\alpha_{33} = \frac{47{\cdot}3\times 0{\cdot}9579-63{\cdot}4\times 0{\cdot}0173}{0{\cdot}9407} = 47{\cdot}0\times 10^{-6}\,{}^\circ\mathrm{C}^{-1}.$$

Q.3.3. Applying the equation (2.6), we have

$$66{\cdot}3 = \alpha_{11}\,0{\cdot}5388+\alpha_{22}\,0{\cdot}3636+\alpha_{33}\,0{\cdot}0999, \quad (1)$$

$$65{\cdot}3 = \alpha_{22}\,0{\cdot}7832+\alpha_{33}\,0{\cdot}2172, \quad (2)$$

$$20{\cdot}0 = \alpha_{33}. \quad (3)$$

Inserting the value of α_{33} into the first two equations, we get

$$64{\cdot}302 = \alpha_{11}\,0{\cdot}5388+\alpha_{22}\,0{\cdot}3636, \quad (4)$$

$$60{\cdot}956 = \alpha_{22}\,0{\cdot}7832. \quad (5)$$

Hence

$$\alpha_{22} = \frac{60{\cdot}956}{0{\cdot}7832} = 77{\cdot}8.$$

Inserting this value in the fourth equation, we have

$$36{\cdot}003 = \alpha_{11}\,0{\cdot}5388,$$

$$\alpha_{11} = \frac{36{\cdot}003}{0{\cdot}5388} = 66{\cdot}8.$$

Finally,

$$\alpha_{11} = 66{\cdot}8,\ \alpha_{22} = 77{\cdot}8,\ \alpha_{33} = 20{\cdot}0\times 10^{-6}\,{}^\circ\mathrm{C}^{-1}.$$

Q.3.4. (1) The direction cosines of the line OM are

$$-0{\cdot}32814,\ 0,\ 0{\cdot}94463.$$

The direction cosines of the line ON are

$$1{\cdot}000,\ 0,\ 0.$$

(2) For OM, remembering that $r_{31} = 2 \tan 19{\cdot}156° = 0{\cdot}69475$, we get

$$u_1 = r_{31}\, x_3 = 0{\cdot}69475 \times 0{\cdot}94463$$

$$= 0{\cdot}65628.$$

Hence, after gliding,

$$x_1' = -0{\cdot}32814 + 0{\cdot}65628 = 0{\cdot}32814,$$

$$x_2' = 0,$$

$$x_3' = 0{\cdot}94463.$$

For ON, $u_1 = 0$ and

$$x_1' = 1{\cdot}000,$$

$$x_2' = 0,$$

$$x_3' = 0.$$

(3) After gliding, the normal to face (a) has direction cosines

$$0{\cdot}9446,\ 0,\ -0{\cdot}3281$$

and the normal to face (b) is unchanged, so that its direction cosines are

$$0,\ 0{\cdot}6081,\ 0{\cdot}7939.$$

4

Tensors of stress and strain

4.1. The strain tensor

IN discussing the glide-twinning of calcite, we met an example where there was only one component representing the strain. This is not characteristic of the treatment of the small strains that occur during piezoelectric or mechanical deformation of crystals. For such deformations it is usual to write

$$r_{ik} = r_{ki}. \tag{4.1}$$

We shall now examine the meaning of equation (4.1). Consider the expression for the component u_1 of the displacement of a point having coordinates x_1, x_2, x_3, which is

$$u_1 = r_{11}x_1 + r_{12}x_2 + r_{13}x_3. \tag{4.2}$$

If $x_2 = x_3 = 0$, i.e. we consider a point on the X_1 axis, then

$$u_1 = r_{11}x_1 \quad \text{or} \quad r_{11} = \frac{u_1}{x_1}.$$

Similarly, for a point on the X_2 axis,

$$r_{22} = \frac{u_2}{x_2},$$

and for a point on the X_3 axis,

$$r_{33} = \frac{u_3}{x_3}.$$

These rs give a change of length per unit length and define the expansion or contraction parallel to the axes X_1, X_2, X_3.

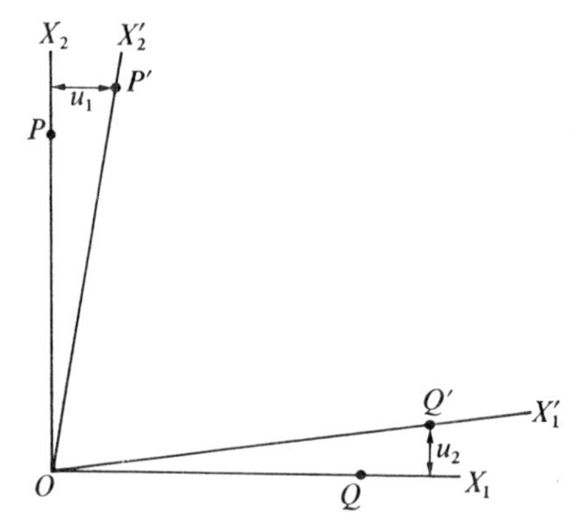

FIG. 4.1. Diagram showing the changes in the directions of the axes when a crystal is subjected to homogeneous deformation.

Now for a point P on the X_2 axis (Fig. 4.1), having coordinates $0, x_2, 0$,

$$u_1 = r_{12}x_2 \quad \text{and} \quad r_{12} = \frac{u_1}{x_2}.$$

When the quantities r_{ik} are small, u_1/x_2 is just the change of angle $P\widehat{O}P'$ experienced by the X_2 axis as a result of the application of the strain.

Similarly, the point $Q(x_1, 0, 0)$ goes to Q' as a result of the deformation u_2, where

$$u_2 = r_{21}x_1,$$

and

$$r_{21} = \frac{u_2}{x_1} = Q\widehat{O}Q'.$$

The convention is usually adopted that makes

$$r_{12} = r_{21}.$$

This also applies to the other planes containing the pairs of axes X_2, X_3 and X_3, X_1. In this way we may write

$$r_{ik} = r_{ki}.$$

This is not the same kind of equality as we have met in connection with thermal conductivity, where the equality of the corresponding coefficients was a consequence of the nature of thermal and electrical conduction. Using this convention, we may apply to the strain tensor the same treatment that we have applied to other second-order tensors.

If the axes of reference are the principal axes of the strain ellipsoid then, as for other second-order tensors,

$$r'_{33} = c_{31}^2 r_{11} + c_{32}^2 r_{22} + c_{33}^2 r_{33}.$$

The representation surface may be a tri-axial ellipsoid or a hyperboloid, since the rs may be either positive or negative (1, p. 113; 13, p. 93; 15, p. 19).

4.2. The stress tensor

The forces acting on a body may be, like the force of gravity, proportional to its volume, or they may be compressive, stretching or shearing forces, which are defined in relation to forces per unit area. The former are called body forces and will not be further discussed here. The second kind are those which we shall consider in relation to the piezoelectric and mechanical properties of crystals.

In all of the following discussion we shall consider that the stress is homogeneous, i.e. that the conditions of stress in every volume element of the body are the same. At first we shall also assume that there is no movement and that all parts of the solid are in mechanical equilibrium.

At first sight it would seem that the stress is a vector, since in common usage the word 'stress' is similar in meaning to 'force'. However, in the work we are doing here the stress tensor relates the vector force to the vector normal to the plane across which the force is applied.

To obtain the relation between the components of force and the components of the unit vector normal to the area under consideration, we consider a tetrahedron within the body represented by $OABC$ (Fig. 4.2). The faces OBC, OAC, OAB are normal to the Cartesian axes X_1, X_2, X_3, and OP is normal to the plane ABC. We suppose equilibrium to be maintained by a force k applied to the face ABC, which need not be perpendicular to that plane, and by forces, normal and tangential, applied to each of the other three faces.

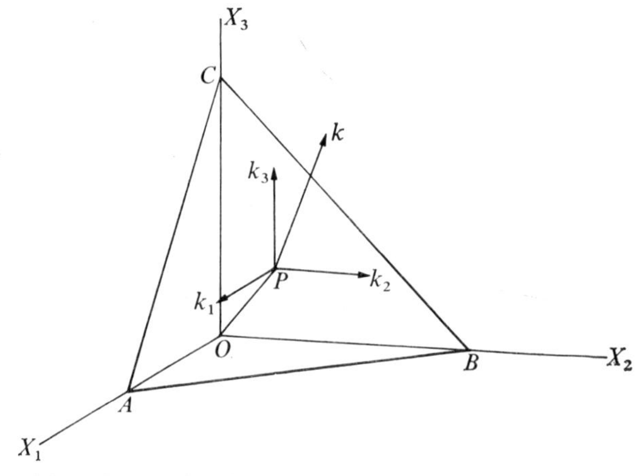

FIG. 4.2. Diagram showing the relations between the components of a force (per unit area) k applied to a face of a piece of crystal $OABC$.

These latter forces will be denoted t_{ik}. The subscript k refers to the normal to the face and the subscript i to the axis to which the force is parallel. Thus t_{11} is a force per unit area acting on face OBC in the direction of X_1, and t_{21} is the force per unit area acting on the same face in the direction of X_2. Under equilibrium conditions the total of forces acting in the direction OX_1 must just balance the total of forces acting in the opposite direction. Thus we may write

$$k_1 \times \text{area } ABC = t_{11} \times \text{area } BOC + t_{12} \times \text{area } COA + \\ + t_{13} \times \text{area } AOB.$$

The direction cosines l_i of the normal to ABC are given by

$$l_1 = \frac{\text{area } OBC}{\text{area } ABC}; \quad l_2 = \frac{\text{area } OAC}{\text{area } ABC}; \quad l_3 = \frac{\text{area } OAB}{\text{area } ABC}.$$

Hence

$$k_1 = t_{11}l_1 + t_{12}l_2 + t_{13}l_3.$$

Similar expressions may be written for k_2 and k_3, so that the general expression becomes

$$k_m = t_{mn}l_n.$$

Thus t_{nm} is a second-order tensor.

The condition that

$$t_{nm} = t_{mn}$$

is one of equilibrium, as may be seen from Fig. 4.3. A tangential force t_{21} acting on face *ABFE* of a unit cube in the direction OX_2 must be balanced by a force of the same magnitude but oppositely directed acting on the face *ODCG*. Similarly, tangential forces t_{12} acting on faces *ABCD* and *OEFG* must be equal and oppositely

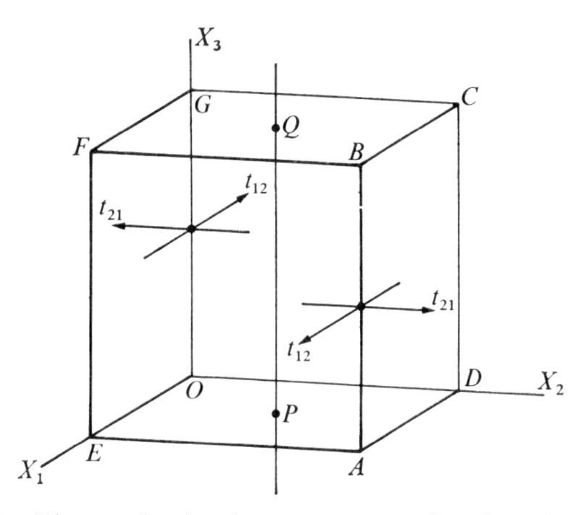

FIG. 4.3. Diagram showing that to prevent rotation about the line *PQ*, t_{12} must equal t_{21}.

directed. If we take moments about an axis *PQ* passing through the centres of the faces *OEAD* and *BCGF*, then the turning moment due to the tangential forces acting on these two faces is zero. Equilibrium thus requires that the two couples t_{12} and t_{21} shall be equal. Similar arguments establish the equality $t_{ik} = t_{ki}$ for all values of i and k.

Thus in general the t_{ik}s transform as all second-order tensors, i.e.

$$t'_{33} = c_{31}^2 t_{11} + c_{32}^2 t_{22} + c_{33}^2 t_{33} + 2c_{31}c_{32}t_{12} + 2c_{32}c_{33}t_{23} + \\ + 2c_{33}c_{31}t_{31},$$

but if the axes of reference are chosen parallel to the principal axes of the quadric, which this equation represents, then

$$t'_{33} = c_{31}^2 t_{11} + c_{32}^2 t_{22} + c_{33}^2 t_{33}.$$

The tensor representing stress has no connection with crystal symmetry and depends simply on how the forces are applied.

Q.4.1. A bar of rectangular section 2×3 cm and 8 cm long is stretched uniformly by a force of 10 kg directed along its length. It increases in length by 20 μm and decreases in the direction of the 2 cm side by 1 μm and in the direction of the 3 cm side by 3μm.

Write down the components of stress and strain, taking the axes X_1, X_2, X_3 parallel to the sides of length 8, 2, and 3 cm respectively.

4.3. Special cases of the stress tensor

When a body is stretched uniformly in one direction we may choose the axis OX_1 to coincide with this direction. In this case, of the six possible t_{ik} values only t_{11} is non-zero. If the force is applied parallel to OX_2, only t_{22} is non-zero, and if parallel to OX_3 only t_{33} is non-zero.

If the body is subjected to a shearing force about the OX_1 axis, then only t_{23} is non-zero. Similar relations apply for shearing about the other axes OX_2 and OX_3.

If the body is subjected to uniform hydrostatic pressure π then

$$t_{11} = t_{22} = t_{33} = \pi, \text{ and } t_{12} = t_{23} = t_{31} = 0,$$

since a liquid is incapable of exerting a tangential force on the body when there is no movement of the liquid.

Q.4.2. A rectangular bar of the same dimensions as that described in Q.4.1 is immersed in a liquid and subjected to a pressure of 1000 kg cm^{-2}. The volume of the bar decreases by 1 mm^3.

Write down the components of stress and strain if the bar is assumed elastically isotropic.

The practical realization of a homogeneous shearing force, constant throughout the whole volume, may be achieved by a combination of compressive and stretching forces.

If a stress component t_{11} is applied along the X_1 axis and an equal stress component but of opposite sign is applied along X_2, then in any arbitrary direction X_1' in the X_1X_2 plane, making an angle θ with X_1 and an angle $\pi/2-\theta$ with $-X_2$, we have

$$t_{22} = -t_{11}$$

and

$$\begin{aligned} t'_{12} &= c_{11}c_{21}t_{11}+c_{12}c_{22}t_{22} \\ &= (+\cos\theta\sin\theta+\sin\theta\cos\theta)t_{11} \\ &= \sin 2\theta t_{11}. \end{aligned}$$

When $\theta = 45°$ this gives

$$t'_{12} = t_{11}.$$

Also we have

$$t'_{11} = c_{11}c_{11}t_{11} + c_{12}c_{12}t_{22}$$
$$= (\cos^2\theta - \sin^2\theta)t_{11},$$

and when $\theta = 45°$

$$t'_{11} = 0.$$

It may be shown similarly that $t'_{22} = 0$. Thus only the shearing force is present under these circumstances (1, p. 118; 13, p. 89; 15, p. 231).

Q.4.3. A hexagonal crystal occurs in the form of a trigonal prism 5 cm long and having each prism face 1 cm wide. The end faces are normal to the length of the prism, which is parallel to the hexad axis. The crystal is subjected to a hydrostatic pressure of 1000 kg cm^{-2} and expands by 5 μm along its length and contracts by 1 μm in the width of the prism faces.

Write down the components of stress and strain. Take the length of the prism as the direction X_3 and a line in one prism face as X_1.

Q.4.4. A block 5×5×2 cm is compressed with a force of 10 kg on two opposite 5×2 cm faces and stretched with an equal force on the other pair of 5×2 cm faces. Taking axes X_1, X_2, X_3 parallel to the 5, 5, and 2 cm edges respectively, write down the components of stress. The reduction of length in the X_1 direction is 2 μm and the increase in length in the X_2 direction is also 2 μm. Write down the components of strain.

Transform to axes X'_1, X'_2, X'_3 such that X'_1 is at 45° to both X_1 and X_2, X'_3 coincides with X_3. Determine relative to the dashed axes the components of stress and strain. Treat compressive stresses as positive and stretching stresses as negative.

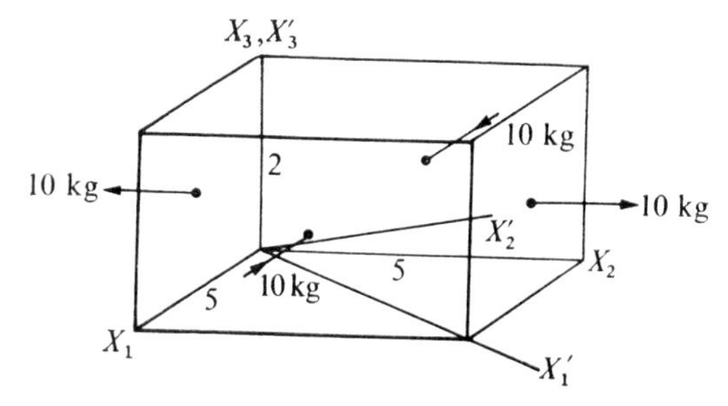

FIG. 4.4. Diagram showing the way in which the forces are applied to the block mentioned in Q.4.4.

4.4. Magnetic Induction

When placed in a magnetic field all crystals react mechanically to a greater or lesser extent. The three types of crystals called ferromagnetic, paramagnetic, and diamagnetic react differently. The ferromagnetic materials are attracted to a magnetic pole from which lines of force are diverging, and the reaction is relatively strong. Paramagnetic crystals are also attracted to stronger parts of the magnetic field, but with a mechanical force that is very small in comparison with the forces experienced by ferromagnetic materials. Diamagnetic crystals are repelled from such a magnetic pole and tend to move towards regions of lower field strength. The forces obtained with diamagnetic materials are also very small in comparison with those given by ferromagnetic materials.

The simplest way of treating this subject is to regard the magnetic field as inducing a magnetic moment of a certain magnitude per unit volume and then to consider the reaction of the applied field on the induced magnetic moment. Diamagnetic materials must be regarded as having constituent atoms that become magnetized so that the N–S direction is just opposite to that produced in ferromagnetic materials. Many organic crystals are anisotropic in a magnetic field, and the coefficients, called susceptibilities, that express the relation between the applied field and the induced magnetic moment may have very different magnitudes in different directions in a given crystal.

The susceptibilities are second-order tensors because the induced magnetic moment **M** is related to the applied field **H** by the equation

$$M_i = \chi_{ik} H_k \tag{4.3}$$

where χ_{ik} represents the susceptibilities. The same symmetry relations apply to the χs as to the thermal conductivities, and also

$$\chi_{ik} = \chi_{ki}.$$

This is not self-evident but is a consequence of thermodynamical reasoning. The determination of the principal susceptibilities χ_{11}, χ_{22}, χ_{33} from the measured values can be carried out by the same kind of analysis that was described in connection with other second-order tensors (**1, p. 164; 13, p. 53; 15, p. 94; 16, p. 34**).

4.5. Electric induction

When placed in an electric field, crystals develop an electric moment within each element of volume, and this is known as dielectric polarization. If the electric field is represented by a vector **E** and the induced electric moment by a vector **P**, then the relation between these quantities is expressed by the usual second-order tensor equation

$$P_m = k_{mn}E_n.$$

The symmetry relations of the ks are the same as for other second-order tensors. The study of dielectric induction is made difficult by the leakage of the charge associated with the polarization. Unlike magnetic poles, which remain constant in a constant field, electric charges tend to neutralize charges of the opposite sign more or less quickly. For most crystals it is impossible to make measurements of the electric induction unless the direction of the electric field is reversed many times per second, (1, p. 157; 13, p. 68; 15, p. 117; 16, p. 65).

Answers

Q.4.1. $$t_{11} = \frac{10\times 10^3\times 981}{6} = 0{\cdot}164\times 10^7 \text{ dyne cm}^{-2}.$$

All other ts are zero.

$$r_{11} = \frac{20\times 10^{-4}}{8} = 2{\cdot}5\times 10^{-4},$$

$$r_{22} = \frac{-10^{-4}}{2} = -0{\cdot}5\times 10^{-4},$$

$$r_{33} = \frac{-3\times 10^{-4}}{3} = -10^{-4}.$$

Since the bar remains rectangular

$$r_{12} = r_{23} = r_{31} = 0.$$

Q.4.2. $$t_{11} = t_{22} = t_{33} = 1000\times 1000\times 981 \text{ dyne cm}^{-2}$$

$$= 0{\cdot}981\times 10^9 \text{ dyne cm}^{-2}.$$

$$t_{12} = t_{23} = t_{31} = 0.$$

$$r_{11} = r_{22} = r_{33}; \qquad r_{12} = r_{23} = r_{31} = 0.$$

The change of volume per unit volume is given by $3r_{11}$ and

$$3\,r_{11} = \frac{10^{-3}}{8\times 2\times 3},$$

$$r_{11} = 0{\cdot}70\times 10^{-5}.$$

Q.4.3. Because hydrostatic pressure is used,

$$t_{11} = t_{22} = t_{33} = 1000 \text{ kg cm}^{-2}, \quad t_{12} = t_{23} = t_{31} = 0.$$

$$r_{33} = \frac{5\times 10^{-4}}{5} = 10^{-4}.$$

Since the shape of the cross-section remains unchanged after the application of the hydrostatic pressure,

$$r_{11} = r_{22}, r_{12} = r_{23} = r_{31} = 0, \text{and } r_{11} = \frac{-10^{-4}}{1} = -10^{-4}.$$

Q.4.4. See Fig. 4.4.

$$t_{11} = \frac{10}{10}\text{ kg cm}^{-2} = 1 \text{ kg cm}^{-2},$$

$$t_{22} = \frac{-10}{10} = -1 \text{ kg cm}^{-2},$$

$$t_{33} = 0,$$

$$t_{12} = t_{23} = t_{31} = 0.$$

$$r_{11} = \frac{-2\times 10^{-4}}{5} = -0{\cdot}4\times 10^{-4},$$

$$r_{22} = \frac{2\times 10^{-4}}{5} = 0{\cdot}4\times 10^{-4},$$

$$r_{33} = r_{12} = r_{23} = r_{31} = 0.$$

	X_1	X_2	X_3
X_1'	$\frac{1}{\sqrt{2}}$	$\frac{1}{\sqrt{2}}$	0
X_2'	$\frac{-1}{\sqrt{2}}$	$\frac{1}{\sqrt{2}}$	0
X_3'	0	0	1

$$t'_{11} = c_{11}c_{11}t_{11}+c_{12}c_{12}t_{22}$$
$$= \tfrac{1}{2}t_{11}+\tfrac{1}{2}t_{22} = 0,$$

$$t'_{22} = c_{21}c_{21}t_{11}+c_{22}c_{22}t_{22}$$
$$= \tfrac{1}{2}t_{11}+\tfrac{1}{2}t_{22} = 0,$$

$$t'_{33} = c_{31}c_{31}t_{11}+c_{32}c_{32}t_{22} = 0,$$

$$t'_{12} = c_{11}c_{21}t_{11}+c_{12}c_{22}t_{22}$$
$$= -\tfrac{1}{2}t_{11}+\tfrac{1}{2}t_{22} = -t_{11},$$

$$t'_{13} = t'_{23} = 0.$$

$$r'_{11} = c_{11}c_{11}r_{11}+c_{12}c_{12}r_{22}$$
$$= \tfrac{1}{2}r_{11}+\tfrac{1}{2}r_{22} = 0 = r'_{22} = r'_{33},$$

$$r'_{12} = c_{11}c_{21}r_{11}+c_{12}c_{22}r_{22}$$
$$= -\tfrac{1}{2}r_{11}+\tfrac{1}{2}r_{22} = 0{\cdot}4\times 10^{-4} = r'_{21},$$

$$r'_{13} = r'_{23} = 0.$$

5

Piezoelectricity: the direct effect

5.1. Introduction

THE initial discovery of the phenomenon of piezoelectricity dates from 1880. The discoverers were J. and P. Curie. The first observation was that certain crystals developed electric charges on their surfaces when subjected to mechanical stresses. Positive charge appeared at one point on a crystal and negative charge at another. This is called the 'direct' piezoelectric effect. They also showed that all these crystals had less than the full symmetry of the crystal system to which they belonged. The same scientists demonstrated the existence of the 'converse' effect, namely that when electric fields are applied in various ways to these crystals, changes in shape occur. These changes in shape are directly related to the changes in shape that are produced by the mechanical stresses in the direct effect. Since that time, the applications of piezoelectricity have been very many. The production of a change of shape by application of an electric field has perhaps had the greatest importance. If an alternating electric field is applied to a piezoelectric crystal in an appropriate way it expands and contracts or twists backwards and forwards. Now every solid bar has a natural frequency of vibration and, if subjected to an alternating force of the same frequency, it goes into resonance and executes a vibration of large amplitude. This principle is applied to piezoelectric crystals by application of an alternating electric field having the same frequency as their natural frequency of mechanical vibration. Crystals such as quartz when excited in this way can control the frequency of the applied electric field. This frequency control is used in radio to keep the frequency of transmitting stations constant, and in clocks that need to keep time, correct to a few seconds per year. The other great application of this effect is to the production of ultrasonic vibrations. The crystal is made to vibrate when in contact with liquids or solids and it sends mechanical waves into the medium with which it is in contact. These waves may be used for detecting cracks in solid bars, or for studying the bed of the sea, or for locating submarines. They may also be used for rapid agitation and find application both in laundering and in the ageing of whisky. The direct effect is used in gramophone pickups, where the movement of a stylus produces a pressure on the crystal, which, in turn, produces the alternating electric charges. These charges are amplified and finally produce the sound in the loudspeaker.

5.2. Third-order tensors

We shall at first consider only the direct effect, i.e. the liberation of electric charge by the application of mechanical forces. We have to deal here with the relation between a first-order tensor and a second-order tensor. The first-order tensor is the electric moment per unit volume, denoted P. Experimentally we measure electric charge per unit area, but this is the same as the electric moment per unit volume. Suppose we have a bar of length l and area A, and let the charge liberated on the end faces of the bar be Q. Then we have

$$\text{charge per unit area of end faces} = Q/A,$$

$$\text{electric moment per unit volume} = Ql/Al,$$

and these two quantities have the same value. The stress is represented, as we have seen, by the tensor t_{kl}. There is a linear relation between P and t_{kl} such that each component of P is proportional to all the components of t. The coefficients of proportionality are called piezoelectric moduli and are denoted d. We can therefore write

$$P_i = d_{ikl}t_{kl}. \tag{5.1}$$

The ds transform in going from one set of axes to another according to the equation

$$d'_{ikl} = c_{im}c_{kn}c_{lo}d_{mno}, \tag{5.2}$$

and in the inverse sense

$$d_{ikl} = c_{mi}c_{nk}c_{ol}d'_{mno}.$$

5.3. The effect of symmetry on the piezoelectric moduli

Centre of symmetry

The first and most important relation between the piezoelectric moduli and crystal symmetry is connected with a centre of symmetry. It requires no analysis to prove that the moduli must all be zero if a centre of symmetry is present. If there were positive charges produced at one point in a crystal having a centre of symmetry the effect of this centre would be to produce an equal positive charge at an opposite point in the crystal. This is physically impossible. Thus no net charge could be liberated. This can be shown analytically in the following way. The transformation scheme for a centre of symmetry is as follows.

	X_1	X_2	X_3			
X'_1	-1	0	0	c_{11}	c_{12}	c_{13}
X'_2	0	-1	0	c_{21}	c_{22}	c_{23}
X'_3	0	0	-1	c_{31}	c_{32}	c_{33}.

Thus only c_{11}, c_{22}, and c_{33} differ from zero, but in every equation for the transformation of the moduli we have

$$d'_{ikl} = c_{im}c_{kn}c_{lo}d_{mno}.$$

The *c*s are each -1, so that

$$d'_{ikl} = -d_{ikl}$$

and, from Neumann's Principle, this means that all the *d*s must be zero.

One two-fold axis

We shall assume that a two-fold axis is parallel to X_3. In this case the transformation scheme is

	X_1	X_2	X_3			
X'_1	-1	0	0	c_{11}	c_{12}	c_{13}
X'_2	0	-1	0	c_{21}	c_{22}	c_{23}
X'_3	0	0	1	c_{31}	c_{32}	c_{33}.

The expression (5.2) for d'_{ikl} shows that only terms involving c_{11}, c_{22}, and c_{33} can give coefficients equal to $+1$. This is the case for $c_{11}c_{11}c_{33}$, $c_{22}c_{22}c_{33}$, $c_{33}c_{33}c_{33}c_{11}c_{22}c_{33}$. Thus the moduli containing one or three 3s can be non-zero. The following example will show this.

$$d'_{113} = c_{11}c_{11}c_{33}d_{113}$$
$$= -1\times -1\times 1d_{113} = d_{113},$$

so that d_{113} is not necessarily equal to zero. But

$$d'_{133} = c_{11}c_{33}c_{33}d_{133}$$
$$= -1\times 1\times 1d_{133} = -d_{133} = 0,$$

and d_{133} must be zero.

The moduli for d_{1kl} can be set out as follows

$$\begin{matrix} 0 & 0 & d_{113} \\ & 0 & d_{123} \\ & & 0 \end{matrix}$$

(d_{131} and d_{132} are also non-zero since $d_{1kl} = d_{1lk}$).

Q.5.1. What schemes of moduli apply to charge components P_2, P_3 respectively?

Q.5.2. The transformation of axes corresponding to a plane of symmetry perpendicular to the X_2 axis is

	X_1	X_2	X_3			
X'_1	1	0	0	c_{11}	c_{12}	c_{13}
X'_2	0	-1	0	c_{21}	c_{22}	c_{23}
X'_3	0	0	1	c_{31}	c_{32}	c_{33}

Find which piezoelectric moduli are non-zero.

For the monoclinic system there are eight non-zero moduli. To reduce the number of digits to be written, a convention is sometimes adopted as shown on the two following lines.

Full symbols	11	22	33	23, 32	31, 13	12, 21
Reduced	1	2	3	4	5	6

The scheme of piezoelectric moduli for a diad axis parallel to X_3 can thus be written

$$\begin{matrix} 0 & 0 & 0 & d_{14} & d_{15} & 0 \\ 0 & 0 & 0 & d_{24} & d_{25} & 0 \\ d_{31} & d_{32} & d_{33} & 0 & 0 & d_{36}. \end{matrix}$$

Thus

$$d_{311} = d_{31} \text{ but } d_{24} = d_{223}+d_{232}.$$

d_{223} and d_{232} can never be separated from one another, and so it is usual to write $d_{24} = 2d_{223}$.

Orthorhombic system—class 222

The symmetry here consists of three mutually perpendicular diad axes. As we have seen in the monoclinic system, a diad axis parallel to X_3 makes all *d*s zero that have no 3s or two 3s. In this system, by the same argument, a diad axis parallel to X_1 must make all *d*s zero that have no 1s or two 1s. Finally the diad axis parallel to X_2 makes equal to zero all *d*s having no 2s or two 2s. The only *d*s that satisfy these requirements are d_{123}, d_{231}, and d_{312}, so that the scheme for charge P_1 becomes

$$\begin{matrix} 0 & 0 & 0 \\ & 0 & d_{123} \\ & & 0 \end{matrix}$$

Q.5.3. What are the schemes of moduli for P_2 and P_3?

Q.5.4. An orthorhombic crystal may have a diad axis parallel to X_3 and two planes of symmetry perpendicular to X_1 and X_2 respectively. Find the non-zero d's.

Tetragonal system—class 4

The transformation of the axes corresponding to a clockwise rotation of 90° about X_3 is

	X_1	X_2	X_3			
X_1'	0	-1	0	c_{11}	c_{12}	c_{13}
X_2'	1	0	0	c_{21}	c_{22}	c_{23}
X_3'	0	0	1	c_{31}	c_{32}	c_{33}

The only products of *c*s that will be non-zero will be those containing c_{12}, c_{21}, and c_{33}. A second rotation about the axis X_3 through 90° introduces the limitations imposed by a diad axis. There are, therefore, only the eight non-zero *d*s applicable to a diad axis parallel to X_3 to be considered. For example,

$$d'_{113} = c_{12}c_{12}c_{33}d_{223} = d_{223} = d_{113}.$$

Q.5.5. What are the corresponding statements for

$$d'_{123},\ d'_{213},\ \text{and}\ d'_{223}?$$

The following *d*s are related to P_3

$$\begin{aligned} d'_{311} &= c_{33}c_{12}c_{12}d_{322} = d_{322} = d_{311}, \\ d'_{312} &= c_{33}c_{12}c_{21}d_{321} = -d_{321} = -d_{312} = d_{312} = 0, \\ d'_{322} &= c_{33}c_{21}c_{21}d_{311} = d_{311} = d_{322}, \\ d'_{333} &= c_{33}c_{33}c_{33}d_{333} = d_{333}. \end{aligned}$$

Thus the scheme of moduli may be written

$$\begin{array}{ccccccccc} 0 & 0 & d_{113} & 0 & 0 & -d_{123} & d_{311} & 0 & 0 \\ & 0 & d_{123} & & 0 & d_{113} & & d_{311} & 0 \\ & & 0 & & & 0 & & & d_{333}. \end{array}$$

There are thus four values of the piezoelectric moduli.

Cubic system

The same method of reducing the number of piezoelectric moduli in accordance with the symmetry of the crystal may be applied to all the classes of symmetry that lack a centre of symmetry. Only one of them has all its coefficients equal to zero and that is class 432.

Q.5.6. The moduli for class 4 are given above. How are they affected in class 432 when the axes X_1, X_2, X_3 become interchangeable?

Q.5.7. A cubic crystal in class $\bar{4}3m$ has the following scheme of piezoelectric moduli

$$\begin{array}{ccccccccc} 0 & 0 & 0 & 0 & 0 & d_{123} & 0 & d_{123} & 0 \\ & 0 & d_{123} & & 0 & 0 & & 0 & 0 \\ & & 0 & & & 0 & & & 0. \end{array}$$

Plates parallel to (100) and (110) planes are prepared and each plate in turn is compressed in a direction normal to its major faces. Write down the charge developed in each case in terms of d_{123} and the components of stress.

Quartz

Quartz is a very important piezoelectric crystal, and its moduli may be written according to the scheme

$$\begin{array}{ccccccccc}
d_{111} & 0 & 0 & 0 & -d_{111} & -d_{123} & 0 & 0 & 0 \\
 & -d_{111} & d_{123} & & 0 & 0 & 0 & & 0 \\
 & & 0 & & & 0 & & & 0.
\end{array}$$

We shall consider the result of applying various stresses to a quartz crystal. Suppose a rectangular block is cut having sides of length a, b, c parallel to a diad axis called X_1, an axis X_2, and a triad axis called X_3 respectively. The block is uniformly compressed by a load T acting in the direction X_1. We require to find the charges developed on the three pairs of parallel faces. The stress components are

$$t_{11} = T/bc \text{ (all other } t\text{s zero).}$$

The components of the electric moment developed are

$$P_1 = d_{111}t_{11},$$

$$P_2 = d_{211}t_{11},$$

$$P_3 = d_{311}t_{11}.$$

Now d_{211} and d_{311} are both zero and only P_1 is non-zero. The charge developed per unit area of the face is equal to the component P_1, so the total charge Q is given by

$$Q = P_1bc = d_{111}t_{11}bc = d_{111}T.$$

Thus the charge developed is independent of the size of the block of quartz, and depends only on the total load.

We now consider the effect of applying a force T to the faces that are perpendicular to the X_2 axis. In this case there is again only one stress component, namely

$$t_{22} = T/ac.$$

The components of the charge are therefore

$$P_1 = d_{122}t_{22} = -d_{111}t_{22} = -d_{111}\frac{\mathrm{T}}{ac} = \frac{\mathrm{Q}}{bc},$$

where Q is the total charge developed on the bc face.
Thus

$$Q = -d_{111}\frac{b}{a}, T$$

$$P_2 = d_{222}t_{22} = 0 \quad \text{since } d_{222} = 0,$$

$$P_3 = d_{322}t_{22} = 0 \quad \text{since } d_{322} = 0.$$

The charge developed in this mode of compression is b/a times greater than when T is applied along X_1.

We now suppose a block of quartz to be cut with the side of length a parallel to the diad axis and the side of length c making 45° with the triad axis (Fig. 5.1). The compression P is applied

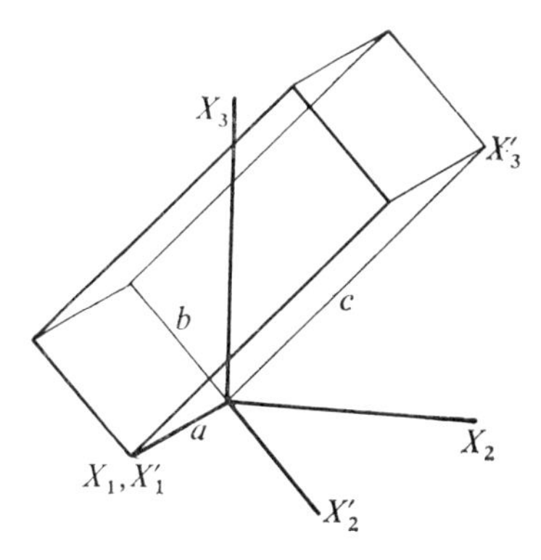

FIG. 5.1. Diagram of a quartz block cut with one edge parallel to the electric axis X_1 and the other two edges inclined at 45° to X_2 and X_3.

normal to the largest faces, which have edges of length c and a. The Cartesian axes are chosen X_1' parallel to the diad axis, X_3' parallel to the edge of length c (Fig. 5.1). There is only one stress component, namely t_{22}', where

$$t_{22}' = T/ac.$$

The direction cosines are given by

	X_1	X_2	X_3			
X_1'	1	0	0	c_{11}	c_{12}	c_{13}
X_2'	0	$\frac{1}{\sqrt{2}}$	$-\frac{1}{\sqrt{2}}$	c_{21}	c_{22}	c_{23}
X_3'	0	$\frac{1}{\sqrt{2}}$	$\frac{1}{\sqrt{2}}$	c_{31}	c_{32}	c_{33}.

Now the stress component t'_{22} can be transformed to the original axes by the equation

$$t_{ik} = c_{2i}c_{2k}t'_{22}.$$

Since of the c_{2i} only c_{22} and c_{23} are non-zero, we have

$$\begin{aligned} t_{22} &= c_{22}c_{22}t'_{22} = \tfrac{1}{2}t'_{22}, \\ t_{23} &= c_{22}c_{23}t'_{22} = -\tfrac{1}{2}t'_{22}, \\ t_{33} &= c_{23}c_{23}t'_{22} = \tfrac{1}{2}t'_{22}, \end{aligned}$$

and all other t_{ik}s are zero. Hence, using the scheme of moduli for quartz and remembering to include t_{32} as well as t_{23} in equation (5.1), we obtain

$$\begin{aligned} P_1 &= -d_{111}t_{22}+2d_{123}t_{23} \\ &= -\tfrac{1}{2}d_{111}t'_{22}-2\times\tfrac{1}{2}d_{123}t'_{22} \\ &= -\tfrac{1}{2}(d_{111}+2d_{123})t'_{22}, \\ P_2 &= 0, \\ P_3 &= 0. \end{aligned}$$

Hence the charge is produced only on the faces bounded by the edges of length b and c.

The total charge liberated is Q where

$$\begin{aligned} Q &= P_1 bc \\ &= -\tfrac{1}{2}(d_{111}+2d_{123})(T/ac)bc \\ &= -\tfrac{1}{2}(d_{111}+2d_{123})Tb/a \end{aligned}$$

(1, p. 159; 13, p. 110; 15, p. 188; 16, p. 72).

Answers

Q.5.1.

$$P_2 \quad \begin{array}{ccc} 0 & 0 & d_{213} \\ & 0 & d_{223} \\ & & 0 \end{array} \qquad\qquad P_3 \quad \begin{array}{ccc} d_{311} & d_{312} & 0 \\ & d_{322} & 0 \\ & & d_{333}. \end{array}$$

Q.5.2. $c_{11} = 1$, $c_{22} = -1$, $c_{33} = 1$, all other cs are zero.
Hence only those ds that contain no 2s or two 2s can be non-zero.
Hence the non-zero ds are

$$\begin{array}{ccc} d_{111} & 0 & d_{113} \\ & d_{122} & 0 \\ & & d_{133} \end{array} \qquad \begin{array}{ccc} 0 & d_{212} & 0 \\ & 0 & d_{223} \\ & & 0 \end{array} \qquad \begin{array}{ccc} d_{311} & 0 & d_{313} \\ & d_{322} & 0 \\ & & d_{333}. \end{array}$$

Q.5.3.

$$P_2 \quad \begin{array}{ccc} 0 & 0 & d_{213} \\ & 0 & 0 \\ & & 0 \end{array} \qquad\qquad P_3 \quad \begin{array}{ccc} 0 & d_{312} & 0 \\ & 0 & 0 \\ & & 0. \end{array}$$

Q.5.4. A diad axis parallel to X_3 requires that *d's containing no* 3*s or two* 3*s shall be zero*. This gives a scheme

$$\begin{array}{ccc} 0 & 0 & d_{113} \\ & 0 & d_{123} \\ & & 0 \end{array} \qquad \begin{array}{ccc} 0 & 0 & d_{213} \\ & 0 & d_{223} \\ & & 0 \end{array} \qquad \begin{array}{ccc} d_{311} & d_{312} & 0 \\ & d_{322} & 0 \\ & & d_{333}. \end{array}$$

A plane perpendicular to X_1 has the transformation

$$\begin{array}{ccc} -1 & 0 & 0 \\ 0 & 1 & 0 \\ 0 & 0 & 1 \end{array} \qquad\qquad \begin{array}{ccc} c_{11} & c_{12} & c_{13} \\ c_{21} & c_{22} & c_{23} \\ c_{31} & c_{32} & c_{33}. \end{array}$$

Those d's that contain one 1 *or three* 1*s will therefore be zero.*

This makes d_{123}, d_{213}, d_{312} zero. A plane perpendicular to X_2 has the transformation

$$\begin{array}{ccc} 1 & 0 & 0 \\ 0 & -1 & 0 \\ 0 & 0 & 1 \end{array} \qquad\qquad \begin{array}{ccc} c_{11} & c_{12} & c_{13} \\ c_{21} & c_{22} & c_{23} \\ c_{31} & c_{32} & c_{33}. \end{array}$$

Those d's that contain one 2 *or three* 2*s will therefore be zero.*

This leaves the following *d*s.

$$\begin{matrix} 0 & 0 & d_{113} \\ & 0 & 0 \\ & & 0 \end{matrix} \qquad \begin{matrix} 0 & 0 & 0 \\ & 0 & d_{223} \\ & & 0 \end{matrix} \qquad \begin{matrix} d_{311} & 0 & 0 \\ & d_{322} & 0 \\ & & d_{333}. \end{matrix}$$

Q.5.5. $d'_{123} = c_{12}c_{21}c_{33}d_{213} = -d_{213} = d_{123}$ (Neumann's Principle),

$d'_{213} = c_{21}c_{12}c_{33}d_{123} = -d_{123} = d_{213}$,

$d'_{223} = c_{21}c_{21}c_{33}d_{113} = d_{113} = d_{223}$.

Q.5.6. If X_1 is interchanged with X_2, X_2 with X_3, and X_3 with X_1, then

$d_{113} = d_{221} = d_{212} = 0$ (from table of moduli for class 4),

$d_{123} = d_{231} = d_{213} = -d_{123} = 0$ (from Neumann's Principle),

$d_{311} = d_{122} = 0$ (from table of moduli for class 4),

$d_{333} = d_{111} = 0$.

Thus all *d*s in the point group 432 are zero.

Q.5.7. For a (100) plate, the only stress component is t_{11}.
Hence $P_1 = d_{111}t_{11} = 0$ (from table of moduli),

$P_2 = d_{211}t_{11} = 0$,

$P_3 = d_{311}t_{11} = 0$.

Hence no charge is developed.

For a (110) plate, we take X'_1 in the direction of the normal to the plate, X'_2 in the plane of the plate, and X'_3 parallel to a tetrad axis (also in the plane of the plate).

Then the only stress component is t'_{11}. Now transform to the axes of the cubic crystal

	X_1	X_2	X_3			
X'_1	$\frac{1}{\sqrt{2}}$	$\frac{1}{\sqrt{2}}$	0	c_{11}	c_{12}	c_{13}
X'_2	$-\frac{1}{\sqrt{2}}$	$\frac{1}{\sqrt{2}}$	0	c_{21}	c_{22}	c_{23}
X'_3	0	0	1	c_{31}	c_{32}	c_{33}.

$t_{ik} = c_{1i}c_{1k}t'_{11}$.

When $k = 3$ the components $t_{ik} = 0$ (since $c_{13} = 0$). Thus the non-zero stress components are

$$t_{11} = c_{11}\,c_{11}\,t'_{11} = \tfrac{1}{2}\,t'_{11},$$

$$t_{12} = c_{11}\,c_{12}\,t'_{11} = \tfrac{1}{2}\,t'_{11},$$

$$t_{22} = c_{12}\,c_{12}\,t'_{11} = \tfrac{1}{2}\,t'_{11}.$$

Inserting these values into the expression for P_i we have $P_1 = 0$,

$P_2 = 0$ (since only $d_{213} = d_{123}$ is non-zero),

$P_3 = 2d_{312}t_{12} = 2d_{123} \times \tfrac{1}{2}t'_{11}$ (t_{21} and t_{12} must be considered)

$\quad = d_{123}t'_{11}$.

6

Piezoelectricity (2)

6.1. Representation surfaces

It is a help in understanding the piezoelectric properties of crystals to construct models representing various relations between the applied force and the electric moment produced. One of these models is defined in the following way. We imagine a parallel-sided plate cut from a crystal. This is compressed and the charge developed is measured. The charge per unit area per unit pressure is obtained for that plate. More plates are imagined cut from the crystal so that a complete survey of the variation of this ratio of charge to the compressing force is obtained for all directions. Finally a solid figure is constructed by marking off lines from a given point as origin. Each line is drawn in the direction of the normal to the plate and its length is proportional to the corresponding ratio of charge to compressive force. Such a solid figure is known as the longitudinal piezoelectric surface.

We may illustrate this by considering quartz. The normal to the compressed plate is regarded as the X'_3 axis and has direction cosines c_{31}, c_{32}, c_{33} defined relative to X_1 (diad axis), X_2 perpendicular to X_1 and X_3, and X_3 (triad axis). If the electric moment of the plate is denoted P'_3 then

$$P'_3 = P_1c_{31}+P_2c_{32}+P_3c_{33}. \qquad (6.1)$$

The scheme of moduli for quartz is as follows.

	P_1			P_2			P_3	
d_{111}	0	0	0	$-d_{111}$	$-d_{123}$	0	0	0
	$-d_{111}$	d_{123}		0	0		0	0
		0			0			0

Hence

$$\left.\begin{aligned} P_1 &= d_{111}t_{11} - d_{111}t_{22} + 2d_{123}t_{23} \\ P_2 &= -2d_{111}t_{12} - 2d_{123}t_{13} \\ P_3 &= 0 \end{aligned}\right\}. \tag{6.2}$$

We now have to find the t_{ik}s.

Since pressure is applied perpendicular to the plane of the plate,

$$t'_{33} = -\pi \text{ (all other } t'_{ik}\text{s} = 0),$$

where π is the pressure (per unit area) applied to the plate. Hence (see equation (2.4))

$$t_{ik} = c_{3i}c_{3k}t'_{33},$$

$$t_{11} = -c_{31}^2\pi, \qquad t_{22} = -c_{32}^2\pi,$$

$$t_{12} = -c_{31}c_{32}\pi, \qquad t_{23} = -c_{32}c_{33}\pi,$$

$$t_{13} = -c_{31}c_{33}\pi, \qquad t_{33} = -c_{33}^2\pi.$$

Inserting these values into equation (6.2), we have

$$\left.\begin{aligned} -P_1 &= \pi(d_{111}c_{31}^2 - d_{111}c_{32}^2 + 2d_{123}c_{32}c_{33}) \\ -P_2 &= \pi(-2d_{111}c_{31}c_{32} - 2d_{123}c_{31}c_{33}) \\ -P_3 &= 0 \end{aligned}\right\}. \tag{6.3}$$

If we now insert these values for P_i into equation (6.1) we obtain

$$\begin{aligned} P'_3 &= -\pi(d_{111}c_{31}^2 - d_{111}c_{32}^2 + 2d_{123}c_{32}c_{33})c_{31} - \\ &\qquad -\pi(-2d_{111}c_{31}c_{32} - 2d_{123}c_{31}c_{33})c_{32} \\ &= -\pi d_{111}(c_{31}^2 - 3c_{32}^2)c_{31}. \end{aligned} \tag{6.4}$$

To determine the solid figure by equation (6.4), consider the sections containing the axes X_1, X_2 and X_1, X_3 respectively. In the former section let

$$c_{31} = \cos\phi, \quad c_{32} = \sin\phi, \quad c_{33} = 0.$$

Now,

$$(\cos^2\phi - 3\sin^2\phi)\cos\phi = \cos 3\phi.$$

Hence

$$P'_3 = -\pi d_{111}\cos 3\phi.$$

The form of this section is shown in Fig. 6.1.

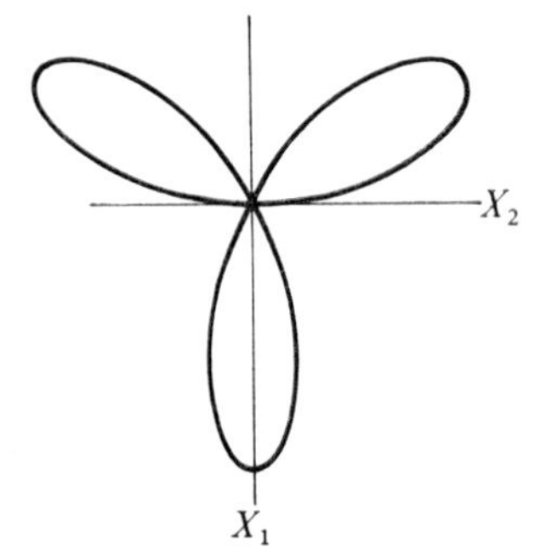

FIG. 6.1. A section of the surface representing the longitudinal piezoelectric effect in quartz. The direction X_1 is that of a diad axis, also called an electric axis. The direction X_2 is normal both to X_1 and to the triad axis.

In the second section let the angle $\widehat{X_3' X_3} = \rho$; then

$$c_{31} = \sin \rho, \qquad c_{32} = 0, \qquad c_{33} = \cos \rho.$$

Then

$$P_3' = -\pi d_{111} \sin^3 \rho,$$

and the form of the section is shown in Fig. 6.2. Thus the whole

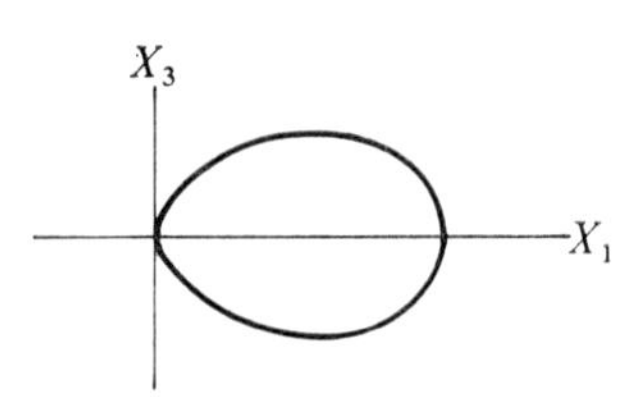

FIG. 6.2. A section containing the diad, X_1, and triad, X_3, axes of the longitudinal piezoelectric surface of quartz.

solid figure consists of three lobes, roughly oval in shape, arranged about the triad axis so that their diad axes coincide with the diad axes of the crystal.

The same problem may be tackled in another way. Instead of using undashed quantities on the right-hand side of equation (5.1) it is possible to use dashed quantities. However, this requires the evaluation of the dashed piezoelectric modulus d'_{333} (equation 5.2) for every direction considered. In the case of quartz we have

$$d'_{333} = c_{31}c_{31}c_{31}d_{111}+c_{31}c_{32}c_{32}d_{122}+2c_{31}c_{32}c_{33}d_{123}+$$
$$+2c_{32}c_{31}c_{32}d_{212}+2c_{32}c_{31}c_{33}d_{213}.$$

The 2s must be inserted because $d_{123} = d_{132}$ and both are required in the expression for d'_{333}. From the scheme of moduli for quartz given above we have

$$d_{122} = -d_{111},$$
$$d_{212} = -d_{111},$$
$$d_{213} = -d_{123}.$$

Inserting these values we obtain

$$\begin{aligned} d'_{333} &= c_{31}^3 d_{111}-c_{31}c_{32}^2 d_{111}+2c_{31}c_{32}c_{33}d_{123}- \\ &\qquad -2c_{31}c_{32}^2 d_{111}-2c_{31}c_{32}c_{33}d_{123} \\ &= d_{111}(c_{31}^2-3c_{32}^2)c_{31}. \end{aligned}$$

It will be seen that this is the same expression as equation (6.4) and leads to the same longitudinal piezoelectric surface, since

$$d'_{333} = P'_3/t'_{33}$$

and all dashed ts other than t'_{33} are zero.

Either of these two methods may be used; sometimes one is more convenient than the other but this depends on the particular problem (15, p. 214).

Q.6.1. The scheme of piezoelectric moduli for zinc blende ZnS is as follows.

$$\begin{array}{ccccccccc} 0 & 0 & 0 & 0 & 0 & d_{123} & 0 & d_{123} & 0 \\ & 0 & d_{123} & & 0 & 0 & & 0 & 0 \\ & & 0 & & & 0 & & & 0 \end{array}$$

Find the general expression for P'_3, the component of the electric moment developed normal to an arbitrarily orientated plate.

Consider a section of the longitudinal piezoelectric surface containing the directions [001] and [111]. Make the angle $\widehat{X_3 X_3} = \theta$ and use the general expression for P'_3 to find the form of this section of the surface.

6.2. The converse piezoelectric effect

When an electric field is applied to a piezoelectric crystal changes of shape are, in general, produced. The relation between the components of strain, r_{ik}, and the components of the electric field strength (a vector **E**) is as follows

$$r_{ik} = d_{lik}E_l, \tag{6.5}$$

where the d_{lik} are the same piezoelectric moduli that we had in connection with the direct effect. This equality of the coefficients in the two effects is a thermodynamical consequence.

The actual changes in size produced by reasonable voltages are very small. For instance, we may calculate the change in thickness of a plate of quartz cut perpendicular to the diad axis, denoted X_1. If a potential V is applied across a plate of thickness S then the electric field strength E is given by

$$E = \frac{V}{S}.$$

In this example $E_1 = E$ and $E_2 = E_3 = 0$. Applying equation (6.5) we have

$$r_{11} = d_{111}E_1 = d_{111}\frac{V}{S}.$$

The value of d_{111} is $6{\cdot}8 \times 10^{-8}$ c.g.s. e.s.u. V may be put equal to 10 000 V and $S = 0{\cdot}1$ cm. The value of r_{11} is then given by

$$r_{11} = 6{\cdot}8 \times 10^{-8} \times \frac{10^4}{300 \times 0{\cdot}1} \quad (300 \text{ V} = 1 \text{ e.s.u. of potential})$$

$$= 2{\cdot}27 \times 10^{-5}.$$

The actual change of thickness is thus $r_{11} \times 0{\cdot}1$, i.e. $0{\cdot}217 \times 10^{-5}$ cm. The corresponding change for a Rochelle salt crystal would be several thousand times greater, i.e. the change of thickness would be several tens of μm.

6.3. The change of shape of a plate having an arbitrary orientation due to the converse effect

We shall consider a rectangular parallelepiped having edges parallel to Cartesian axes X_1', X_2', X_3' of length *a*, *b*, *c* respectively. The length *c* is supposed small in comparison with *a* and *b*, so that we have a thin plate. The major faces are coated with a conducting film, and a potential difference V is applied between them. The electric field strength is V/c and this is denoted E'_3. We may also write

$$E'_1 = E'_2 = 0.$$

We shall study the conversion of the rectangular plate into a parallelogram by determining the component of strain r'_{12}. From equation (6.5) we have

$$r'_{12} = d'_{312}E'_3.$$

The value of d'_{312} can be determined as described in Chapter 5. We may write

$$d_{312} = c_{3i}c_{1k}c_{2l}d_{ikl}. \tag{6.6}$$

In equation (6.6), all non-zero values of d_{ikl} must be inserted. Finally, since r'_{12} represents half the change of angle between the X'_1 and X'_2 axes, the angle between X'_1 and X'_2 after application of the electric field will be $2r'_{12}$.

Q.6.2. Rochelle salt belongs to point group 222, for which the scheme of piezoelectric moduli is

$$\begin{array}{ccccccccc}
0 & 0 & 0 & 0 & 0 & d_{213} & 0 & d_{312} & 0 \\
 & 0 & d_{123} & & 0 & 0 & & 0 & 0 \\
 & & 0 & & & 0 & & & 0
\end{array}$$

The value of d_{123} is $10\,000 \times 10^{-8}$ c.g.s. e.s.u., and for our present purpose we can neglect the small values of d_{213} and d_{312}. A rectangular plate is cut in three orientations I, II, III defined by the direction cosines of the edges a, b, c of the plate, given below.

	Edge $a = X'_1$			Edge $b = X'_2$			Edge $c = X_3$		
	X_1	X_2	X_3	X_1	X_2	X_3	X_1	X_2	X_3
I	0	1	0	0	0	1	1	0	0
II	$\frac{1}{\sqrt{2}}$	$\frac{1}{\sqrt{2}}$	0	0	0	1	$\frac{1}{\sqrt{2}}$	$-\frac{1}{\sqrt{2}}$	0
III	$-\frac{1}{\sqrt{2}}$	$\frac{1}{\sqrt{2}}$	0	$-\sqrt{\frac{1}{6}}$	$-\sqrt{\frac{1}{6}}$	$\sqrt{\frac{2}{3}}$	$\frac{1}{\sqrt{3}}$	$\frac{1}{\sqrt{3}}$	$\frac{1}{\sqrt{3}}$

Find the change of angle between the edges a, b when an electric field of 3000 V cm^{-1} is applied in the c direction.

The side a is 1 cm long. Find the change in length of that side in the cases I, II, III (13, p. 115).

Answers

Q.6.1. The transformation of the piezoelectric moduli is obtained from the equation (5.2), namely,

$$d'_{ikl} = c_{im}c_{kn}c_{lo}d_{mno}.$$

The non-zero piezoelectric moduli are given in the scheme presented in the question. There is only one modulus, namely d_{123}, but the scheme shows that

$$d_{123} = d_{213} = d_{312}.$$

In an arbitrary direction X'_3 defined by the direction cosines, the transformation equation for d'_{ikl} gives

$$d'_{333} = 2c_{31}c_{32}c_{33}d_{123}+2c_{32}c_{31}c_{33}d_{213}+2c_{33}c_{31}c_{32}d_{312}$$

(the factors 2 must be inserted because $d_{123} = d_{132}$), so that

$$d'_{333} = 6c_{31}c_{32}c_{33}d_{123}.$$

If a plate is cut normal to X'_3 the charge density developed, P'_3, is proportional to d'_{ikl} so that the general expression required is

$$6c_{31}c_{32}c_{33}d_{123}.$$

In the section by the plane containing [001] and [111], $c_{31} = c_{32}$.

Further, since $c_{31}^2+c_{32}^2+c_{33}^2 = 1$,

$$2c_{31}^2 = 1-c_{33}^2.$$

If the angle between the direction of the applied pressure and the X_3 axis is θ, then

$$c_{33} = \cos\theta, \qquad c_{31} = \frac{1}{\sqrt{2}}\sin\theta.$$

The equation to the curve of this section of the longitudinal piezoelectric surface is thus

$$3\sin^2\theta\cos\theta$$

(see Fig. 6.3).

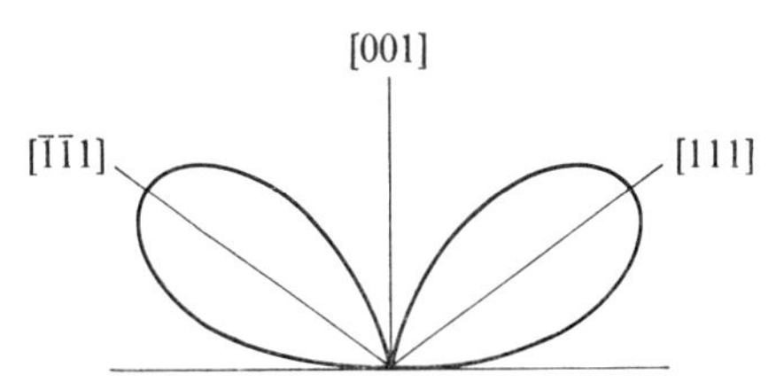

FIG. 6.3. Diagram representing a section of the longitudinal piezoelectric surface for zinc blende, ZnS (Q.6.1).

Q.6.2. $r'_{12} = d'_{312}E'_3.$

$$d'_{312} = c_{31}c_{12}c_{23}d_{123}+c_{31}c_{13}c_{22}d_{132}.$$

$$\text{I} \quad = (1\times1\times1\times10^4+1\times0\times0\times10^4)\times10^{-8}$$

$$= 10^4\times10^{-8}.$$

$$\text{II} \quad = \left(\frac{1}{\sqrt{2}}\times\frac{1}{\sqrt{2}}\times1\times10^4+\frac{1}{\sqrt{2}}\times0\times0\times10^4\right)\times10^{-8}$$

$$= \tfrac{1}{2}\times10^4\times10^{-8}.$$

$$\text{III} \quad = \left(\frac{1}{\sqrt{3}}\times\frac{1}{\sqrt{2}}\times\sqrt{\frac{2}{3}}\times10^4+\frac{1}{\sqrt{3}}\times0\times\left(-\frac{1}{\sqrt{6}}\right)\times10^4\right)\times10^{-8}$$

$$= \tfrac{1}{3}\times10^4\times10^{-8}.$$

$E'_3 = 10$ e.s.u. cm^{-1} in all three cases.

$2r'_{12} = 2\times10^{-4}\times10 = 2\times10^{-3}$ rad in case I,

$= 2\times0{\cdot}5\times10^{-3}$ rad in case II,

$= 2\times0{\cdot}33\times10^{-3}$ rad in case III.

The change in length of side a is given by ar'_{11}.

$$r'_{11} = d'_{311}E'_3.$$

$$d'_{311} = 2c_{31}c_{12}c_{13}d_{123}+2c_{32}c_{11}c_{13}d_{213}+2c_{33}c_{11}c_{12}d_{312}.$$

Case I $= 0 + 0 + 0 = 0$,

Case II $= 0 + 0 + 0 = 0$,

Case III $= 0 + 0 + 2\dfrac{1}{\sqrt{3}}\times-\dfrac{1}{\sqrt{2}}\times\dfrac{1}{\sqrt{2}}\,d_{312}$

$$= -\frac{1}{\sqrt{3}}\,d_{312}.$$

The change in length $= -\dfrac{a}{\sqrt{3}}\times d_{312}\times10.$

7

Elastic properties of crystals

7.1. Introduction

THE study of the elastic properties of crystals is concerned with the relation between the forces applied to the crystal and the consequent changes in shape. Only those small deformations from which the crystal recovers completely on removing the stress are considered here. Further, only those changes in length or angle between lines drawn in the crystal that are proportional to the applied stress come into consideration. In short, we are studying only homogeneous deformation. The elastic properties of crystals are generally anisotropic. Even in most cubic crystals some elastic properties are strongly anisotropic. The majority of substances in common use are polycrystalline and more nearly isotropic than cubic crystals, but even among ordinary metals or wood the preferential orientation of certain crystalline directions often makes them anisotropic. In this chapter we shall deal only with static deformation of crystals. In the next chapter we shall consider the properties of crystals under alternating stresses, such as those caused by the passage of elastic waves through the crystals.

Crystal elasticity is a subject about which there is still some controversy. This arises because some of the assumptions made are not obviously entirely justified. However, the theory developed here has been applied over a long time, and no universally accepted departures from the predictions of the theory have been established. In spite of this, some caution should be exercised in applying the theory because it may be found in the future that there are some small exceptions to its general conclusions.

7.2. Elastic constants

Under homogeneous deformation, there is a linear relation between each component of the stress tensor and each component of the strain tensor. This may be expressed by the equations

$$t_{ik} = c_{ikmn}r_{mn}$$

$$r_{ik} = s_{ikmn}t_{mn}, \tag{7.1}$$

where the ts and rs are components of the stress and strain tensors respectively, and the cs and ss are constants defining the elastic properties. In older books the cs are called constants, but in modern literature they are usually called stiffnesses. The ss used to be called elastic moduli but the more modern name is 'compliances'. The names 'stiffness' and 'compliance' refer to the fact that for a crystal that requires a large stress to produce a small strain c is large and s is small. (It is helpful to notice the alphabetical order of r, s, and t.) When the expressions (7.1) are written out in full we obtain, for example,

$$\begin{aligned} t_{11} &= c_{1111}r_{11}+c_{1122}r_{22}+c_{1133}r_{33}+ \\ &\quad +2c_{1112}r_{12}+2c_{1123}r_{23}+2c_{1113}r_{13} \\ t_{12} &= c_{1211}r_{11}+c_{1222}r_{22}+c_{1233}r_{33}+ \\ &\quad +2c_{1212}r_{12}+2c_{1223}r_{23}+2c_{1213}r_{13}. \end{aligned} \tag{7.2}$$

The reason for the insertion of the factor 2 before terms involving rs with unequal suffixes is that

$$r_{ik} = r_{ki},$$

as we have seen in Chapter 4. In the writing out of the full expression in (7.2), the terms in both r_{12} and r_{21} must be included, and the same applies to r_{23} and r_{13}. Similar expansions can be given for r_{ik} in terms of s_{ikmn} and t_{mn}, and the factor 2 appears in the same way because, in equilibrium,

$$t_{ik} = t_{ki}.$$

7.3. Abbreviations and the interchangeability of suffixes

For many purposes, especially practical ones, it is better to replace the two suffixes i, k or m, n by a single letter or digit, according to the scheme already given in Chapter 4, which is for convenience repeated here, namely

ik	11	22	33	23, 32	31, 13	12, 21
i	1	2	3	4	5	6 .

With these substitutions, two of the equations (7.2) become

$$t_1 = c_{11}r_1+c_{12}r_2+c_{13}r_3+2c_{16}r_6+2c_{14}r_4+2c_{15}r_5$$

$$t_6 = c_{61}r_1+c_{62}r_2+c_{63}r_3+2c_{66}r_6+2c_{64}r_4+2c_{65}r_5. \qquad (7.3)$$

Similarly we may write

$$r_1 = s_{11}t_1+s_{12}t_2+s_{13}t_3+2s_{16}t_6+2s_{14}t_4+2s_{15}t_5$$

$$r_6 = s_{61}t_1+s_{62}t_2+s_{63}t_3+2s_{66}t_6+2s_{64}t_4+2s_{65}t_5. \qquad (7.4)$$

The expressions (7.3) and (7.4) look very similar but there is an important difference between them. Whereas the total change of angle between the axes X_1, X_2 is $r_{12}+r_{21} = 2r_6$ (see Chapter 4), the corresponding component of shear stress is t_6. It is, of course, true that $t_{12} = t_{21} = t_6$, but no factor 2 is required in this case. It is this difference between the components of strain and of stress that is responsible for the introduction of 2s and 4s at various stages in what follows.

There is a further relation that must be considered, affecting the order of the suffixes of c_{ik} and s_{ik} in equation (7.4). The potential energy due to a mechanical deformation can be denoted Φ. This is determined by the forces applied and the deformation they produce. We have seen in Chapter 4 that r_1 (using the abbreviated symbols) is the change in length per unit length in the direction of the X_1 axis. Also t_1 is the force per unit area acting on the crystal in the X_1 direction that has contributed to this deformation. We suppose the crystal to be in equilibrium under the forces applied to it, and in this condition to possess an amount of mechanical potential energy Φ. The crystal is supposed to be further deformed by an amount $\mathrm{d}r_1$ in the X_1 direction. The change of potential energy is equal to the work done or released, according to the sign of $\mathrm{d}r_1$, i.e.

$$\mathrm{d}\Phi = t_1\mathrm{d}r_1.$$

Similar expressions apply for each of the components of the strain tensor, so that the total change of potential energy is given by the expression

$$\mathrm{d}\Phi = t_1\mathrm{d}r_1+t_2\mathrm{d}r_2+t_3\mathrm{d}r_3+2t_4\mathrm{d}r_4+2t_5\mathrm{d}r_5+2t_6\mathrm{d}r_6. \qquad (7.5)$$

The factor 2 occurs in the last three terms because the r_4, r_5, r_6 correspond to only half the total change of angle caused by the shear stresses t_4, t_5, t_6. All the six components of the strain tensor are independent of one another. (For example, an extension corresponding to r_1 is not necessarily accompanied by a strain r_4. Given certain stress components, r_1 may be accompanied by a strain r_4, but it is not necessary that it should be.) We can therefore state that the total change in potential energy is equal to the sum of the six terms, each of which is the contribution from a particular strain component. Each separate contribution is equal to the rate of change of Φ with the particular r, multiplied by the amount of change in r. This may be written as follows.

$$\mathrm{d}\Phi = \frac{\partial \Phi}{\partial r_1}\mathrm{d}r_1 + \frac{\partial \Phi}{\partial r_2}\mathrm{d}r_2 + \ldots + \frac{\partial \Phi}{\partial r_6}\mathrm{d}r_6. \tag{7.6}$$

Comparing equations (7.5) and (7.6), we see that

$$t_1 = \frac{\partial \Phi}{\partial r_1}, \quad t_2 = \frac{\partial \Phi}{\partial r_2}, \quad \ldots, \quad 2t_6 = \frac{\partial \Phi}{\partial r_6}. \tag{7.7}$$

Differentiation of (7.3) partially with respect to r gives the equations

$$\frac{\partial t_1}{\partial r_1} = c_{11}, \quad \frac{\partial t_1}{\partial r_2} = c_{12}, \quad \frac{\partial t_1}{\partial r_6} = 2c_{16}, \quad \text{etc.}$$

$$\frac{\partial t_2}{\partial r_1} = c_{21}, \quad \frac{\partial t_2}{\partial r_2} = c_{22}, \quad \frac{\partial t_2}{\partial r_6} = 2c_{26}, \quad \text{etc.} \tag{7.8}$$

Combining the values of t_i given by equation (7.7) with (7.8), we obtain

$$c_{12} = \frac{\partial t_1}{\partial r_2} = \frac{\partial^2 \Phi}{\partial r_1 \partial r_2}, \qquad c_{21} = \frac{\partial t_2}{\partial r_1} = \frac{\partial^2 \Phi}{\partial r_2 \partial r_1},$$

$$2c_{16} = \frac{\partial t_1}{\partial r_6} = \frac{\partial^2 \Phi}{\partial r_1 \partial r_6}, \qquad 2c_{61} = 2\frac{\partial t_6}{\partial r_1} = \frac{\partial^2 \Phi}{\partial r_6 \partial r_1}.$$

Thus, in general,

$$c_{ik} = c_{ki}.$$

In a similar manner it may be shown that

$$s_{ik} = s_{ki}.$$

(In Chapter 10 we shall consider a fourth-order tensor in which the pairs of suffixes (when using four suffixes) cannot be interchanged.)

To summarize these conclusions, we may say that in the fourth-order tensors of elasticity, c_{iklm} and s_{iklm}, the first two suffixes may be interchanged with one another. Likewise the last two suffixes may be interchanged. Finally, the pairs may be interchanged without affecting the value of the component.

7.4. Relations between compliances and stiffnesses

Equations (7.3) may be rewritten so as to produce a distribution of c_{ik}s that forms a matrix symmetrical about its diagonal. To achieve this, t_4, t_5, and t_6 must be multiplied by 2, when we obtain

$$\begin{aligned}
t_1 &= c_{11}\, r_1 + c_{12}\, r_2 + c_{13}\, r_3 + 2c_{14} r_4 + 2c_{15} r_5 + 2c_{16} r_6 \\
t_2 &= c_{21}\, r_1 + c_{22}\, r_2 + c_{23}\, r_3 + 2c_{24} r_4 + 2c_{25} r_5 + 2c_{26} r_6 \\
t_3 &= c_{31}\, r_1 + c_{32}\, r_2 + c_{33}\, r_3 + 2c_{34} r_4 + 2c_{35} r_5 + 2c_{36} r_6 \\
2t_4 &= 2c_{41} r_1 + 2c_{42} r_2 + 2c_{43} r_3 + 4c_{44} r_4 + 4c_{45} r_5 + 4c_{46} r_6 \\
2t_5 &= 2c_{51} r_1 + 2c_{52} r_2 + 2c_{53} r_3 + 4c_{54} r_4 + 4c_{55} r_5 + 4c_{56} r_6 \\
2t_6 &= 2c_{61} r_1 + 2c_{62} r_2 + 2c_{63} r_3 + 4c_{64} r_4 + 4c_{65} r_5 + 4c_{66} r_6
\end{aligned} \qquad (7.9)$$

Bearing in mind that $c_{ik} = c_{ki}$, we can see that the matrix of cs is diagonally symmetrical. The six simultaneous equations of (7.9) can be solved to give rs in terms of ts and cs. Thus,

$$r_1 = (\delta_{11(c)} t_1 + \delta_{12(c)} t_2 + \delta_{13(c)} t_3 + \delta_{14(c)} 2t_4 + \delta_{15(c)} 2t_5 + \\ + \delta_{16(c)} 2t_6)/\Delta_c$$

where Δ_c is the determinant formed from the c_{ik}s and their 2s and 4s in expression (7.9), and $\delta_{ik(c)}$ is the minor formed by omission of the ith row and kth column from the matrix. If this is compared with the corresponding expression (7.4),

$$r_1 = s_{11} t_1 + s_{12} t_2 + s_{13} t_3 + 2t_4 s_{14} + 2t_5 s_{15} + 2t_6 s_{16},$$

we see that

$$s_{ik} = \delta_{ik(c)}/\Delta_c. \qquad (7.10)$$

In an exactly analogous manner it can be shown that

$$c_{ik} = \delta_{ik(s)}/\Delta_s, \qquad (7.11)$$

where Δ_s and $\delta_{ik(s)}$ have similar meanings to the corresponding Δ_c and $\delta_{ik(c)}$.

Throughout our analysis we have adhered strictly to the relations such as equation (7.1) and have assumed that the subscripts stand for 1, 2, 3 taken in all possible ways. Now the number of ways of writing four numbers depends on what they are. Thus 1111 can be written in only one way. The same is true of 2222 and 3333. However, 1122 can also be written 2211 and when the numbers are subscripts of *c*s or *s*s the physical values of the constants are the same. There are four ways of writing 1123, namely 1123, 1132, 2311, 3211; and eight ways of writing 1223, namely 1223, 1232, 2123, 2132, 2312, 3212, 2321, 3221. The physical values associated with c_{64} and c_{46}, or s_{64} and s_{46}, to which these eight sets of subscripts correspond, are all equal.

The compliances (s_{ik}) correspond with the strict tensor components when they have two suffixes that are 1, 2, or 3. If one suffix is 4, 5, or 6, a factor 2 has generally been introduced, and if both suffixes are 4, 5, or 6 a factor 4. This observation does not apply to the stiffnesses (c_{ik}). The reason for this difference can be seen by comparison of the expressions (7.3) and (7.4). In (7.3), t_1 is related to $2r_4$, the change of angle between the axes X_2 and X_3, by c_{14}. In (7.4), r_1 is related to the shear component t_4 by $2s_{14}$. Similarly, in (7.3), t_6 is related to $2r_4$ by c_{64} but $2r_6$ is related to t_4 in an expression similar to (7.4) by $4s_{64}$. Tables of s_{ik}s usually give the true value of the tensor when i and k are 1, 2, or 3, twice the true value when either i or k is 4, 5, or 6, and four times the true value when both i and k are 4, 5, or 6. It is very important to bear in mind that only true tensors transform according to the rules we have used, and therefore in dealing with plates or bars of general orientation the true tensors must be employed and not the practical values commonly quoted in physical tables.

7.5. Limitations imposed on the elastic properties by crystal symmetry

Triclinic system

When no symmetry is present, the matrix of *c*s given in expression (7.9) has no zero terms. Addition of a centre of symmetry to this does not make any difference, because the transformation given by

$$c'_{iklm} = \alpha_{ip}\alpha_{kq}\alpha_{lr}\alpha_{ms}c_{pqrs} \tag{7.12}$$

(αs are used for direction cosines to avoid confusion with stiffnesses) must have each α equal to -1 (§1.3). The product of the αs is thus $+1$ and the transformed c is equal to the original c.

Monoclinic system

A diad axis parallel to X_3 requires

$$\alpha_{11} = -1, \qquad \alpha_{22} = -1, \qquad \alpha_{33} = 1,$$

$$\text{and all other} \quad \alpha_{ik}\text{s} = 0\,(\S\,1.5).$$

Thus in the equation (7.12) the product of the αs will be -1 when there is only one 3 or three 3s among *iklm*. The cs having one or three 3s in their suffixes must be zero. Thus the matrix (7.9) becomes

$$\begin{matrix} c_{11} & c_{12} & c_{13} & 0 & 0 & 2c_{16} \\ & c_{22} & c_{23} & 0 & 0 & 2c_{26} \\ & & c_{33} & 0 & 0 & 2c_{36} \\ & & & 4c_{44} & 4c_{45} & 0 \\ & & & & 4c_{55} & 0 \\ & & & & & 4c_{66}. \end{matrix}$$

Q.7.1. If the diad axis is parallel to X_2 what will be the scheme of elastic stiffnesses?

Orthorhombic system

When there are three mutually perpendicular diad axes, a non-zero c_{ik} will be obtained only when there is not one or three α_{11}, α_{22}, α_{33} in the product of the four αs. This results in the following scheme.

$$\begin{matrix} c_{11} & c_{12} & c_{13} & 0 & 0 & 0 \\ & c_{22} & c_{23} & 0 & 0 & 0 \\ & & c_{33} & 0 & 0 & 0 \\ & & & 4c_{44} & 0 & 0 \\ & & & & 4c_{55} & 0 \\ & & & & & 4c_{66}. \end{matrix}$$

Tetragonal system—point group 4

The matrix for the transformation of the axes corresponding to a rotation of 90° is

	X_1	X_2	X_3
X_1'	0	1	0
X_2'	−1	0	0
X_3'	0	0	1

$$\begin{matrix} c_{11} & c_{12} & c_{13} \\ c_{21} & c_{22} & c_{23} \\ c_{31} & c_{32} & c_{33}. \end{matrix}$$

Since two rotations of 90° are equal to one rotation of 180°, the non-zero cs cannot include any that are zero in the monoclinic system (diad axis parallel to X_3). Taking the cs for the monoclinic system in order and applying the above transformation, we obtain

$$c'_{11} = \alpha_{12}\alpha_{12}\alpha_{12}\alpha_{12}c_{2222} = c_{22} = c_{11} \text{ (by Neumann's Principle)},$$

$$c'_{12} = \alpha_{12}\alpha_{12}\alpha_{21}\alpha_{21}c_{2211} = 1\times 1\times -1\times -1\times c_{21} = c_{21} = c_{12},$$

$$c'_{13} = \alpha_{12}\alpha_{12}\alpha_{33}\alpha_{33}c_{2233} = 1\times 1\times 1\times 1\times c_{23} = c_{13},$$

$$c'_{16} = \alpha_{12}\alpha_{12}\alpha_{12}\alpha_{21}c_{2221} = 1\times 1\times 1\times -1\times c_{26} = -c_{26} = c_{16},$$

$$c'_{22} = c_{11},$$

$$c'_{23} = c_{13},$$

$$c'_{26} = -c_{16},$$

$$c'_{33} = \alpha_{33}\alpha_{33}\alpha_{33}\alpha_{33}c_{3333} = c_{33},$$

$$c'_{36} = \alpha_{33}\alpha_{33}\alpha_{12}\alpha_{21}c_{3321} = 1\times 1\times 1\times -1\times c_{36} = -c_{36} = c_{36}$$
$$= 0,$$

$$c'_{44} = \alpha_{21}\alpha_{33}\alpha_{21}\alpha_{33}c_{1313} = -1\times 1\times -1\times 1\times c_{55} = c_{55} = c_{44},$$

$$c'_{45} = \alpha_{21}\alpha_{33}\alpha_{12}\alpha_{33}c_{1323} = -1\times 1\times 1\times 1\times c_{54} = -c_{45} = c_{45}$$
$$= 0,$$

$$c'_{66} = \alpha_{12}\alpha_{21}\alpha_{12}\alpha_{21}c_{2121} = 1\times -1\times 1\times -1\times c_{66} = c_{66}.$$

The scheme of cs is therefore as follows.

$$\begin{matrix} c_{11} & c_{12} & c_{13} & 0 & 0 & 2c_{11} \\ & c_{11} & c_{13} & 0 & 0 & -2c_{11} \\ & & c_{33} & 0 & 0 & 0 \\ & & & 4c_{44} & 0 & 0 \\ & & & & 4c_{44} & 0 \\ & & & & & 4c_{66}. \end{matrix}$$

Q.7.2. What scheme of cs applies to point group $4/m$?

Q.7.3. Which of the cs become zero when a diad axis is present, e.g. in point group 422?

Cubic system

In this system, the axes X_1, X_2, X_3 cannot be distinguished from one another. Thus if we change 1 into 2, 2 into 3, and 3 into 1, no change can result. Applying this to particular cs, we have

$$c_{1122} = c_{2233} = c_{3311} \text{ or } c_{12} = c_{23} = c_{13},$$

$$c_{2323} = c_{3131} = c_{1212} \text{ or } c_{44} = c_{55} = c_{66}.$$

Insertion of this requirement into the orthorhombic scheme gives the following set of cs,

$$\begin{matrix} c_{11} & c_{12} & c_{12} & 0 & 0 & 0 \\ & c_{11} & c_{12} & 0 & 0 & 0 \\ & & c_{11} & 0 & 0 & 0 \\ & & & 4c_{44} & 0 & 0 \\ & & & & 4c_{44} & 0 \\ & & & & & 4c_{44}. \end{matrix}$$

The relation between c's and s's in the cubic system

From equations (7.10) and (7.11) we can work out the ss in terms of cs or the cs in terms of the ss. Thus, missing out the first row and the first column of the matrix of cs for the cubic system, we have

$$s_{11}\Delta_c = \begin{vmatrix} c_{11} & c_{12} & 0 & 0 & 0 \\ c_{12} & c_{11} & 0 & 0 & 0 \\ 0 & 0 & 4c_{44} & 0 & 0 \\ 0 & 0 & 0 & 4c_{44} & 0 \\ 0 & 0 & 0 & 0 & 4c_{44} \end{vmatrix},$$

where Δ_c is the determinant derived from the whole c matrix. This reduces to

$$s_{11} = \frac{\begin{vmatrix} c_{11} & c_{12} \\ c_{12} & c_{11} \end{vmatrix}}{\begin{vmatrix} c_{11} & c_{12} & c_{12} \\ c_{12} & c_{11} & c_{12} \\ c_{12} & c_{12} & c_{11} \end{vmatrix}} = \frac{(c_{11}+c_{12})}{(c_{11}-c_{12})(c_{11}+2c_{12})}.$$

Proceeding in the same way, omitting the first row and the second column from the cubic c scheme, we have

$$s_{12}\Delta_c = -\begin{vmatrix} c_{12} & c_{12} & 0 & 0 & 0 \\ c_{12} & c_{11} & 0 & 0 & 0 \\ 0 & 0 & 4c_{44} & 0 & 0 \\ 0 & 0 & 0 & 4c_{44} & 0 \\ 0 & 0 & 0 & 0 & 4c_{44} \end{vmatrix}$$

$$s_{12} = \frac{-(c_{11}c_{12}-c_{12}^2)}{(c_{11}-c_{12})^2(c_{11}+2c_{12})} = \frac{-c_{12}}{(c_{11}-c_{12})(c_{11}+2c_{12})}.$$

Finally, omitting the fourth row and the fourth column, we have

$$s_{44} = \frac{(4c_{44})^2}{(4c_{44})^3} = \frac{1}{4c_{44}}.$$

It must be remembered that this s_{44} is a true tensor and that the practical s_{44} is four times greater.

7.6. Representation surfaces

One representation surface can be constructed to represent the change in length per unit length in any arbitrary direction due to a stretching force in the same direction. If we take the arbitrary direction as X_3' then we have r_{33}' as the change in length per unit length and t_{33}' as the only stress component.

Thus we may write

$$r_{33}' = s_{3333}' t_{33}'.$$

A surface, called a longitudinal elastic surface, can be constructed having each radius vector described along the direction of stretching and the length proportional to s_{33}'. We shall work out for a cubic crystal the type of surface obtained. We know that the moduli correspond to the scheme

$$\begin{matrix} s_{11} & s_{12} & s_{12} & 0 & 0 & 0 \\ & s_{11} & s_{12} & 0 & 0 & 0 \\ & & s_{11} & 0 & 0 & 0 \\ & & & 4s_{44} & 0 & 0 \\ & & & & 4s_{44} & 0 \\ & & & & & 4s_{44}. \end{matrix}$$

Hence

$$s'_{33} = s'_{3333} = \alpha_{3i}\alpha_{3k}\alpha_{3m}\alpha_{3n}s_{ikmn}. \tag{7.12}$$

Taking X'_3 as lying in the plane (100), and making an angle θ with X_3, we get

$$\alpha_{31} = 0, \qquad \alpha_{32} = \sin\theta, \qquad \alpha_{33} = \cos\theta.$$

Inserting these values in equation (7.12) and using only non-zero ss in the cubic scheme, we obtain,

$$\begin{aligned} s'_{3333} &= \alpha_{32}\alpha_{32}\alpha_{32}\alpha_{32}s_{2222}+\alpha_{33}\alpha_{33}\alpha_{33}\alpha_{33}s_{3333}+ \\ &\quad +\alpha_{32}\alpha_{32}\alpha_{33}\alpha_{33}s_{2233}+\alpha_{33}\alpha_{33}\alpha_{32}\alpha_{32}s_{3322}+ \\ &\quad +\alpha_{32}\alpha_{33}\alpha_{32}\alpha_{33}s_{2323}+\alpha_{32}\alpha_{33}\alpha_{33}\alpha_{32}s_{2332}+ \\ &\quad +\alpha_{33}\alpha_{32}\alpha_{32}\alpha_{33}s_{3223}+\alpha_{33}\alpha_{32}\alpha_{33}\alpha_{32}s_{3232}. \\ &= (\alpha_{32}^4+\alpha_{33}^4)s_{1111}+2\alpha_{32}^2\alpha_{33}^2s_{1122}+4\alpha_{32}^2\alpha_{33}^2s_{2323}. \end{aligned}$$

Substituting $\alpha_{32} = \sin\theta$, $\quad \alpha_{33} = \cos\theta$, $\quad$ we obtain

$$(\alpha_{32}^2+\alpha_{33}^2)^2 = 1 = \alpha_{32}^4+\alpha_{33}^4+2\alpha_{32}^2\alpha_{33}^2,$$

$$\begin{aligned} s'_{3333} &= s_{1111}-2\alpha_{32}^2\alpha_{33}^2(s_{1111}-s_{1122}-2s_{2323}) \\ &= s_{1111}-2\sin^2\theta\cos^2\theta(s_{1111}-s_{1122}-2s_{2323}) \\ &= s_{1111}-\tfrac{1}{2}\sin^2 2\theta(s_{1111}-s_{1122}-2s_{2323}). \end{aligned} \tag{7.13}$$

For the case of sodium chloride this leads to a curve of the form shown in Fig. 7.1. Another commonly used representation surface gives s'_{44} in all directions radiating from a centre.

7.7. Examples of particular applications of stress

Simple stretching

We suppose a rod cut in an arbitrary direction X'_3 from a crystal, and only one stress component, namely t'_{33} to be applied. We can then discuss a compliance or modulus of elasticity but not a stiffness or constant of elasticity. In the expressions for the components of stress and strain

$$r'_{33} = s'_{3333}t'_{33}, \tag{7.14}$$

$$t'_{33} = c'_{33ik}r'_{ik}. \tag{7.15}$$

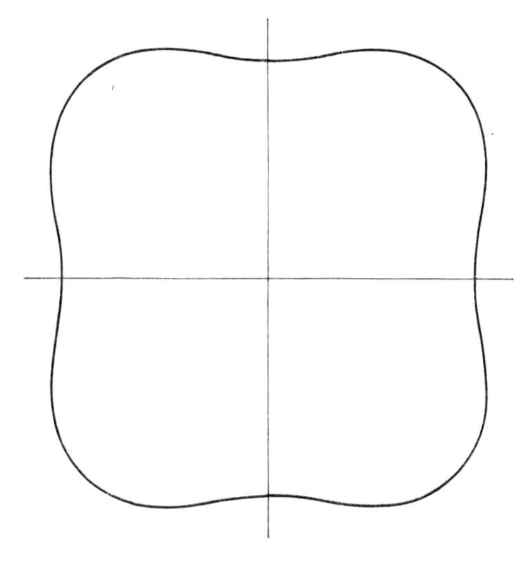

FIG. 7.1. Diagram of a central section, parallel to (100), of the surface representing s'_{33}, for rock salt, NaCl.

We have a single term in expression (7.14) but an unknown number of terms not exceeding six in expression (7.15). We can define the stress as t'_{33} but we do not know, without more investigation, what are all the components of strain called into being by that stress. However, the first equation relates the change in shape we are really concerned with, namely r'_{33}, to the stress component t'_{33} producing it. The value of t'_{33} may be put equal to the tension per unit area of the rod, and r'_{33} is the change of length per unit length. Young's modulus, denoted Y, is by definition t'_{33}/r'_{33} and is therefore equal to $1/s'_{33}$. To illustrate the calculation of the stretching, let us suppose the bar is cut along the [011] direction of a cubic crystal. Then, as we have seen (equation (7.13)),

$$s'_{3333} = s_{1111} - \tfrac{1}{2}\sin^2 2\theta\,(s_{1111} - s_{1122} - 2s_{2323}),$$

and since $2\theta = 90°$

$$s'_{3333} = \tfrac{1}{2}(s_{1111} + s_{1122} + 2s_{2323}) = r'_{33}/t'_{33}.$$

Simple shear

An application of compression along, say, [011] and stretching along a perpendicular direction, say [0$\bar{1}$1], results in a shear tending to change the angle between the axes X_2, X_3. We choose axes $X_1' = X_1$, $X_2' = [011]$, $X_3' = [0\bar{1}1]$, for which the direction-cosine scheme is

	X_1	X_2	X_3			
X_1'	1	0	0	α_{11}	α_{12}	α_{13}
X_2'	0	$\frac{1}{\sqrt{2}}$	$\frac{1}{\sqrt{2}}$	α_{21}	α_{22}	α_{23}
X_3'	0	$\frac{-1}{\sqrt{2}}$	$\frac{1}{\sqrt{2}}$	α_{31}	α_{32}	α_{33}.

The stress components present are t'_{22} and t'_{33}, all other t'_{ik}s being zero. We may write

$$t'_{22} = +\pi, \qquad t'_{33} = -\pi.$$

From this we obtain

$$t_{11} = \alpha_{21}\alpha_{21}t'_{22}+\alpha_{31}\alpha_{31}t'_{33} = 0,$$

$$t_{12} = \alpha_{21}\alpha_{22}t'_{22}+\alpha_{31}\alpha_{32}t'_{33} = 0,$$

$$t_{13} = \alpha_{21}\alpha_{23}t'_{22}+\alpha_{31}\alpha_{33}t'_{33} = 0,$$

$$t_{22} = \alpha_{22}\alpha_{22}t'_{22}+\alpha_{32}\alpha_{32}t'_{33} = (+\tfrac{1}{2}-\tfrac{1}{2})t'_{22} = 0,$$

$$t_{23} = \alpha_{22}\alpha_{23}t'_{22}+\alpha_{32}\alpha_{33}t'_{33} = (+\tfrac{1}{2}+\tfrac{1}{2})t'_{22} = +\pi,$$

$$t_{33} = \alpha_{23}\alpha_{23}t'_{22}+\alpha_{33}\alpha_{33}t'_{33} = (+\tfrac{1}{2}-\tfrac{1}{2})t'_{22} = 0.$$

Thus only the stress component t_{23} is non-zero, and the corresponding strain component is given by

$$r_{23} = s_{2323}t_{23}+s_{2332}t_{32} = 2s_{44}t_{23} = 2s_{44}\pi.$$

Note that what is measured by the change of angle is $2r_{23}$.

Hydrostatic pressure

When the crystal is subjected to hydrostatic pressure,

$$t_{11} = t_{22} = t_{33} = \pi, \qquad t_{12} = t_{23} = t_{31} = 0.$$

The strain components are then

$$\begin{aligned}
r_{11} &= s_{1111}t_{11}+s_{1122}t_{22}+s_{1133}t_{33},\\
r_{12} &= s_{1211}t_{11}+s_{1222}t_{22}+s_{1233}t_{33},\\
r_{13} &= s_{1311}t_{11}+s_{1322}t_{22}+s_{1333}t_{33},\\
r_{22} &= s_{2211}t_{11}+s_{2222}t_{22}+s_{2233}t_{33},\\
r_{23} &= s_{2311}t_{11}+s_{2322}t_{22}+s_{2333}t_{33},\\
r_{33} &= s_{3311}t_{11}+s_{3322}t_{22}+s_{3333}t_{33}.
\end{aligned}$$

In a cubic crystal this leads to

$$\begin{aligned}
r_{11} &= \pi(s_{1111}+2s_{1122}) = r_{22} = r_{33},\\
r_{12} &= 0 = r_{13} = r_{23}
\end{aligned}$$

(1, p. 124; 13, p. 131; 15, p. 231; 16, p. 101).

Q.7.4. An orthorhombic crystal is subjected to hydrostatic pressure. Write down the values of all six components of strain. The scheme of elastic moduli is as follows.

$$\begin{matrix}
s_{11} & s_{12} & s_{13} & 0 & 0 & 0\\
 & s_{22} & s_{23} & 0 & 0 & 0\\
 & & s_{33} & 0 & 0 & 0\\
 & & & 4s_{44} & 0 & 0\\
 & & & & 4s_{55} & 0\\
 & & & & & 4s_{66}.
\end{matrix}$$

Answers

Q.7.1. If one 2 or three 2s are present in the subscripts of a c_{iklm} that constant is zero. Thus the scheme of *c*s is as follows.

$$\begin{matrix}
c_{11} & c_{12} & c_{13} & 0 & 2c_{15} & 0\\
 & c_{22} & c_{23} & 0 & 2c_{25} & 0\\
 & & c_{33} & 0 & 2c_{35} & 0\\
 & & & 4c_{44} & 0 & 4c_{46}\\
 & & & & 4c_{55} & 0\\
 & & & & & 4c_{66}.
\end{matrix}$$

Q.7.2. If a centre of symmetry is added to point group 4 it becomes point group $4/m$. The transformation for a centre of symmetry has

$$\alpha_{11} = \alpha_{22} = \alpha_{33} = -1, \text{and all other } \alpha_{ik}\text{s} = 0.$$

The product of four -1s is $+1$ and hence the scheme of cs is unchanged by addition of a centre of symmetry.

Q.7.3. A diad axis parallel to X_1 requires that if one 1 or three 1s are present the c is zero. The only constant thus made zero is c_{16}.

Q.7.4. $t_{11} = t_{22} = t_{33} = \pi, \qquad t_{12} = t_{23} = t_{31} = 0.$

$$r_{11} = s_{1111}t_{11} + s_{1122}t_{22} + s_{1133}t_{33}$$

$$= \pi(s_{11} + s_{12} + s_{13}),$$

$$r_{12} = 0,$$

$$r_{13} = 0,$$

$$r_{22} = s_{2211}t_{11} + s_{2222}t_{22} + s_{2233}t_{33}$$

$$= \pi(s_{12} + s_{22} + s_{23}),$$

$$r_{23} = r_{31} = 0,$$

$$r_{33} = s_{3311}t_{11} + s_{3322}t_{22} + s_{3333}t_{33}$$

$$= \pi(s_{13} + s_{23} + s_{33}).$$

8

The transmission of elastic waves through crystals

8.1. Introduction

IN Chapter 7 the elastic properties of crystals were considered in relation to stretching, shearing, or hydrostatic compression by steadily applied forces. Many of the early measurements of the constants and moduli were made in this way. Most modern observations are made by sending waves into the crystal and measuring the wavelength associated with a particular frequency of vibration. Electronic devices can measure the velocity with which elastic waves travel along crystal bars, and some of the stiffnesses can be obtained in this way. Rectangular or cylindrical bars have natural frequencies of vibration, and, provided these can be measured and associated with certain of the dimensions of the vibrating bar or plate, one or more elastic stiffnesses can be obtained. Piezoelectric crystals, such as quartz, are very useful in exciting particular modes of vibration in any crystal bar to which they may be attached. When such vibrations pass through a transparent crystal there are usually changes of refractive index caused by the ultrasonic waves. If a light beam is sent through such a vibrating crystal, the spatial alternation of the refractive index gives a diffraction effect from which one or more of the elastic stiffnesses can be obtained. Ultrasonic vibrations in solids and liquids are of great importance in many technical applications for which a thorough understanding of the passage of elastic waves through crystals is necessary.

8.2. The differential equation governing wave propagation

Plane waves having a wave normal in the direction X_2 (Fig. 8.1) are supposed to be travelling through the crystal. Two planes AB, CD are normal to the axis X_2 and are separated by a distance dx_2.

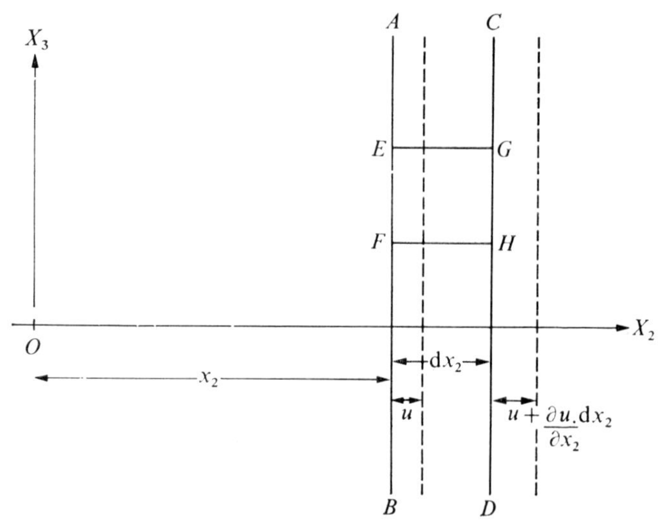

FIG. 8.1. Diagram showing the displacements associated with a plane wave travelling in the direction OX_2.

On plane *AB* there is a unit area *EF* and on plane *CD* a unit area *GH*, such that the lines *EG*, *FH* are parallel to X_2. We suppose the displacement due to the wave motion in the direction OX_2 of the plane *AB* from its rest position to be u. The plane *CD* is then displaced by a distance $u+(\partial u/\partial x_2)\mathrm{d}x_2$. The component of strain, i.e. change of length per unit length, of the element *EFGH* in the direction X_2 is $\partial u/\partial x_2$. The force acting across unit area of this element is the stress component t_{22}, where

$$t_{22} = cr_{22} = c\frac{\partial u}{\partial x_2}$$

(see § 7.2).

If this is taken as the tension or compression across plane *EF* then the corresponding tension or compression across plane *GH* is

$$c\left(\frac{\partial u}{\partial x_2}+\frac{\partial^2 u}{\partial x_2^2}\mathrm{d}x_2\right).$$

Thus the resultant force per unit volume acting on this element *EFGH* is

$$c\,\frac{\partial^2 u}{\mathrm{d}x_2^2}.$$

This force must be equal to the product of the mass, ρ, and the acceleration, $\partial^2 u/\partial t^2$. Thus we have

$$\rho\frac{\partial^2 u}{\partial t^2} = c\frac{\partial^2 u}{\partial x_2^2}. \tag{8.1}$$

This one-dimensional case can be generalized. The displacement u may be written

$$u_m = r_{ml}x_l \tag{8.2}$$

(see § 4.1), where x_l are the coordinates of a point suffering displacements u_m. The components of stress are given by

$$t_{pq} = c_{pqml}r_{ml} \tag{8.3}$$

(see § 7.2). From equation (8.2) we have

$$\frac{\partial u_m}{\partial x_l} = r_{ml}$$

(where m and l now have particular values). Substituting this value of r_{ml} into equation (8.3), we obtain

$$t_{pq} = c_{pqml}\frac{\partial u_m}{\partial x_l}. \tag{8.4}$$

The force acting in a given direction on a volume element is equal, as we have seen above, to the rate of change of the stress component in that direction. For example, if $p = q = 2$,

$\frac{\partial t_{22}}{\partial x_2}\,dx_2 =$ resultant force acting in the direction of the X_2 axis due to forces parallel to the X_2 axis.

If $p = 1, q = 2$ (Fig. 8.2),

$\frac{\partial t_{12}}{\partial x_2}\,dx_2 =$ resultant force acting on the volume element in the X_1 direction due to tangential forces acting on the face normal to the X_2 axis.

Thus, in general,

$\frac{\partial t_{pq}}{\partial x_q}\,dx_q =$ resultant force acting on the volume element of unit area and thickness dx_q in the X_p direction due to the stress component t_{pq}.

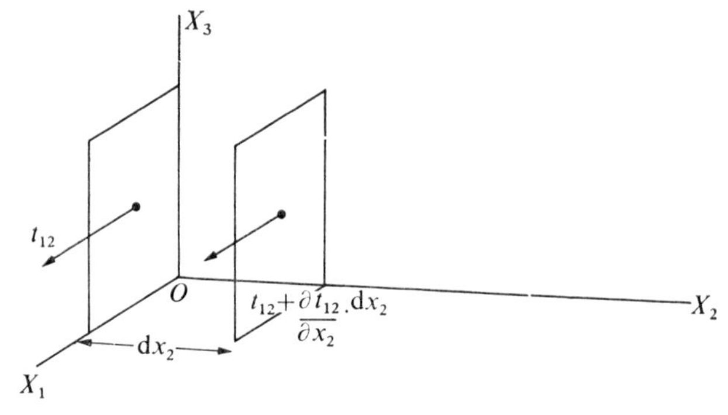

FIG. 8.2. Diagram showing the change of shear stress component t_{12} along the direction OX_2 in which the plane wave is travelling.

Using equations (8.4) and (8.1), we may therefore write

$$\frac{\partial t_{pq}}{\partial x_q} = c_{pqml}\frac{\partial^2 u_m}{\partial x_q \partial x_l} = \frac{\rho \partial^2 u_p}{\partial t^2}. \tag{8.5}$$

This is the differential equation on which all elastic wave propagation in crystals is based.

8.3. Christoffel's determinant

We suppose a simple harmonic wave to be travelling through the crystal. The wave vector is denoted $\mathbf{K}^*_{(i)}$, the frequency ν, and amplitude $\xi_{(i)}$. The subscript (i) has the values 1, 2, and 3 in turn and refers to the three waves having a common direction of travel. The numerical value of $\mathbf{K}^*_{(i)}$, namely $|K^*_{(i)}|$, is equal to $1/\lambda_{(i)}$. An asterisk is used in connection with $\mathbf{K}^*_{(i)}$ since it is normal to certain lattice planes and is therefore best referred to the reciprocal lattice. The reciprocal lattice vectors are usually distinguished from the Bravais lattice vectors by addition of the asterisk. The usual equation for a wave motion† permits us to write the relation between the displacement $\mathbf{u}$ and $\mathbf{K}^*$, ν, and ξ as follows,

† A simple harmonic wave is often described by the equation

$$u = \xi \cos 2\pi\,(x/\lambda - t/\tau).$$

In equation (8.6) the exponential form gives a general expression for such a wave; $\mathbf{K}^*\cdot\mathbf{r}$, being the projection of $\mathbf{r}$ on $\mathbf{K}^*$, takes the place of x/λ, and νt replaces t/τ.

$$u_p = \xi_{(i)p}\exp\{2\pi i(\mathbf{K}^*_{(i)}\cdot\mathbf{r}-\nu t)\} \tag{8.6}$$

where $\mathbf{r}$ is a vector from the origin to the point which has the displacement $\mathbf{u}$.

In this analysis we shall find the velocity with which elastic waves travel in directions radiating from a given origin. We can therefore restrict $\mathbf{r}$ to the same direction as $\mathbf{K}^*$. From equation (8.6)

$$\frac{\partial u_p}{\partial t} = -2\pi i\nu u_p$$

and so

$$\frac{\partial^2 u_p}{\partial t^2} = -4\pi^2\nu^2 u_p. \tag{8.7}$$

In the Bravais lattice we may choose unit vectors $\mathbf{n}$, $\mathbf{o}$, $\mathbf{p}$ parallel to the axes X_1, X_2, X_3. Then the vector $\mathbf{r}$ is given by

$$\mathbf{r} = x_1\mathbf{n}+x_2\mathbf{o}+x_3\mathbf{p},$$

where x_1, x_2, x_3 are the coordinates of $\mathbf{r}$. In the reciprocal lattice there are corresponding unit vectors $\mathbf{n}^*$, $\mathbf{o}^*$, $\mathbf{p}^*$, and the vector $\mathbf{K}^*_{(i)}$ can be expressed by the equation

$$\mathbf{K}^*_{(i)} = (f_1\mathbf{n}^*+f_2\mathbf{o}^*+f_3\mathbf{p}^*)\,|K^*_{(i)}|,$$

where f_1, f_2, and f_3 are the direction cosines of $\mathbf{K}^*_{(i)}$. In equation (8.6) there occurs the scalar product of $\mathbf{K}^*_{(i)}$ and $\mathbf{r}$. This is given by

$$\mathbf{K}^*_{(i)}\cdot\mathbf{r} = (f_1x_1+f_2x_2+f_3x_3)\,|K^*_{(i)}| = f_qx_q\,|K^*_{(i)}|,$$

since, by definition of the reciprocal lattice,

$$\mathbf{n}^*\cdot\mathbf{n} = \mathbf{o}^*\cdot\mathbf{o} = \mathbf{p}^*\cdot\mathbf{p} = 1$$

and

$$\mathbf{n}^*\cdot\mathbf{o} = 0, \text{ etc.}$$

Differentiating $\mathbf{K}^*_{(i)}\cdot\mathbf{r}$ with respect to x_q only, we have

$$\frac{\partial(\mathbf{K}^*_{(i)}\cdot\mathbf{r})}{\partial x_q} = |K^*_{(i)}|f_q.$$

From equation (8.6) it follows that

$$\frac{\partial u_m}{\partial x_q} = 2\pi i\,|K^*_{(i)}|\,f_q u_m,$$

and

$$\frac{\partial^2 u_m}{\partial x_q \partial x_l} = -4\pi^2 \left|K^*_{(i)}\right|^2 f_q f_l u_m. \tag{8.8}$$

Combining equations (8.5), (8.6), (8.7), and (8.8), we obtain

$$\rho v^2 \xi_{(i)p} = c_{pqml} f_q f_l \left|K^*_{(i)}\right|^2 \xi_{(i)m}. \tag{8.9}$$

The quantities $c_{pqml} f_q f_l$ are all determined by the elastic constants and the direction cosines of the wave normal. We may abbreviate the expression by writing

$$A_{pm} = c_{pqml} f_q f_l. \tag{8.10}$$

To illustrate the meaning of these A_{pm}s the following example is written out in full, using the common abbreviations of the suffixes of the cs.

$$A_{11} = c_{11} f_1^2 + c_{66} f_2^2 + c_{55} f_3^2 + 2c_{56} f_2 f_3 + 2c_{15} f_3 f_1 + 2c_{16} f_1 f_2. \tag{8.11}$$

If $V_{(i)}$ is the velocity of the elastic wave (i) ($i = 1, 2$, or 3),

$$V_{(i)}^2 = v^2 / \left|K^*_{(i)}\right|^2.$$

Substituting in equation (8.9) we obtain

$$\rho V_{(i)}^2 \xi_{(i)p} = A_{pm} \xi_{(i)m}. \tag{8.12}$$

The expansion of equation (8.12) gives three equations when p has the values 1, 2, 3 in turn and m the values 1, 2, and 3 in each equation. These equations are as follows.

$$\xi_{(i)1}(A_{11} - \rho V_{(i)}^2) + \xi_{(i)2} A_{12} + \xi_{(i)3} A_{13} = 0, \tag{8.13a}$$

$$\xi_{(i)1} A_{21} + \xi_{(i)2}(A_{22} - \rho V_{(i)}^2) + \xi_{(i)3} A_{23} = 0, \tag{8.13b}$$

$$\xi_{(i)1} A_{31} + \xi_{(i)2} A_{32} + \xi_{(i)3}(A_{33} - \rho V_{(i)}^2) = 0. \tag{8.13c}$$

The equations (8.13) have a real solution only if the determinant formed by the coefficients of the $\xi_{(i)p}$s is equal to zero. This determinant is named after Christoffel, and it plays a great part in the study of the propagation of elastic waves through crystals.

Since in equation (8.10) the subscripts of the fs must take the values 1, 2, and 3 in all possible combinations, it is clear that $A_{pm} = A_{mp}$. The Christoffel determinant is therefore diagonally symmetrical. When this determinant is expanded, it gives a cubic expression involving powers up to the third of $\rho V_{(i)}^2$. Since this determinant is in real cases equal to zero, there are three roots of the equation, which correspond to the ρV^2 values of the longitudinal and two transverse wave motions respectively.

To illustrate the use of the equations (8.13) we shall consider a wave normal directed along [110] of a cubic crystal. The direction cosines f_p of the wave normal $\mathbf{K}_{(i)}^*$ are $1/\sqrt{2}$, $1/\sqrt{2}$, 0. These values and the cubic elastic stiffnesses c_{11}, c_{12}, and c_{44} (see § 7.5) are inserted in the expression (8.10) for A_{pq}, namely

$$A_{pq} = c_{prqs} f_r f_s.$$

From this we obtain

$$\begin{aligned}
A_{11} &= c_{1111} f_1^2 + c_{1212} f_2^2 = \tfrac{1}{2}(c_{11}+c_{44}),\\
A_{22} &= c_{2222} f_2^2 + c_{2121} f_1^2 = \tfrac{1}{2}(c_{11}+c_{44}),\\
A_{33} &= c_{3131} f_1^2 + c_{3232} f_2^2 = c_{44},\\
A_{12} &= c_{1122} f_1 f_2 + c_{1221} f_2 f_1 = \tfrac{1}{2}(c_{12}+c_{44}),\\
A_{13} &= c_{1231} f_2 f_1 + c_{1132} f_1 f_2 = 0,\\
A_{23} &= 0.
\end{aligned}$$

The Christoffel determinant in this case is

$$\begin{vmatrix}
\rho V_{(i)}^2 - \tfrac{1}{2}(c_{11}+c_{44}) & \tfrac{1}{2}(c_{12}+c_{44}) & 0\\
\tfrac{1}{2}(c_{12}+c_{44}) & \rho V_{(i)}^2 - \tfrac{1}{2}(c_{11}+c_{44}) & 0\\
0 & 0 & \rho V_{(i)}^2 - c_{44}
\end{vmatrix} = 0.$$

On expanding this determinant we obtain the cubic equation

$$[\{\rho V_{(i)}^2 - \tfrac{1}{2}(c_{11}+c_{44})\}^2 - \{\tfrac{1}{2}(c_{12}+c_{44})\}^2](\rho V_{(i)}^2 - c_{44}) = 0. \quad (8.14)$$

The last factor of this expression gives one root of the cubic equation, and this will be denoted $V_{(3)}$. Then we have

$$\rho V_{(3)}^2 = c_{44}.$$

The other two roots, which are denoted $V_{(1)}$ and $V_{(2)}$ respectively, are obtained from the expression in the square brackets of the equation (8.14). Factorizing this difference of two squares gives the two roots, namely

$$\rho V_{(1)}^2 = \tfrac{1}{2}(c_{11}+c_{12}+2c_{44}),$$

$$\rho V_{(2)}^2 = \tfrac{1}{2}(c_{11}-c_{12}).$$

In this way the three wave velocities for travel along the direction [110] are obtained.

Q.8.1. Find by means of Christoffel's determinant and the elastic stiffnesses c_{11}, c_{12}, c_{44} for a cubic crystal the velocities of waves having [100] as wave normal.

8.4. The polarization vectors of elastic waves

After the determination of the three wave velocities that are, in general, associated with any given direction of travel of the waves, it is necessary to consider the direction of vibration, or polarization vector as it is often called, for each wave. If the wave normal is parallel to certain symmetry axes, the longitudinal vibration direction will coincide with the wave-normal direction, but in general it is inclined to it at a small angle. The other two waves, called 'transverse', have vibration directions that are in every case perpendicular to one another and to the longitudinal vibration direction. Each such vector is taken to be of unit length. The longitudinal vector is denoted $\mathbf{e}_{(1)}$ and the transverse vectors are denoted $\mathbf{e}_{(2)}$ and $\mathbf{e}_{(3)}$ respectively. A further subscript is used to denote the Cartesian component of the polarization vector. Thus $e_{(i)k}$ stands for the kth component of the ith polarization vector. Since the vectors are of unit length, the values of $e_{(i)k}$ are direction cosines of the polarization vectors. It should be noted that these direction cosines also apply to the amplitude vectors $\xi_{(i)}$ of equation (8.6).

When the values of the wave velocities $V_{(i)}$ have been found as described in § 8.3, it is possible to use equation (8.12) to find the corresponding values of $e_{(i)k}$. Substituting $e_{(i)k}$ for $\xi_{(i)k}$ in equation (8.12), we obtain

$$\rho V_{(i)}^2 e_{(i)p} = A_{pm} e_{(i)m}. \tag{8.15}$$

We now take particular values of i, p, and m to work out the components of the vibration directions of the three waves discussed at the end of § 8.3. Putting $i = 1$, we have

$$\rho V_{(1)}^2 e_{(1)1} = A_{11}e_{(1)1}+A_{12}e_{(1)2}+A_{13}e_{(1)3},$$

$$\tfrac{1}{2}(c_{11}+c_{12}+2c_{44})e_{(1)1} = \tfrac{1}{2}(c_{11}+c_{44})e_{(1)1}+\tfrac{1}{2}(c_{12}+c_{44})e_{(1)2},$$

$$(c_{12}+c_{44})e_{(1)1} = (c_{12}+c_{44})e_{(1)2},$$

$$\text{or } e_{(1)1} = e_{(1)2}.$$

Further, since $A_{33} = c_{44}$ and $A_{31} = A_{32} = 0$,

$$\rho V_{(1)}^2 e_{(1)3} = A_{31}e_{(1)1}+A_{32}e_{(1)2}+A_{33}e_{(1)3}$$

$$= c_{44}e_{(1)3},$$

since $\rho V_{(1)}^2 \neq c_{44}$, $\quad e_{(1)3} = 0.$

Combining this result with the equality of $e_{(1)1}$ and $e_{(1)2}$, we get

$$e_{(1)1} = e_{(1)2} = 1/\sqrt{2}.$$

We now consider the components of the second wave.

$$\rho V_{(2)}^2 e_{(2)1} = A_{11}e_{(2)1}+A_{12}e_{(2)2}+A_{13}e_{(2)3},$$

$$\tfrac{1}{2}(c_{11}-c_{12})e_{(2)1} = \tfrac{1}{2}(c_{11}+c_{44})e_{(2)1}+\tfrac{1}{2}(c_{12}+c_{44})e_{(2)2},$$

$$(c_{11}-c_{12}-c_{11}-c_{44})e_{(2)1} = (c_{12}+c_{44})e_{(2)2},$$

$$\text{or } e_{(2)1} = -e_{(2)2}.$$

Further

$$\rho V_{(2)}^2 e_{(2)3} = A_{31}e_{(2)1}+A_{32}e_{(2)2}+A_{33}e_{(2)3}$$

and

$$\tfrac{1}{2}(c_{11}-c_{12})e_{(2)3} = c_{44}e_{(2)3}.$$

Since c_{44} is not, except by accident, equal to $\tfrac{1}{2}(c_{11}-c_{12})$, this equation requires that $e_{(2)3} = 0$, and hence

$$e_{(2)1} = 1/\sqrt{2} \text{ and } e_{(2)2} = -1/\sqrt{2}.$$

Finally, we consider the third wave.

$$\rho V_{(3)}^2 e_{(3)1} = A_{11}e_{(3)1}+A_{12}e_{(3)2}+A_{13}e_{(3)3},$$

$$e_{(3)1}\{c_{44}-\tfrac{1}{2}(c_{11}+c_{44})\} = \tfrac{1}{2}(c_{12}+c_{44})e_{(3)2},$$

$$e_{(3)1}\{-\tfrac{1}{2}(c_{11}-c_{44})\} = e_{(3)2}\tfrac{1}{2}(c_{12}+c_{44}).$$

The only values of $e_{(3)1}$ and $e_{(3)2}$ that will satisfy the above equation are zero, since the cs are physical constants. This makes $e_{(3)3}$ unity, which is consistent with the following equation.

$$\rho V_{(3)}^2 e_{(3)3} = A_{31}e_{(3)1}+A_{32}e_{(3)2}+A_{33}e_{(3)3}$$

$$= c_{44}\, e_{(3)3}.$$

The polarization vectors of the three waves are orthogonal, and their components are

Wave 1	$\frac{1}{\sqrt{2}}$	$\frac{1}{\sqrt{2}}$	0,
Wave 2	$\frac{1}{\sqrt{2}}$	$\frac{-1}{\sqrt{2}}$	0,
Wave 3	0	0	1.

The vibration direction of the longitudinal wave (1) coincides with the wave normal direction [110]; one transverse wave, (3), has a vibration direction parallel to [001], and the other, (2), is parallel to $[1\bar{1}0]$.

8.5. Degenerate modes of vibration

Waves travelling in the directions of symmetry axes may have an indeterminate transverse vibration direction. Consider, for example, waves travelling along the tetrad axis [100] of a cubic crystal. It may be shown that the three wave velocities are

$$V_{(1)} = \sqrt{\frac{c_{11}}{\rho}}, \qquad V_{(2)} = \sqrt{\frac{c_{44}}{\rho}}, \qquad V_{(3)} = \sqrt{\frac{c_{44}}{\rho}}.$$

To determine the vibration directions we have the equation

$$\rho V_{(i)}^2 e_{(i)p} = A_{pm} e_{(i)m}.$$

When $i = 1$,

$$\rho V_{(1)}^2 e_{(1)1} = A_{11} e_{(1)1} + A_{12} e_{(1)2} + A_{13} e_{(1)3}.$$

From equation (8.10) are obtained the values

$$A_{11} = c_{11}, \qquad A_{22} = A_{33} = c_{44}, \qquad A_{12} = A_{23} = A_{31} = 0;$$

hence

$$c_{11} e_{(1)1} = c_{11} e_{(1)1}.$$

This equation does not help to define $e_{(1)1}$. We have further that

$$\rho V_{(1)}^2 e_{(1)2} = A_{21} e_{(1)1} + A_{22} e_{(1)2} + A_{23} e_{(1)3}$$

and

$$c_{11} e_{(1)2} = c_{44} e_{(1)2}.$$

This equation requires that $e_{(1)2} = 0$.

Similarly it may be shown that $e_{(1)3} = 0$, and thus the direction cosines of the polarization vector of wave (1) are 1, 0, 0.

For wave (2) we have

$$\rho V_{(2)}^2 e_{(2)p} = A_{pm} e_{(2)m}.$$

Thus

$$\begin{aligned} c_{44} e_{(2)1} &= A_{11} e_{(2)1} + A_{12} e_{(2)2} + A_{13} e_{(2)3} \\ &= c_{11} e_{(2)1}, \end{aligned}$$

which requires that $e_{(2)1} = 0$.

Then

$$\begin{aligned} c_{44} e_{(2)2} &= A_{21} e_{(2)1} + A_{22} e_{(2)2} + A_{23} e_{(2)3} \\ &= c_{44} e_{(2)2}, \end{aligned}$$

but this gives no information about $e_{(2)2}$. Lastly,

$$\begin{aligned} c_{44} e_{(2)3} &= A_{31} e_{(2)1} + A_{32} e_{(2)2} + A_{33} e_{(2)3} \\ &= c_{44} e_{(2)3}, \end{aligned}$$

and again we have no information about $e_{(2)3}$.

Thus although the polarization vector of wave (2) must be normal to the wave vector, its vibration direction in the plane perpendicular to the wave vector is indeterminate.

The same result is obtained for wave (3). Thus we have the waves (2) and (3) which are called degenerate and cannot be distinguished from one another.

Q.8.2. The elastic-constant scheme for an orthorhombic crystal is as follows

$$\begin{matrix} c_{11} & c_{12} & c_{13} & 0 & 0 & 0 \\ & c_{22} & c_{23} & 0 & 0 & 0 \\ & & c_{33} & 0 & 0 & 0 \\ & & & 4c_{44} & 0 & 0 \\ & & & & 4c_{55} & 0 \\ & & & & & 4c_{66}. \end{matrix}$$

Find the velocities of travel of the longitudinal waves along the directions [100], [010], [001] respectively.

Q.8.3. Find the direction cosines of the transverse vibrations for a wave travelling along [100] in the case given in Q.8.2.

Answers

Q.8.1. For the wave normal along [100],

$$A_{pm} = c_{pqml} f_q f_l, \qquad f_1 = 1, \qquad f_2 = 0 = f_3.$$

Hence

$$A_{11} = c_{1111} = c_{11},$$

$$A_{22} = c_{2121} = c_{44},$$

$$A_{33} = c_{3131} = c_{44},$$

$$A_{12} = c_{1121} = 0 = A_{13} = A_{23}.$$

The Christoffel determinant is therefore

$$\begin{vmatrix} \rho V^2 - c_{11} & 0 & 0 \\ 0 & \rho V^2 - c_{44} & 0 \\ 0 & 0 & \rho V^2 - c_{44} \end{vmatrix} = 0.$$

Hence the three waves have velocities

$$\sqrt{\frac{c_{11}}{\rho}}, \quad \sqrt{\frac{c_{44}}{\rho}}, \quad \sqrt{\frac{c_{44}}{\rho}} \quad \text{respectively.}$$

Q.8.2. For the wave normal along [100], $f_1 = 1$, $f_2 = 0 = f_3$,

$$A_{pm} = c_{pqml} f_q f_l.$$

Hence

$$A_{11} = c_{1111} = c_{11},$$

$$A_{22} = c_{2121} = c_{66},$$

$$A_{33} = c_{3131} = c_{55},$$

$$A_{12} = c_{1121} = c_{16} = 0,$$

$$A_{23} = c_{2131} = c_{65} = 0,$$

$$A_{31} = c_{3111} = c_{15} = 0.$$

Hence the Christoffel determinant is

$$\begin{vmatrix} \rho V^2 - c_{11} & 0 & 0 \\ 0 & \rho V^2 - c_{66} & 0 \\ 0 & 0 & \rho V^2 - c_{55} \end{vmatrix} = 0.$$

If we take $\rho V^2 = c_{11}$ to define the wave (1), then

$$\begin{aligned} c_{11}e_{(1)1} &= A_{11}e_{(1)1}+A_{12}e_{(1)2}+A_{13}e_{(1)3} \\ &= c_{11}e_{(1)1}, \\ c_{11}e_{(1)2} &= A_{21}e_{(1)1}+A_{22}e_{(1)2}+A_{23}e_{(1)3} \\ &= \quad 0 \quad +c_{66}e_{(1)2}+0. \end{aligned}$$

Hence $e_{(1)2} = 0$, and also $e_{(1)3} = 0$. Thus $e_{(1)1} = 1$, and c_{11} gives the longitudinal wave with velocity $\sqrt{(c_{11}/\rho)}$.

Wave normal along [010]. $\quad f_1 = 0, \quad f_2 = 1, \quad f_3 = 0.$

$$\begin{aligned} A_{11} &= c_{1212} = c_{66}, & A_{12} &= c_{1222} = c_{26} = 0, \\ A_{22} &= c_{2222} = c_{22}, & A_{13} &= c_{1232} = c_{46} = 0, \\ A_{33} &= c_{3232} = c_{44}, & A_{23} &= c_{2232} = c_{24} = 0. \end{aligned}$$

Hence the Christoffel determinant is

$$\begin{vmatrix} \rho V^2-c_{66} & 0 & 0 \\ 0 & \rho V^2-c_{22} & 0 \\ 0 & 0 & \rho V^2-c_{44} \end{vmatrix} = 0.$$

Consider the wave given by

$$\rho V^2 = c_{22}.$$

Applying the relation

$$\rho V^2_{(i)}e_{(i)p} = A_{pm}e_{(i)m},$$

we have, putting $i = 7$

$$\begin{aligned} c_{22}e_{(1)1} &= A_{11}e_{(1)1}+A_{12}e_{(1)2}+A_{13}e_{(1)3} \\ &= c_{66}e_{(1)1}, \end{aligned}$$

so $\qquad e_{(1)1} = 0.$

$$\begin{aligned} c_{22}e_{(1)2} &= A_{21}e_{(1)1}+A_{22}e_{(1)2}+A_{23}e_{(1)3} \\ &= c_{22}e_{(1)2}, \\ c_{22}e_{(1)3} &= A_{31}e_{(1)1}+A_{32}e_{(1)2}+A_{33}e_{(1)3} \\ &= c_{44}e_{(1)3}, \end{aligned}$$

and $\qquad e_{(1)3} = 0.$

Thus $_{(}e_{1)2} = 1, \qquad e_{(1)1} = e_{(1)3} = 0,$

and this is the longitudinal wave travelling along [010], having velocity $\sqrt{(c_{22}/\rho)}$.

Similarly the wave given by

$$\rho V^2 = c_{33}$$

is the longitudinal wave travelling along [001].

Q.8.3. The two transverse vibrations have associated velocities given by the Christoffel determinant, namely

$$\rho V_{(2)}^2 = c_{66}, \qquad \rho V_{(3)}^2 = c_{55}.$$

For wave (2),

$$c_{66}e_{(2)1} = A_{11}e_{(2)1}+A_{12}e_{(2)2}+A_{13}e_{(2)3}.$$

From Q.8.2 we have

$$A_{11} = c_{11}, \qquad A_{22} = c_{66}, \qquad A_{33} = c_{55},$$

all other A_{ik}s $= 0$.

Hence,

$$c_{66}e_{(2)1} = c_{11}e_{(2)1} \quad \text{and} \quad e_{(2)1} = 0.$$

Further,

$$c_{66}e_{(2)2} = A_{21}e_{(2)1}+A_{22}e_{(2)2}+A_{23}e_{(2)3}$$
$$= c_{66}e_{(2)2},$$

which gives no information about $e_{(2)2}$. Finally,

$$c_{66}e_{(2)3} = A_{31}e_{(2)1}+A_{32}e_{(2)2}+A_{33}e_{(2)3}$$
$$= c_{55}e_{(2)3}$$

and $e_{(2)3} = 0$.

Hence the direction cosines for wave normal (2) are 0, 1, 0.
For wave (3),

$$c_{55}e_{(3)1} = A_{11}e_{(3)1} = c_{11}e_{(3)1} \quad \text{and} \quad e_{(3)1} = 0,$$
$$c_{55}e_{(3)2} = A_{22}e_{(3)2} = c_{66}e_{(3)2} \quad \text{and} \quad e_{(3)2} = 0.$$

Hence the direction cosines for wave normal (3) are 0, 0, 1.

9

Polar and axial tensors

9.1. Introduction

UP to this point the chapters have dealt with the relations between like tensors. Among the second-order tensors, for example, thermal expansion relates the change in the length and orientation of a line drawn in the crystal to the length and orientation of that line. The quantities related in thermal expansion are called polar vectors. The second-order tensor of diamagnetic susceptibility also relates two like quantities, namely, magnetic moment and magnetic field. These are both called axial vectors, because not only can they be represented by a line of a certain length drawn in a certain direction, but also a sense of rotation is associated with the direction of each vector. When an electric current circulates in a coil, a magnetic field is produced along the axis of the coil, and, if the positive carriers of current are circulating in a clockwise sense as seen by an observer looking in a direction at right-angles to the plane of the coil, the magnetic lines of force are directed through the coil away from the observer. A magnetic moment in a solid body is produced by the cumulative effect of atoms in which the circulating electrons are rotating round the nucleus in the same sense. Thus the vectors of both magnetic field and magnetic moment have an associated sense of rotation as well as a magnitude, and a direction which is determined by the sense of the rotation. The equations representing thermal expansion and diamagnetic susceptibility are of the same form, namely

$$h_i = k_{ij} l_j \tag{9.1}$$

where $\mathbf{h}$ and $\mathbf{l}$ are the like vectors and k is the second-order polar tensor.

The representation surface for k_{ij} is in the general case either a tri-axial ellipsoid or hyperboloidal sheets, and the expression for the k_{ij} in an arbitrary direction is

$$k'_{33} = c_{3i}c_{3j}k_{ij}, \tag{9.2}$$

where the cs are the direction cosines of the k'_{33} relative to the coordinate axes (see Chapter 2).

A different relation exists in connection with optical activity. When plane-polarized light passes through certain crystals in certain directions, the plane of polarization on emergence from the crystal is rotated with respect to the corresponding plane on entry. The incident light beam can be represented by a unit vector, which is polar, since it has no sense of rotation. The plane of polarization of the light at emergence, however, has been rotated, and this change must be represented by an axial vector. In this physical property, a polar and an axial vector are related to one another. The coefficients connecting the two vectors form a second-order axial tensor, and, if the rotation per unit distance traversed, or rotatory power, in an arbitrary direction X'_3, is represented by ρ'_{33} and the direction cosines of the direction of travel of the light are c_{3i}, then a second-order tensor requires an expression of the form

$$\rho'_{33} = c_{3i}c_{3j}g_{ij}, \tag{9.3}$$

where the g_{ij} terms are the coefficients for rotatory polarization. The equations (9.2) and (9.3) are formally similar, but there is an important difference between them. The coefficients that are made zero by the operation of the symmetry and the application of Neumann's Principle are different according to whether the tensor is polar or axial.

9.2. The relation between polar and axial vectors and axes, planes and centres of symmetry

A two-fold axis of symmetry parallel to X_3 changes the components of a polar vector from x_1, x_2, x_3 to $\bar{x}_1$, $\bar{x}_2$, x_3 (Fig. 9.1(a)). The action of the same two-fold axis on an axial vector can be seen from Fig. 9.1(b). The vector OP is in this figure provided with an arc ABC which indicates a clockwise sense of rotation (as seen by an observer at the origin O looking towards P). The diad axis X_3 rotates OP to OP' and changes the points ABC to the points $A'B'C'$. The sense of rotation given by $A'B'C'$ (as seen by an observer at O

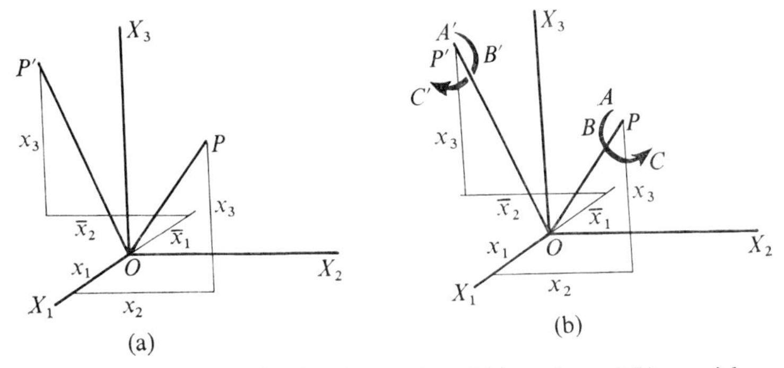

FIG. 9.1. Diagrams showing the rotation of (a) a polar and (b) an axial vector about a diad axis OX_3.

looking towards P') is also clockwise. The positive direction along OP' is therefore from O to P'. Thus a diad axis acts in the same way on both polar and axial vectors. In the same way it may be shown that three-fold, four-fold, or six-fold axes act in the same way on both types of vectors.

In Fig. 9.2(a), (b) the action of a plane of symmetry X_1OX_2 on the two types of vectors is illustrated. In Fig. 9.2(a) the polar vector

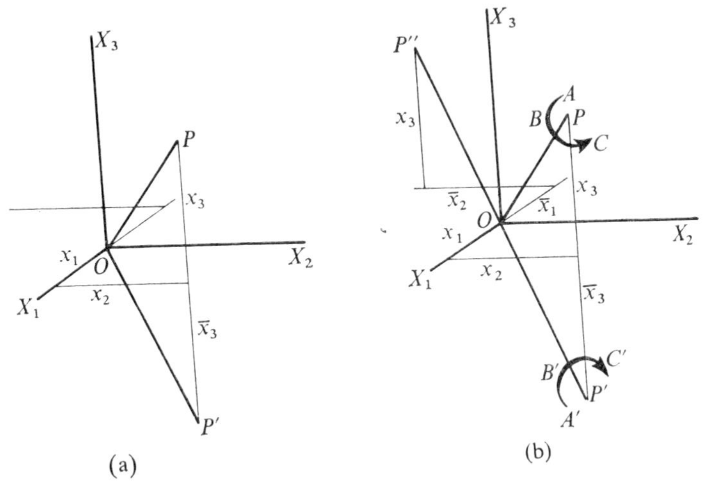

FIG. 9.2. Diagrams showing the reflection in a plane of symmetry X_1OX_2 of (a) a polar and (b) an axial vector.

OP is reflected to OP' and the new components are x_1, x_2, $\bar{x}_3$. In Fig. 9.2(b) the arc ABC is reflected to $A'B'C'$ and the sense of rotation as seen from O looking towards P' is reversed. This means that the positive direction of the reflected axial vector is from O to P'' and the vector has components $\bar{x}_1$, $\bar{x}_2$, x_3. Thus a plane of symmetry acts differently on axial and polar vectors.

In Figs. 9.3(a), (b) the action of a centre of symmetry at O on (a)

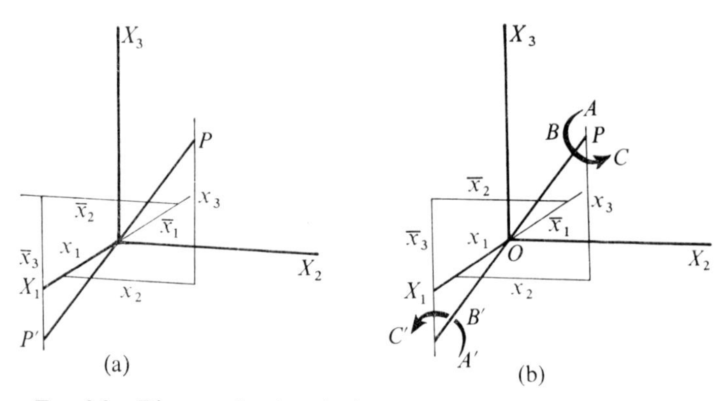

FIG. 9.3. Diagram showing the inversion through a centre of symmetry (a) with a polar vector and (b) with an axial vector.

a polar, and (b) an axial vector is shown. In Fig. 9.3(a) it can be seen that P is inverted to P' and the new components are $\bar{x}_1$, $\bar{x}_2$, $\bar{x}_3$. In Fig. 9.3(b) the arc ABC is inverted to the arc $A'B'C'$, and looking from O the sense of rotation is anticlockwise for $A'B'C'$ and clockwise for ABC. Thus the positive direction for the inverted axial vector is from O to P, i.e. the centre of symmetry has not introduced any change in the components.

In studies of the variation of physical properties, use has been made of the transformation of the axial system X_1, X_2, X_3 into X_1', X_2', X_3'. For the first example discussed above, we have the direction-cosine schemes for a diad axis parallel to X_3,

Polar vector

	X_1	X_2	X_3
X_1'	-1	0	0
X_2'	0	-1	0
X_3'	0	0	1

Axial vector

	X_1	X_2	X_3
X_1'	-1	0	0
X_2'	0	-1	0
X_3'	0	0	1.

For the second example, a plane of symmetry X_1OX_2,

$$\begin{matrix} 1 & 0 & 0 \\ 0 & 1 & 0 \\ 0 & 0 & -1 \end{matrix} \qquad \begin{matrix} -1 & 0 & 0 \\ 0 & -1 & 0 \\ 0 & 0 & 1. \end{matrix}$$

Lastly for the centre of symmetry

$$\begin{matrix} -1 & 0 & 0 \\ 0 & -1 & 0 \\ 0 & 0 & -1 \end{matrix} \qquad \begin{matrix} 1 & 0 & 0 \\ 0 & 1 & 0 \\ 0 & 0 & 1. \end{matrix}$$

Expressed in general terms, we can say that the components p of a vector transform according to the equation

$$p'_j = c_{ji}p_i$$

if the transformation is due to axes of rotation only, no matter whether the vector is polar or axial. But if the transformation involves planes or centres of symmetry, then, for an axial vector,

$$p'_j = -c_{ji}p_i$$

(1, p. 14; 13, p. 38; 16, p. 93).

9.3. Optical activity

When plane-polarized light travels through certain crystals that lack a centre of symmetry, the plane of polarization is rotated. The sense of the rotation is defined as positive when, to an observer looking towards the source of light, the rotation appears clockwise, and as negative when it appears anticlockwise. Quartz is the most carefully studied crystal so far as this property is concerned. In some crystals of quartz, the rotation is positive along the direction of the optic axis and negative in directions perpendicular to that axis, and in others the senses of rotation along the directions specified are reversed. The relation between the rotation ρ and the components $c_{3i}c_{3j}$ corresponding to the direction of travel of the light is given by the equation (9.3).

There is no physical difference between $c_{3i}c_{3j}$ and $c_{3j}c_{3i}$, and so $g_{ij} = g_{ji}$, i.e. g_{ij} is a symmetric tensor which is axial. The transformation of g_{ij} is therefore given by

$$g'_{ij} = \pm c_{ip}c_{jq}g_{pq}, \tag{9.4}$$

the + sign being taken when the rotation axes are involved and the − sign when planes, centres, or inversion–rotation axes are involved.

Relation between coefficients of rotatory power and symmetry

In the triclinic system, no symmetry is present, and hence the scheme of constants in point group 1 is

$$\begin{matrix} g_{11} & g_{12} & g_{13} \\ & g_{22} & g_{23} \\ & & g_{33}. \end{matrix}$$

In class $\bar{1}$ the negative sign must be taken in equation (9.4), and the transformation of the axes is given by

	X_1	X_2	X_3
X'	-1	0	0
Y'	0	-1	0
Z'	0	0	-1.

Thus $g'_{11} = -1 \times c_{11}c_{11}g_{11} = -1 \times -1 \times -1 \times g_{11} = -g_{11} = g_{11}$ (by Neumann's Principle).

Hence $g_{11} = 0$.

Similarly g'_{22}, g'_{33}, and all other g'_{ij} are zero. Thus optical activity cannot exist in any point group having a centre of symmetry.

Point group 2

In the monoclinic system, point group 2, with the diad axis parallel to [010], we have

	X_1	X_2	X_3
X'_1	-1	0	0
X'_2	0	1	0
X'_3	0	0	-1.

Thus

$$g'_{11} = -1 \times -1 \times g_{11} \neq 0,$$

$$g'_{22} = 1 \times 1 \times g_{22} \neq 0,$$

$$g'_{33} = -1 \times -1 \times g_{33} \neq 0,$$

$$g'_{12} = -1\times 1\times g_{12} = 0,$$
$$g'_{13} = -1\times -1\times g_{13} \neq 0,$$
$$g'_{23} = 1\times -1\times g_{23} = 0.$$

Thus the scheme of coefficients has zeros where a single 2 occurs in the suffixes, as shown here.

$$\begin{matrix} g_{11} & 0 & g_{13} \\ & g_{22} & 0 \\ & & g_{33}. \end{matrix}$$

The representation surface is thus given by

$$\rho'_{33} = c_{31}^2 g_{11} + c_{32}^2 g_{22} + c_{33}^2 g_{33} + 2c_{31}c_{33}g_{13}.$$

This surface is a triaxial ellipsoid or triaxial hyperboloid depending on the similarity or dissimilarity in the signs of the g_{ik}s. Only the axis X_2 coincides with a principal axis of the ellipsoid or hyperboloid.

Point group m

If we take the plane of symmetry normal to the X_2 axis the transformation matrix c_{ik} is

$$\begin{matrix} 1 & 0 & 0 \\ 0 & -1 & 0 \\ 0 & 0 & 1. \end{matrix}$$

Since the operation is that of a plane of symmetry, the equation for g_{ik} is

$$g'_{ik} = -c_{il}c_{km}g_{lm}.$$

This gives

$$g'_{11} = -c_{11}c_{11}g_{11} = -1\times 1\times 1\times g_{11} = g_{11} = 0,$$
$$g'_{22} = -c_{22}c_{22}g_{22} = -1\times -1\times -1\times g_{22} = g_{22} = 0,$$
$$g'_{33} = -c_{33}c_{33}g_{33} = -1\times 1\times 1\times g_{33} = g_{33} = 0,$$
$$g'_{12} = -c_{11}c_{22}g_{12} = -1\times 1\times -1\times g_{12} = g_{12} \neq 0,$$
$$g'_{13} = -c_{11}c_{33}g_{13} = -1\times 1\times 1\times g_{13} = g_{13} = 0,$$
$$g'_{23} = -c_{22}c_{33}g_{23} = -1\times -1\times 1\times g_{23} \neq 0.$$

Thus, the g_{ik} scheme is

$$\begin{matrix} 0 & g_{12} & 0 \\ & 0 & g_{23} \\ & & 0. \end{matrix}$$

The representation surface for such a scheme of g_{ik}s is given by the equation

$$\rho'_{33} = 2c_{31}c_{32}g_{12}+2c_{32}c_{33}g_{23}.$$

A central section normal to the X_1 axis has $c_{31} = 0$, $c_{32} = \sin\phi$, and $c_{33} = \cos\phi$, where ϕ is the angle between the axes X'_3 and X_3. For this section

$$\rho'_{33} = 2\sin\phi\cos\phi\, g_{23} = \sin 2\phi\, g_{23}.$$

The section thus has the form shown in Fig. 9.4 with two positive and two negative lobes. A section normal to the axis X_2 has $c_{32} = 0$,

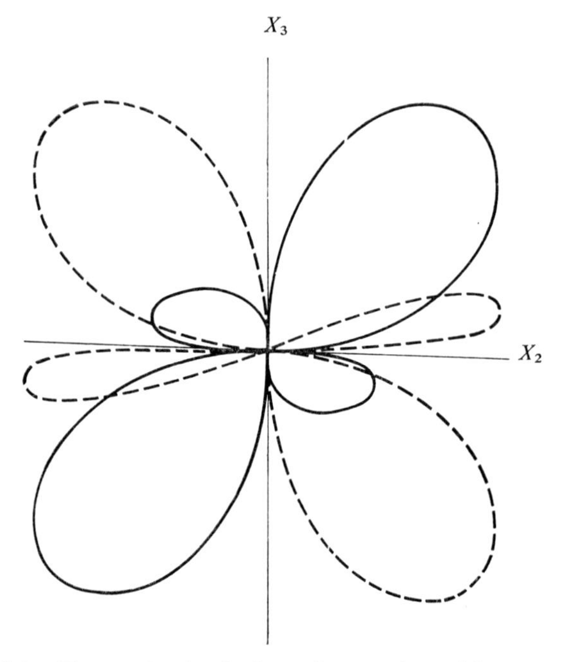

FIG. 9.4. Diagram showing the form of two sections of the representation surface for optical activity in point group *m*.

so that ρ'_{33} is everywhere zero. A section normal to the X_3 axis has $c_{33} = 0$ and $\rho'_{33} = 2c_{31}c_{32}g_{12}$.
This clearly has a form similar to that of the section normal to axis X_1 but changed in scale. Intermediate sections have corresponding positive and negative lobes. The whole representation surface thus consists of two positive and two negative regions, related by a plane of symmetry within which all directions have zero coefficients.

Point group $\bar{4}$

In the tetragonal system, and point group $\bar{4}$, the transformation of the axes may be given by

	X_1	X_2	X_3
X'_1	0	-1	0
X'_2	1	0	0
X'_3	0	0	-1.

In this point group, an inversion axis is present and therefore the factor -1 must be used in equation (9.4). Thus we have

$$g'_{11} = -1\times c_{12}c_{12}g_{22} = -1\times -1\times -1\times g_{22} = -g_{22} = g_{11},$$
$$g'_{33} = -1\times c_{33}c_{33}g_{33} = -1\times -1\times -1\times g_{33} = -g_{33} = 0,$$
$$g'_{12} = -1\times -1\times 1\times g_{21} = g_{12} \neq 0,$$
$$g'_{13} = -1\times -1\times -1\times g_{23} = -g_{23} = g_{13},$$
$$g'_{23} = -1\times 1\times -1\times g_{13} = g_{13} = g_{23},$$

but since $g_{13} = -g_{23}$ both are zero.

Hence the scheme is

$$\begin{matrix} g_{11} & g_{12} & 0 \\ & -g_{11} & 0 \\ & & 0 \end{matrix}$$

(1, p. 187; 13, p. 269).

Q.9.1. What scheme of g_{ik}s applies to point group 422?

9.4. The Hall effect

When an electric current flows through a rod-shaped isotropic conductor, the potential at all points in a plane perpendicular to the direction of current flow is the same. This statement is no longer true when a magnetic field is applied across the conductor. The lines of current flow and the surfaces of the same electric potential are distorted by the magnetic field. The simplest arrangement, and the only one to be discussed here, is that in which the lines of magnetic force are perpendicular to the direction of current flow. The conductor is supposed to be in the form of a rectangular bar of thickness w in the direction of the magnetic field (Fig. 9.5), which is of strength

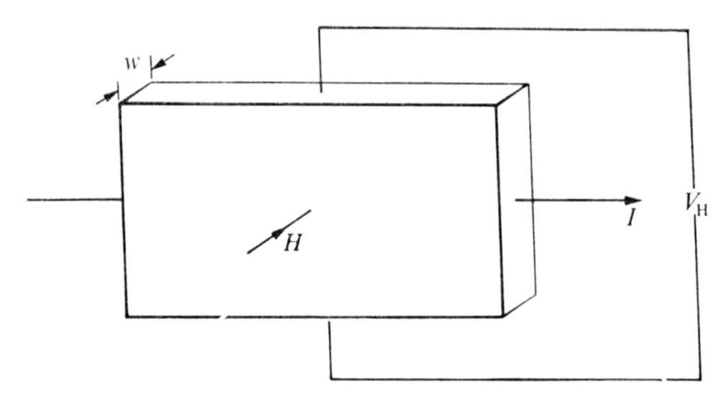

FIG. 9.5. Diagram showing the quantities involved in finding the Hall voltage.

H. The current through the conductor is denoted I and the voltage across the specimen, due to the application of the magnetic field, is V_{H} (Hall voltage). The Hall coefficient is denoted R_{H}, and this is related to the other quantities mentioned by the equation

$$R_{\mathrm{H}} = \frac{V_{\mathrm{H}} w}{HI}. \tag{9.5}$$

This expression could also be written

$$\frac{V_{\mathrm{H}}}{I} = \frac{1}{w} R_{\mathrm{H}} H.$$

Now V_{H}/I represents a resistivity, i.e. a second-order polar tensor, and H represents a magnetic field, which is an axial vector. R_{H} is therefore a third-order axial tensor.

The resistivity associated with the Hall voltage differs from that discussed in Chapter 2. There it was stated that heat flows outwards from a point source in straight lines, in a crystal of any symmetry, and for this reason the resistivity tensor r_{mn} is symmetric, and $r_{mn} = r_{nm}$. In the absence of a magnetic field this is also true of electrical resistance, but when a field is applied the lines of current flow are no longer straight. The magnetic field exerts a force on the charge carriers in a direction perpendicular to their direction of motion. The tensor representing conductivity or resistivity can, therefore, no longer be symmetric. It is always possible to resolve such a tensor into two parts, one fully symmetric and the other antisymmetric. This may be shown as follows. If r_{pq} is any kind of second-order polar tensor, symmetric or non-symmetric, we may write

$$r'_{mn} = c_{mp}c_{nq}r_{pq}.$$

We put $r_{12}+r_{21} = 2r''_{12}$ and $r_{12}-r_{21} = 2a_{12}$, so that $r_{12} = r''_{12}+a_{12}$, $r_{21} = r''_{12}-a_{12}$, etc.

We also put

$$r_{11} = r''_{11},\ r_{22} = r''_{22},\ r_{33} = r''_{33},$$

and

$$a_{11} = a_{22} = a_{33} = 0;$$

then, for example,

$$\begin{aligned}
r'_{12} = {} & c_{11}c_{21}r_{11} && +c_{11}c_{22}r_{12} && +c_{11}c_{23}r_{13}+ \\
& +c_{12}c_{21}r_{21} && +c_{12}c_{22}r_{22} && +c_{12}c_{23}r_{23}+ \\
& +c_{13}c_{21}r_{31} && +c_{13}c_{22}r_{32} && +c_{13}c_{23}r_{33}. \\
= {} & c_{11}c_{21}r''_{11} && +c_{11}c_{22}r''_{12} && +c_{11}c_{23}r''_{13}+ \\
& +c_{12}c_{21}r''_{12} && +c_{12}c_{22}r''_{22} && +c_{12}c_{23}r''_{23}+ \\
& +c_{13}c_{21}r''_{13} && +c_{13}c_{22}r''_{23} && +c_{13}c_{23}r''_{33}+ \\
& +0 && +c_{11}c_{22}a_{12} && +c_{11}c_{23}a_{13}+ \\
& +c_{12}c_{21}(-a_{12})+0 && && +c_{12}c_{23}a_{23}+ \\
& +c_{13}c_{21}(-a_{13})+c_{13}c_{22}(-a_{23})+0.
\end{aligned}$$

Thus we have split the non-symmetric tensor r'_{mn} into two tensors, one symmetric, r''_{mn}, and the other antisymmetric, a_{mn}. Written out in full, these are

$$\begin{matrix} r''_{11} & r''_{12} & r''_{13} \\ r''_{12} & r''_{22} & r''_{23} \\ r''_{13} & r''_{23} & r''_{33} \end{matrix} \qquad \begin{matrix} 0 & a_{12} & a_{13} \\ -a_{12} & 0 & a_{23} \\ -a_{13} & -a_{23} & 0. \end{matrix}$$

We may now express the resistivity, in a magnetic field $r_{mn}(H)$, in a series of increasing order of powers of the magnetic field. Going no further than second-order terms in the magnetic field, we obtain

$$r_{mn}(H) = r^0_{mn} + \rho_{mnp}H_p + \rho_{mnpq}H_pH_q,$$

where r^0_{mn} is the resistivity tensor in the absence of a magnetic field and ρ_{mnp} is the third-order tensor representing the Hall effect. If H_p is changed to $-H_p$ and H_q to $-H_q$, the value of H_pH_q is unchanged. Thus the term involving ρ_{mnpq} must be centrosymmetric. The antisymmetric part of the tensor, $r_{mn}(H)$, is given by the odd-order term. Hence we may write

$$a_{ik} = \rho_{ikp}H_p.$$

The Hall tensor relates the axial vector H_p to the antisymmetric second-order tensor a_{ik}.

Relation between Hall coefficients and symmetry

Point group 2

If the diad axis is parallel to X_3, the transformation matrix is

	X_1	X_2	X_3
X'_1	-1	0	0
X'_2	0	-1	0
X'_3	0	0	1.

Here we are dealing with a rotation axis of symmetry, and therefore

$$\rho'_{ikl} = c_{ip}c_{kq}c_{lr}\rho_{pqr},$$

i.e. the plus sign occurs on the right-hand side of the equation. All terms having the first two suffixes equal are zero, since $a_{ii} = 0$, a_{ii} being an antisymmetric tensor. Similarly,

$$\rho_{ikl} = -\rho_{kil}$$

because $a_{ik} = -a_{ki}$ when $i \neq k$.

We shall consider the possible non-zero terms in order.

$$\begin{aligned}\rho'_{121} &= c_{11}c_{22}c_{11}\rho_{121} = (-1)(-1)(-1)\rho_{121}\\ &= -\rho_{121} = \rho_{121} = 0,\end{aligned}$$

$$\begin{aligned}\rho'_{131} &= c_{11}c_{33}c_{11}\rho_{131} = (-1)\times 1\times(-1)\rho_{131}\\ &= \rho_{131} \neq 0,\end{aligned}$$

$$\begin{aligned}\rho'_{231} &= c_{22}c_{33}c_{11}\rho_{231} = (-1)\times 1\times(-1)\rho_{231}\\ &= \rho_{231} \neq 0.\end{aligned}$$

Only if one 3 is present in the suffixes can the coefficient be non-zero. Thus the first portion of the tensor, relating to component H_1, can be written

$$\begin{matrix} 0 & 0 & \rho_{131} \\ 0 & 0 & \rho_{231} \\ -\rho_{131} & -\rho_{231} & 0. \end{matrix}$$

The second portion, relating to H_2, follows in the same way, and gives

$$\begin{matrix} 0 & 0 & \rho_{132} \\ 0 & 0 & \rho_{232} \\ -\rho_{132} & -\rho_{232} & 0. \end{matrix}$$

The third gives

$$\begin{matrix} 0 & \rho_{123} & 0 \\ -\rho_{123} & 0 & 0 \\ 0 & 0 & 0. \end{matrix}$$

Point group $\bar{3}m$

This is the point group to which bismuth belongs. The Hall effect in bismuth is relatively large, and it is therefore much used in studies on this subject. The point-group symmetry includes a $\bar{3}$ axis parallel to X_3 and a diad axis parallel to X_1. Taking the transformation for the diad axis first, we have the matrix

$$\begin{matrix} 1 & 0 & 0 \\ 0 & -1 & 0 \\ 0 & 0 & -1. \end{matrix}$$

Applying this to the components of the Hall tensor relating to H_1, we have

$$\rho_{121} = c_{11}c_{22}c_{11}\rho_{121} = -\rho_{121} = 0,$$

and it can be seen that only terms having one 1 can be non-zero. Hence the scheme of components is

$$\begin{matrix} & H_1 & \\ 0 & 0 & 0 \\ 0 & 0 & \rho_{231} \\ 0 & -\rho_{231} & 0 \end{matrix} \qquad \begin{matrix} & H_2 & \\ 0 & \rho_{122} & \rho_{132} \\ -\rho_{122} & 0 & 0 \\ -\rho_{132} & 0 & 0 \end{matrix}$$

$$\begin{matrix} & H_3 & \\ 0 & \rho_{123} & \rho_{133} \\ -\rho_{123} & 0 & 0 \\ -\rho_{133} & 0 & 0. \end{matrix}$$

The transformation for the $\bar{3}$ axis has the matrix corresponding to 3 + centre, i.e.

$$\begin{matrix} \frac{1}{2} & -\frac{\sqrt{3}}{2} & 0 \\ \frac{\sqrt{3}}{2} & \frac{1}{2} & 0 \\ 0 & 0 & -1. \end{matrix}$$

The non-zero ρs given by the diad axis will now be considered in turn in relation to the matrix for the $\bar{3}$ axis. (The centre requires minus signs).

$$\begin{aligned} \rho'_{231} &= -c_{21}c_{33}c_{11}\rho_{131} + (-)c_{22}c_{33}c_{11}\rho_{231} + \\ &\quad + (-)\,c_{21}c_{33}c_{12}\rho_{132} + (-)c_{22}c_{33}c_{12}\rho_{232} \\ &= 0 + \tfrac{1}{4}\rho_{231} + (-)\tfrac{3}{4}\rho_{132} + 0 \\ &= \rho_{231} \text{ (by Neumann's Principle).} \end{aligned}$$

Hence $\rho_{231} = -\rho_{132}$,

$$\rho'_{122} = -c_{11}c_{22}c_{22}\rho_{122}+(-)c_{12}c_{21}c_{22}\rho_{212}$$
$$= -\tfrac{1}{8}\rho_{122}-\tfrac{3}{8}\rho_{122} = -\tfrac{1}{2}\rho_{122} = \rho_{122} = 0,$$
$$\rho'_{132} = -c_{12}c_{33}c_{21}\rho_{231}+(-)c_{11}c_{33}c_{22}\rho_{132}$$
$$= -\tfrac{3}{4}\rho_{231}+\tfrac{1}{4}\rho_{132} = \rho_{132},$$
$$\rho_{132} = -\rho_{231},$$
$$\rho'_{123} = -c_{11}c_{22}c_{33}\rho_{123}+(-)c_{12}c_{21}c_{33}\rho_{213}$$
$$= \tfrac{1}{4}\rho_{123}+\tfrac{3}{4}\rho_{123} = \rho_{123} \neq 0,$$
$$\rho'_{133} = -c_{11}c_{33}c_{33}\rho_{133}$$
$$= -\tfrac{1}{2}\rho_{133} = \rho_{133} = 0.$$

Hence the scheme of Hall coefficients for bismuth is

$$H_1 \qquad\qquad H_2$$
$$\begin{matrix} 0 & 0 & 0 \\ 0 & 0 & \rho_{231} \\ 0 & -\rho_{231} & 0 \end{matrix} \qquad \begin{matrix} 0 & 0 & -\rho_{231} \\ 0 & 0 & 0 \\ \rho_{231} & 0 & 0 \end{matrix}$$

$$H_3$$
$$\begin{matrix} 0 & \rho_{123} & 0 \\ -\rho_{123} & 0 & 0 \\ 0 & 0 & 0. \end{matrix}$$

The Hall coefficient for a plate of an arbitrary orientation

The current is assumed to travel along the axis X'_1 (Fig. 9.5), the magnetic field is parallel to X'_3 and the Hall field generated by this field is determined along X'_2. The Hall coefficient is then ρ'_{123}. To find the corresponding coefficients in terms of the known Hall constants we can use the transformation

$$\rho'_{123} = c_{1p}c_{2q}c_{3r}\rho_{pqr},$$

where ρ_{pqr} stands for all the non-zero ρs in turn, and the cs express the transformation of the axes X_1, X_2, X_3.

As an illustration we shall consider a plate of bismuth having its length perpendicular to the trigonal axis, and the normal to its major faces inclined to the trigonal axis at an angle θ. The length of the plate is designated X_1' and the normal to the major faces X_3', so that the transformation matrix can be given as

	X_1	X_2	X_3
X_1'	0	1	0
X_2'	$-\cos\theta$	0	$\sin\theta$
X_3'	$\sin\theta$	0	$\cos\theta$.

We assume that the magnetic field is directed along X_3' so that the relevant Hall coefficient is ρ'_{123}. The variation of this Hall coefficient with θ can be obtained in the following way.

From the ρs for bismuth given above we can write

$$\begin{aligned}\rho'_{123} &= c_{12}c_{23}c_{31}\rho_{231}+c_{12}c_{21}c_{33}\rho_{213}\\ &= \sin^2\theta\,\rho_{231}+(-)\cos^2\theta\,(-\rho_{123})\\ &= \sin^2\theta\,\rho_{231}+\cos^2\theta\,\rho_{123}.\end{aligned}$$

Thus when $\theta = 0°$

$$\rho'_{123} = \rho_{123}$$

and when $\theta = 90°$

$$\rho'_{123} = \rho_{231}$$

(1, p. 197; 16, p. 90).

Q.9.2. A single crystal plate of bismuth is cut parallel to the basal plane (i.e. its normal is X_3). The length of the plate is inclined to the diad axis, X_1, at an angle θ. Find how the Hall voltage depends on θ.

Answers

Q.9.1. In point group 422 there is a tetrad axis parallel to X_3 and a diad axis parallel to X_1, and there is also a diad axis bisecting the angle between X_1 and X_2. Two rotations of 90° about the tetrad axis correspond to the transformation scheme of a diad axis parallel to X_3, namely

$$\begin{matrix}-1 & 0 & 0\\ 0 & -1 & 0\\ 0 & 0 & 1.\end{matrix}$$

Any g_{ij} that contains only one 3 must be zero, just as in § 9.3 we saw that if the diad axis is parallel to X_2 and g_{ij} containing one 2 is zero. Thus the scheme of gs is

$$\begin{matrix} g_{11} & g_{12} & 0 \\ & g_{22} & 0 \\ & & g_{33}. \end{matrix}$$

To the non-zero gs we now apply the transformation for a rotation of 90° about the axis X_3, namely

$$\begin{matrix} 0 & -1 & 0 \\ 1 & 0 & 0 \\ 0 & 0 & 1. \end{matrix}$$

This gives the following relations

$$g'_{11} = c_{12}c_{12}g_{22} = -1\times -1\times g_{22} = g_{22} = g_{11},$$

$$g'_{12} = c_{12}c_{21}g_{21} = -1\times 1\times g_{21} = -g_{21} = g_{12} = 0,$$

$$g'_{33} = c_{33}c_{33}g_{33} = 1\times 1\times g_{33} = g_{33} \neq 0.$$

Thus the g scheme is

$$\begin{matrix} g_{11} & 0 & 0 \\ & g_{11} & 0 \\ & & g_{33}. \end{matrix}$$

The transformation matrix for the diad axis bisecting the angle between axes X_1 and X_2 is

$$\begin{matrix} 0 & 1 & 0 \\ 1 & 0 & 0 \\ 0 & 0 & -1. \end{matrix}$$

Applying this to the above g matrix we obtain

$$g'_{11} = c_{12}c_{12}g_{22} = 1\times 1\times g_{22} = g_{11}\ ,$$

$$g_{33} = c_{33}c_{33}g_{33} = -1\times -1\times g_{33} = g_{33},$$

so that the g matrix is unchanged by the operation of this diad axis.

Q.9.2. The transformation matrix for the axes of the bismuth plate is

	X_1	X_2	X_3
X_1'	$\cos\theta$	$\sin\theta$	0
X_2'	$-\sin\theta$	$\cos\theta$	0
X_3'	0	0	1.

The required Hall constant is ρ'_{123}. The non-zero undashed ρs for bismuth give the following result.

$$\rho'_{123} = c_{11}c_{22}c_{33}\rho_{123} + c_{12}c_{21}c_{33}\rho_{213}$$
$$= \cos^2\theta\ \rho_{123} - \sin^2\theta\ \rho_{213}$$
$$= (\cos^2\theta + \sin^2\theta)\ \rho_{123}$$
$$= \rho_{123}.$$

Thus the Hall constant is independent of θ for a plate normal to X_3.

10

Photoelasticity

10.1. Introduction

When any transparent material is subjected to stress it changes its refractive index. It becomes doubly refracting if originally it was isotropic and, if it was anisotropic, it changes its double refraction. The phenomena of photoelasticity find great application in relation to the design of all kinds of solid objects that have to withstand stress, e.g. wheels, beams, and frameworks. For this purpose a model is usually made in gelatine or some plastic and then viewed between crossed polaroids in white or monochromatic light. The double refraction caused by the applied stress enables the changes in refractive index to be measured and, if the photoelastic properties of the material are known, the stresses applied can be found from the optical pattern. It is also possible to study photoelasticity with an interferometer. A beam of light is split, and one part passes through the specimen under test and the other through a standard reference piece, either of the same or of a different material. The path difference between the two beams is measured by the interference pattern. When the piece under test is loaded there is a change of the interference pattern from which the stress produced at particular points can be evaluated. Although the apparatus required for this is more elaborate than the polarized-light apparatus, the results are easier to interpret.

10.2. The photoelastic tensor

The optical properties of any crystal can be represented by an ellipsoid–triaxial for triclinic, monoclinic, and orthorhombic crystals, an ellipsoid of revolution for trigonal, hexagonal, and tetragonal crystals, and a sphere for cubic crystals. In the unstressed state, we can describe this ellipsoid in the most general case by the expression

$$B_{11}^0 x_1^2 + B_{22}^0 x_2^2 + B_{33}^0 x_3^2 + 2B_{23}^0 x_2 x_3 + 2B_{31}^0 x_3 x_1 + 2B_{12}^0 x_1 x_2 = 1.$$

When stress is applied all these Bs can change and this is denoted by dropping the superscript '0'. Then we can write, using p_{ijkl} to denote strain-optical coefficients (i, j, k, l run from 1 to 3),

$$B_{11} - B_{11}^0 = p_{11kl} r_{kl},$$

$$B_{22} - B_{22}^0 = p_{22kl} r_{kl},$$

$$B_{33} - B_{33}^0 = p_{33kl} r_{kl},$$

$$B_{23} - B_{23}^0 = p_{23kl} r_{kl},$$

$$B_{31} - B_{31}^0 = p_{31kl} r_{kl},$$

$$B_{12} - B_{12}^0 = p_{12kl} r_{kl}, \text{ etc.}$$

or, in abbreviated notation (see § 7.3), $B_i - B_i^0 = p_{ik} r_k$ where i and k run from 1 to 6.

In terms of stress we can write similarly

$$B_i - B_i^0 = -q_{ik} t_k$$

(The negative sign arises if stretching forces are counted positive.)

The strain-optical (sometimes called elasto-optical) p_{ik} and stress-optical (sometimes called piezo-optical) coefficients q_{ik} may be shown to be related by the equations

$$p_{ik} = q_{ij} c_{jk},$$

$$q_{ik} = p_{ij} s_{jk},$$

where c_{jk} and s_{jk} are elastic stiffnesses and compliances respectively.

It was mentioned in § 7.3 that the first pair of suffixes in p_{ijkl} or q_{ijkl} cannot be interchanged with the second pair as is the case with the compliances and stiffnesses. This is clearly so if we take a particular example, e.g. p_{1123}. This coefficient occurs in the equation

$$B_{11} - B_{11}^0 = p_{1123} r_{23} + \text{etc.}$$

and gives the change of the term B_{11} due to a strain that changes the angle between axes X_2 and X_3. The coefficient p_{2311}, however, occurs in the equation

$$B_{23} - B_{23}^0 = p_{2311} r_{11} + \text{etc.}$$

and relates the change in B_{23} to the change in length per unit length in the X_1 direction. There is no necessary connection between p_{1123} and p_{2311}. There are therefore in general 6×6 independent photoelastic coefficients (not 21 as in elasticity). The same necessity arises here to insert 2s in certain of the equations as in the study of elasticity. The *p*s and *q*s having four suffixes are treated as true tensors which transform according to the usual law. Writing the expressions in full we obtain

$$\begin{aligned} B_{11}-B_{11}^0 &= p_{1111}r_{11}+p_{1112}r_{12}+p_{1121}r_{21}+\text{etc}, \\ &= p_{1111}r_{11}+p_{1112}(r_{12}+r_{21})+\text{etc}, \\ &= p_{11}r_{12}+2p_{16}r_6+\text{etc}. \end{aligned}$$

However, $2r_6$ is the total change of angle between the axes X_1 and X_2 and the value given in physical tables relates $B_{11}-B_{11}^0$ to $2r_6$, not to r_6, i.e. it gives p_{16}. Similarly we have

$$B_{23}-B_{23}^0 = 2p_{45}r_5+\text{etc},$$

and $2r_5$ is the total change of angle between the axes X_1 and X_3. The physical tables therefore quote p_{45}. Thus no factor 2 is involved in going from the usually quoted tensor component to the true tensor value.

The relations involving the stress-optical coefficients are a little different. Thus we may write

$$\begin{aligned} B_{11}-B_{11}^0 &= -q_{1112}t_{12}-q_{1121}t_{21}+\text{etc.}, \\ &= -2q_{16}t_6+\text{etc.}, \\ B_{23}-B_{23}^0 &= -2q_{45}t_5+\text{etc.} \end{aligned}$$

Physical tables give the ratio of the difference of the *B*s to the stress component, and thus when 4, 5, or 6 are involved a factor 2 is introduced.

10.3. The photoelastic coefficients in relation to point–group symmetry

It must be noted that, since stress and strain, and also the change of refractive index, are not affected by the presence or absence of a centre of symmetry, all point groups that differ only in the possession of a centre of symmetry must have the same non-zero sets of ps and qs.

Point group mmm

For the transformation of the true four-suffix tensor, we can write

$$p'_{ijkl} = c_{im}c_{jn}c_{kr}c_{ls}p_{mnrs}.$$

The transformation for a diad axis parallel to X_1 is

	X_1	X_2	X_3
X_1'	1	0	0
X_2'	0	−1	0
X_3'	0	0	−1.

Since only c_{11}, c_{22}, and c_{33} are involved, for any p containing one or three 1s the result must be a zero coefficient. In abbreviated notation this gives the result

$$\begin{matrix} p_{11} & p_{12} & p_{13} & p_{14} & 0 & 0 \\ p_{21} & p_{22} & p_{23} & p_{24} & 0 & 0 \\ p_{31} & p_{32} & p_{33} & p_{34} & 0 & 0 \\ p_{41} & p_{42} & p_{43} & p_{44} & 0 & 0 \\ 0 & 0 & 0 & 0 & p_{55} & p_{56} \\ 0 & 0 & 0 & 0 & p_{65} & p_{66}. \end{matrix}$$

For a diad axis parallel to X_2, the coefficients containing one or three 2s must be zero, and for a diad axis parallel to X_3 the coefficients containing one or three 3s must be zero.

Thus, for *mmm* all true tensor four-suffix coefficients containing one or three of any number 1, 2, 3 must be zero. This gives the scheme

$$\begin{matrix} p_{11} & p_{12} & p_{13} & 0 & 0 & 0 \\ p_{21} & p_{22} & p_{23} & 0 & 0 & 0 \\ p_{31} & p_{32} & p_{33} & 0 & 0 & 0 \\ 0 & 0 & 0 & p_{44} & 0 & 0 \\ 0 & 0 & 0 & 0 & p_{55} & 0 \\ 0 & 0 & 0 & 0 & 0 & p_{66}. \end{matrix}$$

Point group m3m

Starting from the non-zero *p*s of point group *mmm*, we make X_1, X_2, X_3 all equivalent. The transformation is

	X_1	X_2	X_3
X_1'	0	1	0
X_2'	0	0	1
X_3'	1	0	0.

Then we have

$$p'_{1111} = c_{12}c_{12}c_{12}c_{12}p_{2222} = p_{2222} = p_{1111} \text{ (by Neumann's Principle)},$$

$$p'_{2222} = c_{23}c_{23}c_{23}c_{23}p_{3333} = p_{3333} = p_{2222} = p_{1111},$$

$$p'_{1122} = c_{12}c_{12}c_{23}c_{23}p_{2233} = p_{2233} = p_{1122},$$

$$p'_{1133} = c_{12}c_{12}c_{31}c_{31}p_{2211} = p_{2211} = p_{1133},$$

$$p'_{2233} = c_{23}c_{23}c_{31}c_{31}p_{3311} = p_{3311} = p_{2233} = p_{1122},$$

$$p'_{2211} = c_{23}c_{23}c_{12}c_{12}p_{3322} = p_{3322} = p_{2211} = p_{1133},$$

$$p'_{3311} = c_{31}c_{31}c_{12}c_{12}p_{1122} = p_{1122} = p_{3311},$$

$$p'_{3322} = c_{31}c_{31}c_{23}c_{23}p_{1133} = p_{1133} = p_{3322}.$$

Hence we have the following scheme for these coefficients.

$$\begin{matrix} p_{11} & p_{12} & p_{13} \\ p_{13} & p_{11} & p_{12} \\ p_{12} & p_{13} & p_{11}. \end{matrix}$$

Further,

$$p'_{2323} = c_{23}c_{31}c_{23}c_{31}p_{3131} = p_{3131} = p_{2323},$$

$$p'_{3131} = c_{31}c_{12}c_{31}c_{12}p_{1212} = p_{1212} = p_{3131},$$

and hence

$$p_{44} = p_{55} = p_{66}.$$

The transformation for the diagonal plane of symmetry corresponding to the terminal *m* in *m*3*m* is given by the matrix

	X_1	X_2	X_3
X_1'	0	1	0
X_2'	1	0	0
X_3'	0	0	1.

Hence,

$$p'_{1122} = c_{12}c_{12}c_{21}c_{21}p_{2211} = p_{2211} = p_{1122}.$$

Thus the final scheme becomes

$$\begin{matrix} p_{11} & p_{12} & p_{12} & 0 & 0 & 0 \\ p_{12} & p_{11} & p_{12} & 0 & 0 & 0 \\ p_{12} & p_{12} & p_{11} & 0 & 0 & 0 \\ 0 & 0 & 0 & p_{44} & 0 & 0 \\ 0 & 0 & 0 & 0 & p_{44} & 0 \\ 0 & 0 & 0 & 0 & 0 & p_{44}. \end{matrix}$$

Point group 32

A diad axis parallel to X_1 gives

$$\begin{matrix} p_{11} & p_{12} & p_{13} & p_{14} & 0 & 0 \\ p_{21} & p_{22} & p_{23} & p_{24} & 0 & 0 \\ p_{31} & p_{32} & p_{33} & p_{34} & 0 & 0 \\ p_{41} & p_{42} & p_{43} & p_{44} & 0 & 0 \\ 0 & 0 & 0 & 0 & p_{55} & p_{56} \\ 0 & 0 & 0 & 0 & p_{65} & p_{66}. \end{matrix}$$

A triad axis parallel to X_3 transforms according to the scheme

$$\begin{matrix} -\frac{1}{2} & \frac{\sqrt{3}}{2} & 0 \\ -\sqrt{\frac{3}{2}} & -\frac{1}{2} & 0 \\ 0 & 0 & 1. \end{matrix}$$

We shall consider the non-zero ps in the above matrix in order.

$$\begin{aligned} p'_{1111} = {} & c_{11}c_{11}c_{11}c_{11}p_{1111} + c_{12}c_{12}c_{12}c_{12}p_{2222} + \\ & + c_{11}c_{11}c_{12}c_{12}p_{1122} + c_{12}c_{12}c_{11}c_{11}p_{2211} + \\ & + c_{11}c_{12}c_{11}c_{12}p_{1212} + c_{12}c_{11}c_{12}c_{11}p_{2121} + \\ & + c_{11}c_{12}c_{12}c_{11}p_{1221} + c_{12}c_{11}c_{11}c_{12}p_{2112} \\ = {} & \tfrac{1}{16}(p_{11} + 9p_{22} + 3p_{12} + 3p_{21} + 12p_{66}) = p_{11}, \end{aligned}$$

i.e. $\quad -15p_{11}+9p_{22}+3p_{12}+3p_{21}+12p_{66} = 0,$

or $\quad -5p_{11}+3p_{22}+p_{12}+p_{21}+4p_{66} = 0. \qquad (10.1)$

$$\begin{aligned}p'_{2222} &= c_{21}c_{21}c_{21}c_{21}p_{1111}+c_{22}c_{22}c_{22}c_{22}p_{2222}+\\&\quad +c_{21}c_{21}c_{22}c_{22}p_{1122}+c_{22}c_{22}c_{21}c_{21}p_{2211}+\\&\quad +c_{21}c_{22}c_{21}c_{22}p_{1212}+c_{22}c_{21}c_{22}c_{21}p_{2121}+\\&\quad +c_{21}c_{22}c_{22}c_{21}p_{1221}+c_{22}c_{21}c_{21}c_{22}p_{2112}\\&= \tfrac{1}{16}(9p_{11}+p_{22}+3p_{12}+3p_{21}+12p_{66}) = p_{22},\end{aligned}$$

i.e. $\quad 9p_{11}-15p_{22}+3p_{12}+3p_{21}+12p_{66} = 0,$

or $\quad 3p_{11}-5p_{22}+p_{12}+p_{21}+4p_{66} = 0. \qquad (10.2)$

Subtracting equation (10.2) from equation (10.1), we obtain

$$-8p_{11}+8p_{22} = 0,$$

or $\quad p_{11} = p_{22}.$

For the next non-zero coefficient,

$$\begin{aligned}p'_{2211} &= c_{21}c_{21}c_{11}c_{11}p_{1111}+c_{21}c_{21}c_{12}c_{12}p_{1122}+\\&\quad +c_{22}c_{22}c_{11}c_{11}p_{2211}+c_{22}c_{22}c_{12}c_{12}p_{2222}+\\&\quad +c_{21}c_{22}c_{11}c_{12}p_{1212}+c_{21}c_{22}c_{12}c_{11}p_{1221}+\\&\quad +c_{22}c_{21}c_{11}c_{12}p_{2112}+c_{22}c_{21}c_{12}c_{11}p_{2121}\\&= \tfrac{1}{16}(3p_{11}+9p_{12}+p_{21}+3p_{22}-3p_{66}-3p_{66}-3p_{66}-3p_{66})\\&= p_{21}.\end{aligned}$$

Replacing p_{22} by p_{11} we obtain

$$6p_{11}-15p_{21}+9p_{12}-12p_{66} = 0. \qquad (10.3)$$

$$\begin{aligned}p'_{1122} &= c_{11}c_{11}c_{21}c_{21}p_{1111}+c_{11}c_{11}c_{22}c_{22}p_{1122}+\\&\quad +c_{12}c_{12}c_{21}c_{21}p_{2211}+c_{12}c_{12}c_{22}c_{22}p_{2222}+\\&\quad +c_{11}c_{12}c_{21}c_{22}p_{1212}+c_{11}c_{12}c_{22}c_{21}p_{1221}+\\&\quad +c_{12}c_{11}c_{21}c_{22}p_{2112}+c_{12}c_{11}c_{22}c_{21}p_{2121}\\&= \tfrac{1}{16}(3p_{11}+p_{12}+9p_{21}+3p_{22}-3p_{66}-3p_{66}-3p_{66}-3p_{66})\\&= p_{12}.\end{aligned}$$

Replacing p_{22} by p_{11} we obtain

$$6p_{11}+9p_{21}-15p_{12}-12p_{66} = 0. \tag{10.4}$$

Subtracting equation (10.3) from equation (10.4), we obtain

$$24p_{21}-24p_{12} = 0$$

or

$$p_{12} = p_{21}.$$

Substituting p_{12} for p_{21} in equation (10.3) we have

$$6p_{11}-6p_{12}-12p_{66} = 0$$

or

$$p_{66} = \tfrac{1}{2}(p_{11}-p_{12}).$$

So

$$\begin{aligned} p'_{1133} &= c_{11}c_{11}c_{33}c_{33}p_{1133}+c_{12}c_{12}c_{33}c_{33}p_{2233} \\ &= \tfrac{1}{4}(p_{13}+3p_{23}) = p_{13}, \end{aligned}$$

i.e.

$$-3p_{13}+3p_{23} = 0,$$

$$p_{13} = p_{23}.$$

Next,

$$\begin{aligned} p'_{1123} &= c_{11}c_{11}c_{22}c_{33}p_{1123}+c_{12}c_{12}c_{22}c_{33}p_{2223}+ \\ &\qquad +2c_{11}c_{12}c_{21}c_{33}p_{1213} \\ &= \tfrac{1}{8}\{-p_{14}+(-3)p_{24}+6p_{65}\} = p_{14}, \end{aligned}$$

or

$$-3p_{14}-p_{24}+2p_{65} = 0. \tag{10.5}$$

$$\begin{aligned} p'_{2223} &= c_{21}c_{21}c_{22}c_{33}p_{1123}+c_{22}c_{22}c_{22}c_{33}p_{2223}+ \\ &\qquad +2c_{21}c_{22}c_{21}c_{33}p_{1213} \\ &= \tfrac{1}{8}(-3p_{14}-p_{24}-6p_{65}) = p_{24}, \end{aligned}$$

i.e.

$$-p_{14}-3p_{24}-2p_{65} = 0. \tag{10.6}$$

Addition of equations (10.5) and (10.6) gives

$$-4p_{14}-4p_{24} = 0$$

or

$$p_{24} = -p_{14}.$$

For p'_{3311} we have

$$p'_{3311} = c_{33}c_{33}c_{11}c_{11}p_{3311}+c_{33}c_{33}c_{12}c_{12}p_{3322}$$
$$= \tfrac{1}{4}(p_{31}+3p_{32}) = p_{31},$$

or

$$-p_{31}+p_{32} = 0,$$

i.e. $\qquad p_{32} = p_{31}.$

Next,

$$p'_{3333} = c_{33}c_{33}c_{33}c_{33}p_{3333} = p_{3333} \neq 0.$$

For p'_{3323} we have

$$p'_{3323} = c_{33}c_{33}c_{22}c_{33}p_{3323}$$
$$= -\tfrac{1}{2}p_{34} = p_{34};$$

hence $p_{34} = 0$.

Similarly $p_{43} = 0$.

For the next coefficient,

$$p'_{2311} = c_{22}c_{33}c_{11}c_{11}p_{2311}+2c_{21}c_{33}c_{11}c_{12}p_{1312}+$$
$$+c_{22}c_{33}c_{12}c_{12}p_{2322}$$
$$= \tfrac{1}{8}(-p_{41}-3p_{42}+6p_{56}) = p_{41},$$

i.e. $$-3p_{41}-p_{42}+2p_{56} = 0. \tag{10.7}$$

We have also

$$p'_{2322} = c_{22}c_{33}c_{21}c_{21}p_{2311}+c_{22}c_{33}c_{22}c_{22}p_{2322}+$$
$$+2c_{21}c_{33}c_{21}c_{22}p_{1312}$$
$$= \tfrac{1}{8}(-3p_{41}-p_{42}-6p_{56}) = p_{42},$$

and so

$$-p_{41}-3p_{42}-2p_{56} = 0. \tag{10.8}$$

Adding equations (10.7) and (10.8) we get

$$-4p_{41}-4p_{42} = 0,$$

and hence

$$p_{42} = -p_{41}.$$

From equation (10.8) we obtain, after putting $p_{42} = -p_{41}$,

$$p_{56} = p_{41}.$$

In the same way we may show that $p_{65} = p_{14}$. Finally,

$$\begin{aligned} p'_{2323} &= c_{21}c_{33}c_{21}c_{33}p_{1313}+c_{22}c_{33}c_{22}c_{33}p_{2323} \\ &= \tfrac{1}{4}(3p_{55}+p_{44}) = p_{44}; \end{aligned}$$

hence $\quad 3p_{55} = 3p_{44}$

or $\quad p_{44} = p_{55}.$

The final matrix is therefore

$$\begin{matrix} p_{11} & p_{12} & p_{13} & p_{14} & 0 & 0 \\ p_{12} & p_{11} & p_{13} & -p_{14} & 0 & 0 \\ p_{31} & p_{31} & p_{33} & 0 & 0 & 0 \\ p_{41} & -p_{41} & 0 & p_{44} & 0 & 0 \\ 0 & 0 & 0 & 0 & p_{44} & p_{41} \\ 0 & 0 & 0 & 0 & p_{14} & \tfrac{1}{2}(p_{11}-p_{12}) \end{matrix}$$

(1, p. 182; 13, p. 245).

Q.10.1. Which are the non-zero strain-optical coefficients in point groups 4, 422, and 23?

This chapter concludes our study of the applications of tensors to the physical properties of crystals. There are a number of tensors of higher order than those we have considered, but the principles governing their use are the same as those already discussed in the preceding chapters.

The symmetry utilized in the discussion up to this point is just that employed in morphological crystallography. This is not as precise or detailed as that employed in group theory, and in Part II of this book we shall see how this more detailed analysis permits a more direct study of the physical constants associated with given physical properties (see Chapter 17). The tensor method and the group theory approach are complementary to one another. Questions involving symmetry are more thoroughly dealt with by group theory, but the actual physical magnitudes can usually be determined only by using a tensor analysis.

Answers

Q.10.1. A tetrad axis parallel to X_3 has the transformation

$$\begin{matrix} 0 & 1 & 0 \\ -1 & 0 & 0 \\ 0 & 0 & 1. \end{matrix}$$

This can be applied to the non-zero ps for a diad axis parallel to X_3, namely

$$\begin{matrix} p_{11} & p_{12} & p_{13} & 0 & 0 & p_{16} \\ p_{21} & p_{22} & p_{23} & 0 & 0 & p_{26} \\ p_{31} & p_{32} & p_{33} & 0 & 0 & p_{36} \\ 0 & 0 & 0 & p_{44} & p_{45} & 0 \\ 0 & 0 & 0 & p_{54} & p_{55} & 0 \\ p_{61} & p_{62} & p_{63} & 0 & 0 & p_{66}. \end{matrix}$$

$$p'_{1111} = c_{12}c_{12}c_{12}c_{12}p_{2222} = p_{22} = p_{11},$$

$$p'_{2222} = c_{21}c_{21}c_{21}c_{21}p_{1111} = p_{11} = p_{22},$$

$$p'_{3333} = c_{33}c_{33}c_{33}c_{33}p_{3333} = p_{33} = p_{33},$$

$$p'_{1122} = c_{12}c_{12}c_{21}c_{21}p_{2211} = p_{21} = p_{12},$$

$$p'_{1133} = c_{12}c_{12}c_{33}c_{33}p_{2233} = p_{23} = p_{13},$$

$$p'_{3311} = c_{33}c_{33}c_{12}c_{12}p_{3322} = p_{32} = p_{31},$$

$$p'_{1112} = c_{12}c_{12}c_{12}c_{21}p_{2221} = 1 \times 1 \times 1 \times -1 \times p_{26} = -p_{26} = p_{16},$$

$$p'_{3312} = c_{33}c_{33}c_{12}c_{21}p_{3321} = 1 \times 1 \times 1 \times -1 \times p_{36} = p_{36} = 0,$$

$$p'_{2323} = c_{21}c_{33}c_{21}c_{33}p_{1313} = p_{55} = p_{44},$$

$$p'_{1212} = c_{12}c_{21}c_{12}c_{21}p_{2121} = p_{66},$$

$$p'_{1211} = c_{12}c_{21}c_{12}c_{12}p_{2122} = -p_{62} = p_{61},$$

$$p'_{1233} = c_{12}c_{21}c_{33}c_{33}p_{2133} = -p_{63} = p_{63} = 0,$$

$$p'_{2331} = c_{21}c_{33}c_{33}c_{12}p_{1332} = -p_{54} = p_{45}.$$

Hence the scheme of ps for point group 4 is as follows.

$$\begin{matrix} p_{11} & p_{12} & p_{13} & 0 & 0 & p_{16} \\ p_{12} & p_{11} & p_{13} & 0 & 0 & -p_{16} \\ p_{31} & p_{31} & p_{33} & 0 & 0 & 0 \\ 0 & 0 & 0 & p_{44} & p_{45} & 0 \\ 0 & 0 & 0 & -p_{45} & p_{44} & 0 \\ p_{61} & -p_{61} & 0 & 0 & 0 & p_{66}. \end{matrix}$$

Point group 422: we add to point group 4 a diad axis parallel to X_1. This has the effect of making zero any coefficient that has single 1s or three 1s in its four suffixes. Thus p_{16}, p_{61}, p_{45} all become zero and the resulting scheme is

$$\begin{matrix} p_{11} & p_{12} & p_{13} & 0 & 0 & 0 \\ p_{12} & p_{11} & p_{13} & 0 & 0 & 0 \\ p_{31} & p_{31} & p_{33} & 0 & 0 & 0 \\ 0 & 0 & 0 & p_{44} & 0 & 0 \\ 0 & 0 & 0 & 0 & p_{44} & 0 \\ 0 & 0 & 0 & 0 & 0 & p_{66}. \end{matrix}$$

Point group 23: in §10.3 the non-zero ps were found for point group $m3$ prior to imposing the diagonal plane of symmetry. Since $m3$ is the same as 23 plus a centre of symmetry, the scheme of ps must be the same for both point groups, namely

$$\begin{matrix} p_{11} & p_{12} & p_{13} & 0 & 0 & 0 \\ p_{13} & p_{11} & p_{12} & 0 & 0 & 0 \\ p_{12} & p_{13} & p_{11} & 0 & 0 & 0 \\ 0 & 0 & 0 & p_{44} & 0 & 0 \\ 0 & 0 & 0 & 0 & p_{44} & 0 \\ 0 & 0 & 0 & 0 & 0 & p_{44}. \end{matrix}$$

Part II

Group theory in relation to point groups and some physical properties of crystals

11

Character tables in point groups of low symmetry

11.1. Introduction

THE symmetry shown by the external faces of crystals has been studied since the seventeenth century. By the middle of the nineteenth century a complete account of this symmetry had been given. It formed a closed subject, and those who developed it did not relate it to the group theory of the mathematicians. During the second half of the last century, the study of crystal symmetry was extended to include the symmetry of the arrangements of atoms and molecules in crystals. The subject up to this time did not greatly suffer through being self-contained, but during the last few decades this situation has changed. The movements, in contrast with the positions, of atoms in crystals have become important. This might be called a dynamic aspect of crystal symmetry as compared with the static aspect previously studied. Formal group theory is essential for the study of this dynamic aspect.

There are a number of difficulties for the crystallographer who takes up the study of group theory. One is the use of certain words to mean different things in crystallography and in group theory. Thus, to the crystallographer, the word 'class' has meant one of the possible combinations of axes, planes of symmetry and centres of symmetry presented by the external faces of crystals. In group theory 'class' has the meaning of a group of symmetry elements that are related to one another by the other symmetry elements in the group, e.g. the three diad axes in quartz form such a group-theory class.

The word 'element' to the crystallographer means just one element of symmetry but in group theory an 'element' is any single symmetry operation. Thus in crystallography a triad axis is simply a triad axis, and only one element of symmetry, but in group theory it comprises two elements, one a rotation about a given axis through 120° anticlockwise and one a rotation about the same axis through 120° clockwise. In the stereographic projection of general equivalent positions in a given point group, each pole corresponds to one 'element' in the group-theoretical sense. It is unusual to regard a given symmetry operation as several distinct symmetry operations, but, in group theory, a tetrad axis, for example, not only becomes two tetrad axes, one right-handed and the other left-handed, but also involves a diad axis.

In crystallographic drawings we have representations of crystals, but a group theory representation for any point group is a selection of the numbers ± 3, ± 2, ± 1, 0. Another difficulty for the crystallographer arises from the symbolism and the fact that different authors writing about group theory use different symbols for the same thing. It is also a hindrance to the crystallographer that the Schoenfliess symbols are often used in preference to the Hermann–Mauguin notation of the *International tables for X-ray crystallography*.

The value of the group-theoretical approach to crystallography arises from its more detailed analysis of the combinations of symmetry elements. This is especially valuable in discussing the physical properties of crystals. On the whole, the discussion of the number of physical constants required to define a physical property of a crystal is more elegantly derived by group-theoretical than by tensor-analysis methods, which have been explained in Chapters 1–10. In fact, historically the number of photoelastic constants in a cubic crystal was first (incorrectly) derived by tensor methods (see Chapter 10), and then after some years this was shown to be in error by group-theoretical methods. The more complicated the physical property becomes, the more useful is the group-theoretical method. The study of vibrations—elastic or optical—is also very well dealt with by group-theoretical methods. Thus the number of infrared absorption lines or Raman spectral lines that can occur in a crystal of given symmetry can be worked out by these methods. The vibrations postulated in wave mechanics, which play such an important part in the theory of conductors and semiconductors, are also amenable to a treatment by group theory. In all of these aspects of the study of crystals, group theory extends the theory of symmetry with which crystallographers normally deal.

11.2. Character tables†

The combinations of symmetry elements that can be arranged about, or through, a given point as origin, in a self-consistent manner, form the thirty-two point groups (see Chapter 1). The results of this study can be expressed in the form of thirty-two stereograms giving the distribution of directions joining the equivalent points to the origin (see Appendix 1). Coordinates x, y, z can be given to these equivalent points, and a set of such coordinates defines the symmetry elements present in the given point group.

Non-centrosymmetric point groups

Point group m (C_s).

In the point group m the equivalent points are x, y, z and $x, y, \bar{z}$ when the plane of symmetry is normal to z. Group theory distinguishes between the different relations that exist between the elements of symmetry and between the coordinates. In the above example there are only two elements of symmetry in the point group, namely E, direct superposition, and σ_h, a plane of symmetry perpendicular to z. Now the square of each of the coordinates of the two equivalent points in the point group m is the same, whereas the first power of the z coordinate is positive for the element E and negative for the element σ_h. This relationship is expressed in what is called a character table as follows (Table 11.1).

TABLE 11.1

Character table of the point group m (C_s)

	E	σ_h	
A'	1	1	x^2, y^2, z^2
A''	1	-1	z

† A complete set of character tables for the thirty-two point groups is given in Appendix 1.

A table of equivalent Hermann–Mauguin and Schoenflies symbols is given in Appendix 2. The Schoenflies symbols are given in brackets following the Hermann–Mauguin symbols.

The symbols A', A'' distinguish between the row in the table which, like z^2, is symmetric, i.e. $+1$, for the σ_h column, and the row which, like z, is antisymmetric, i.e. -1, for this column. The two rows are called *representations* of the point group. The functions of the coordinates in the right-hand column are such that, if under any symmetry element the sign of the coordinate, or function of the coordinate, is changed from what it is under the identity operator E, then a -1 is entered in the column corresponding to that symmetry element. The functions of the coordinates that can be put in the right-hand column are not limited to those given here. For instance, the A' representation could have $x, y, x^2+y^2, x^2+y^2+z^2, xy$ in addition to those given in Table 11.1. Similarly, the representation A'' could have xyz in addition to z. In practice, powers of the coordinates higher than the third are seldom used.

It may seem at first that too much attention is being devoted to a small difference. This is not so, as the subsequent work will show. Tables of this kind may be drawn up for all the point groups and they form the basis for much of the analysis provided by group theory. As will be seen in subsequent chapters, the representations consist of matrices. In most cases these are 1×1, but in some cases they are 2×2 or 3×3 matrices. The sum of the diagonal elements of a matrix is defined as its 'character', and it is these numbers that appear in the character tables. The simplest matrix is 1×1 and it can have the character $+1$ or -1. In the first few point groups that we shall study, the character tables have only $+1$ or -1. In point groups with higher symmetry, representations occur in which the characters are 0, $+2$ or -2, $+3$ or -3.

In addition to the coordinates x, y, z and functions of these coordinates, the quantities R_x, R_y, R_z also appear in the right-hand columns of the tables. The Rs correspond to infinitesimally small rotations about the axes and may be regarded as axial vectors (see Chapter 9), and they are transformed in the same way as polar vectors, e.g. x, y, z, by axes of symmetry but oppositely by planes or centres of symmetry. Thus, in the point groups containing only axes of symmetry R_x, R_y, R_z are placed alongside x, y, z respectively, but in all other point groups, containing planes or centres of symmetry, the Rs and corresponding coordinates are associated with different representations. When, for example, R_x appears at the end of a row in the character table and a $+1$ occurs under any given symmetry element, then an infinitesimally small rotation about the axis X from the original positions of the equivalent points, is not inconsistent with the symmetry of the corresponding element given at the top of the column containing the $+1$. If, however, a -1 occurs in the column, then the rotation is not consistent with that symmetry element. In point group m, for example, a small rotation about the axis z does not contravene the symmetry σ_h, and so R_z occurs in the A' representation. R_x and R_y both contravene this symmetry and therefore occur in the A'' representation.

Point group 222 (D_2).

The elements of symmetry present in the point group 222 are E, C_{2x}, C_{2y}, and C_{2z}. The equivalent points in general positions have coordinates corresponding to these symmetry elements, namely

$$x, y, z; \quad x, \bar{y}, \bar{z}; \quad \bar{x}, y, \bar{z}; \quad \text{and } \bar{x}, \bar{y}, z.$$

The character table is shown below (Table 11.2).

TABLE 11.2

Character table of point group 222 (D_2)

	E	C_{2z}	C_{2y}	C_{2x}		
A	1	1	1	1		x^2, y^2, z^2
B_2	1	-1	1	-1	y, R_y	xz
B_1	1	1	-1	-1	z, R_z	xy
B_3	1	-1	-1	1	x, R_x	yz

Each horizontal row of the table is called a representation of the point group. The first row is denoted A because the characters referring to the E column and to the axes of symmetry are $+1$. These characters are called 'symmetric'. The following three rows are denoted B because the characters in the E column are $+1$ and some of the characters under the axes are -1. The expressions involving x, y, z, R_x, R_y and R_z on the right-hand side of the table are those for which the $+1$ or -1 applies to the particular symmetry element considered. The first representation is associated with second powers of x, y, z, the signs of which are all $+1$. Each symmetry element is therefore represented by $+1$. The second row of the table refers to y or the product of z and x. (First-order and second-order terms involving the coordinates are very important in the study of infrared and Raman spectra and also in other vibrational problems.) The number in this row of the table in any particular column corresponds to the sign of y, or of zx, produced by the operation of the symmetry element at the top of the column. The third and fourth rows similarly give the signs of z, xy and x, yz respectively. The symbols R_x, R_y, R_z stand for very small rotations about the axis denoted by the subscripts. Fig. 11.1 shows that a small rotation in,

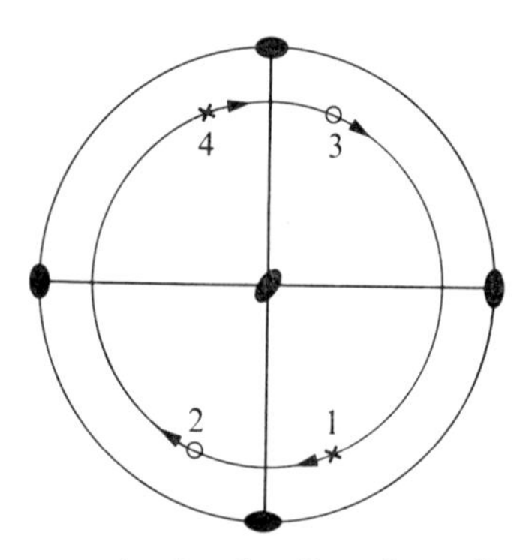

FIG. 11.1. Stereogram showing the effect of a small rotation R_z on the poles corresponding to point group 222 (D_2).

say, a clockwise sense of the poles marked 1, 2, 3, 4 about the diad axis perpendicular to the paper, C_{2z}, conserves the symmetry of that axis. In the third row of the table this corresponds to $+1$ in the column headed C_{2z}. The same small rotations of poles 1 and 2 are not consistent with C_{2x}, nor are the rotations of poles 1 and 3 consistent with C_{2y}. Thus -1 occurs in the columns headed C_{2x} and C_{2y} in the third row of the character table. Similar reasoning requires R_x and R_y to be placed in the fourth and second rows respectively of the table. It should be noted that in this point group all diad axes convert a plus coordinate into either the same plus or a negative coordinate. Consequently, in all except the top row, and in all except the left-hand column, there are equal numbers of $+1$s and -1s. This necessitates as many rows as columns or, in other words, as many representations as there are classes. This result will be found to apply to all the character tables. The coordinates x, y, z are important in considering atomic and molecular vibrations that involve translations, and R_x, R_y, R_z are used in the study of molecular oscillations (see Chapter 18).

*Point group mm*2 (C_{2v}).

The elements of symmetry present in point group $mm2$ (C_{2v}) are E, C_2, σ_x, and σ_y. The C_2 axis is taken parallel to the z axis, and the two planes of symmetry denoted σ_y and σ_x coincide with the coordinate planes xz and yz respectively. The character table is given in Table 11.3.

TABLE 11.3

Character table for point group mm2 (C_{2v})

	E	C_2	σ_y	σ_x		
A_1	1	1	1	1	z	x^2, y^2, z^2
B_1	1	-1	1	-1	x, R_y	xz
A_2	1	1	-1	-1	R_z	xy
B_2	1	-1	-1	1	y, R_x	yz

The As in the first column correspond with $+1$ under C_2; the Bs correspond with -1 under the axis C_2. The coordinates of the equivalent points in this point group and the corresponding elements of symmetry are as follows.

$$x, y, z \qquad E$$
$$\bar{x}, \bar{y}, z \qquad C_2$$
$$x, \bar{y}, z \qquad \sigma_y$$
$$\bar{x}, y, z \qquad \sigma_x.$$

Since z is positive for all the elements of symmetry, it corresponds to the A_1 representation. The signs of x^2, y^2, and z^2 are all positive for the four elements of symmetry and so correspond to the same representation. The sign of x is positive for E and σ_y and negative for C_2 and σ_x. Thus it corresponds to the B_1 representation. Fig. 11.2 shows by the arrows the small rotation corresponding to R_y.

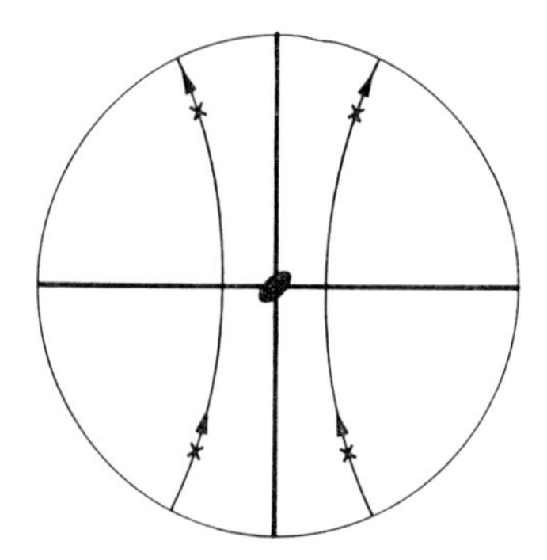

FIG. 11.2. Stereogram showing the effect of a small rotation R_y on the poles corresponding to point group *mm*2 (C_{2v}).

The symmetry of the arrows satisfies σ_y but contravenes the C_2 axis and the σ_x plane. Thus it also belongs to the B_1 representation. Clearly, since z is positive for all four sets of coordinates, xz has the same sign as x and belongs to the same representation, i.e. B_1. The relation of the A_2 representation to R_z and xy follows the same principles, and similarly for the B_2 representation.

Centrosymmetric point groups

Point group $\bar{1}(C_i)$

The elements of symmetry present are E and I, a centre of symmetry, and the coordinates of the equivalent points are x, y, z and $\bar{x}, \bar{y}, \bar{z}$. The character table is as shown in Table 11.4.

TABLE 11.4

Character table of point group $\bar{1}$ (C_i)

	E	I		
A_g	1	1	R_x, R_y, R_z	$x^2, y^2, z^2, xy, xz, yz$
A_u	1	-1	x, y, z	

The representations are A-type because the character in the E column is 1 and there are no axes of symmetry. The subscript 'g' (German *gerade*, meaning 'even') implies a symmetric representation under I, i.e. $+1$ in the I column. The subscript 'u' (German *ungerade*, meaning 'odd') implies an antisymmetric representation under I, i.e. -1 in the I column. Fig. 11.3 shows the small rotation R_x of the

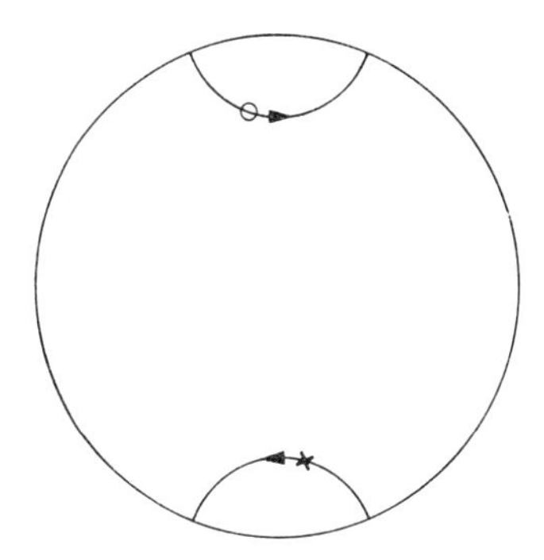

FIG. 11.3. Stereogram showing the effect of a small rotation R_x on the poles corresponding to point group $\bar{1}$ (C_i).

poles corresponding to the two equivalent points. This rotation moves the poles to positions satisfying the centre of symmetry and thus it belongs to the A_g representation. A translation of both equivalent points in the direction of the X axis is not consistent with the centre and hence it belongs to the A_u representation. The rest of the character table follows in the same way. When a centre of symmetry is inserted into a point group which does not possess a centre of symmetry, the character table always consists of four quadrants, three of which are identical (top right, top left, bottom left), and the fourth (bottom right) is the same as the other three but has the signs reversed. The centre acts on every symmetry element, or group of symmetry elements, in the same way as it acts on E.

Point group $2/m$ (C_{2h}).

The character table for point group 2 is similar in the arrangement of 1s to the table for $\bar{1}$, while the character table for $2/m$ (C_{2h}), which has a centre of symmetry, is as follows (Table 11.5).

TABLE 11.5

Character table for point group $2/m$ (C_{2h})

	E	C_2	I	σ_h		
A_g	1	1	1	1	R_z	x^2, y^2, z^2, xy
B_g	1	-1	1	-1	R_x, R_y	xz, yz
A_u	1	1	-1	-1	z	
B_u	1	-1	-1	1	x, y	

It can easily be seen that this character table bears the same relation, so far as the 1s are concerned, to the table for $\bar{1}$ as the latter does to the table for point group 1 (C_1). This relationship is valid for all the point groups, so that it is necessary to work out only the tables for point groups lacking a centre of symmetry, and then the rest are easily written down. (4, p. 475; 5, p. 71; 6, p. 14; 8, pp. 115, 125; 10, p. 72).

Q.11.1. Write down the coordinates of the equivalent points and draw stereograms showing the directions joining these points to the origin of the coordinates for point groups 2(C_2) and *mmm* (D_{2h}).

Q.11.2. List the elements of symmetry present in point group *mmm* (D_{2h}). Regarding *mmm* as being derived from 222 (D_2) by addition of a centre of symmetry, draw up the character table for *mmm*.

Answers

Q.11.1. Coordinates of equivalent points are

Point group 2 (C_2): $x, y, z;\ \bar{x}, \bar{y}, z;$

Point group *mmm* (D_{2h}): $x, y, z;\ x, \bar{y}, z;\ \bar{x}, y, z;\ \bar{x}, \bar{y}, z;$

$x, y, \bar{z};\ x, \bar{y}, \bar{z};\ \bar{x}, y, \bar{z};\ \bar{x}, \bar{y}, \bar{z}.$

The required stereograms are shown in Figs 11.4 and 11.5.

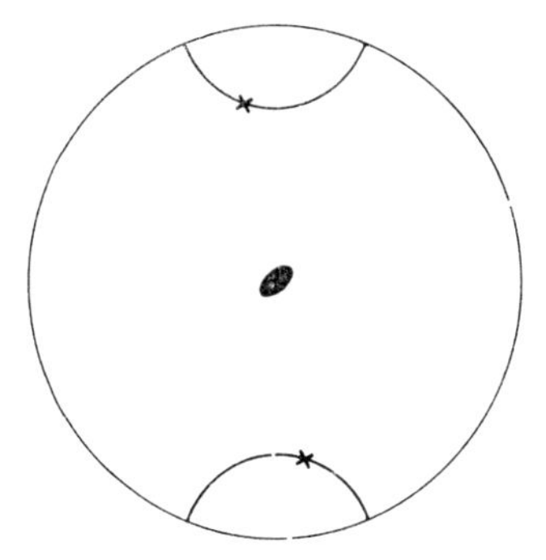

FIG. 11.4. Stereogram showing equivalent directions in point group 2 (C_2).

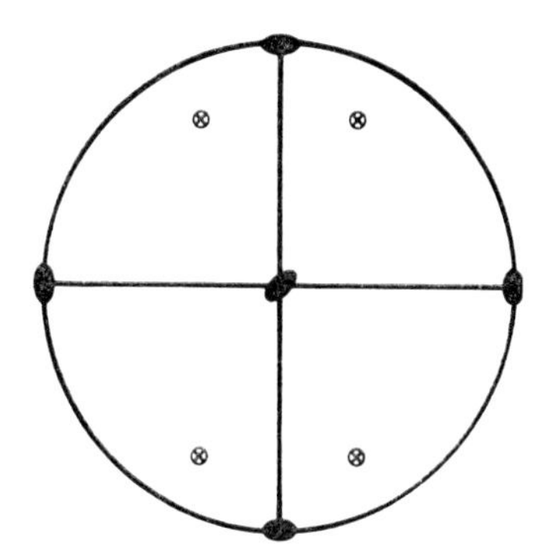

FIG. 11.5. Stereogram showing equivalent directions in point group *mmm* (D_{2h}).

Q.11.2. Elements of symmetry present in point group *mmm* (D_{2h}) are

$$E, C_{2z}, C_{2y}, C_{2x}, I, \sigma_x, \sigma_y, \sigma_z.$$

TABLE 11.6

Character table for point group mmm (D_{2h})

	E	C_{2z}	C_{2y}	C_{2x}	I	σ_z	σ_y	σ_x		
A_g	1	1	1	1	1	1	1	1		x^2, y^2, z^2
B_{2g}	1	−1	1	−1	1	−1	1	−1	R_y	xz
B_{1g}	1	1	−1	−1	1	1	−1	−1	R_z	xy
B_{3g}	1	−1	−1	1	1	−1	−1	1	R_x	yz
A_u	1	1	1	1	−1	−1	−1	−1		xyz
B_{2u}	1	−1	1	−1	−1	1	−1	1	y	
B_{1u}	1	1	−1	−1	−1	−1	1	1	z	
B_{3u}	1	−1	−1	1	−1	1	1	−1	x	

12

Conjugation and multiplication of symmetry elements

12.1. Group–theoretical classes of symmetry elements

THE approach illustrated by the discussion of the character table of point group 222 (D_2) can be applied to those point groups that have only one element of symmetry in each group-theoretical class. Those now to be discussed have, at least in one of the classes in the point group, more than one element of symmetry. Let us consider the point group $4mm$ (C_{4v}), which has the following elements of symmetry, E, C_4, C_4^3, C_2, σ_x, σ_y, σ_d, $\sigma_{d'}$. The stereogram of the equivalent points is shown in Fig. 12.1. The symbol C_4 means a rotation

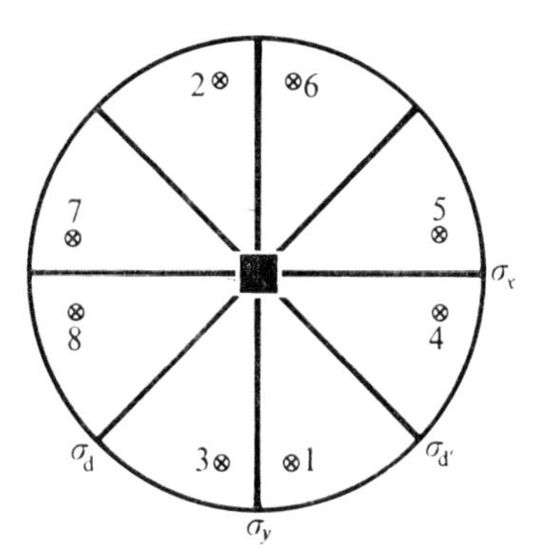

FIG. 12.1. Stereogram of equivalent general positions in point group $4mm$.

of $2\pi/4$ about the z axis in an anticlockwise sense. C_4^3 means a similar rotation but in a clockwise sense (or three rotations of $2\pi/4$ in an anticlockwise sense). Crystallographers have not, within a given point group, distinguished between these elements of symmetry, and this is one reason why group theory goes beyond the usual crystallographic analysis. Two rotations of $2\pi/4$ in an anti-clockwise sense could be represented as C_4^2, but this brings cross 1 (Fig. 12.1) to cross 2 and therefore corresponds to the operation of a diad axis, which is denoted C_2. The diad axis is counted as one of the symmetry elements present even though it arises from the tetrad axis. The symbols σ_x, σ_y, σ_d, $\sigma_{d'}$ stand for planes of symmetry perpendicular respectively to the x, y, [110], and [1$\bar{1}$0] axes.

The combination of symmetry elements within a given point group forms a group in the sense used by mathematicians (see Appendix 4 for the formal definition of a group in the mathematical sense). The elements of symmetry within a given point group are mutually compatible. Taking any stereogram representing one of the thirty-two point groups and starting from any cross or circle, we arrive at another cross or circle by operating with any of the elements of symmetry within the point group. This may be expressed by saying that, if A and B are two symmetry operations within a point group, A multiplied by B results in another symmetry operation C of the same group, or

$$BA = C.$$

In such products as BA we start on the right-hand side and work towards the left-hand side. In general, the order in which the symmetry operations are carried out is important. For example, in Fig. 12.1, the operation of C_4 takes cross 1 to cross 5, and if this is followed by σ_x we come to cross 4. If, however, we start with σ_x, cross 1 goes to cross 6, and then C_4 takes cross 6 to cross 7. Thus we may write

$$\sigma_x C_4 \neq C_4 \sigma_x.$$

Symmetry elements within a given point group which are similar, e.g. C_4 and C_4^3, are said to be in the same *class*. Such elements must satisfy the tests given in §§ 12.2 and 13.1. Certain of the eight elements of symmetry in point group $4mm$ may be shown to belong to the same class. Thus C_4 and C_4^3, σ_x and σ_y, and σ_d and $\sigma_{d'}$ form three classes. The remaining elements E and C_2 belong to separate classes, thus making five classes in all.

12.2. Conjugation of symmetry elements

The reason for placing certain symmetry elements in the same class must now be explained. In Fig. 12.1, the cross marked 1 is taken to the cross marked 5 by the operation C_4. The operation taking cross 5 to cross 1 is called C_4^{-1}, or the inverse of C_4. Similarly the operation σ_x takes cross 1 to cross 6, and σ_x^{-1} takes cross 6 to cross 1, i.e. σ_x^{-1} is a reflection across the plane of symmetry like σ_x. The symbol $\sigma_x^{-1}C_4\sigma_x$ means, reading from right to left, that cross 1 is taken to cross 6 (σ_x), cross 6 is rotated to cross 7 (C_4), and cross 7 is taken to cross 8 (σ_x^{-1}). Thus the whole sequence of operations takes cross 1 to cross 8, which is the same as the operation C_4^3. Thus we may write

$$\sigma_x^{-1}C_4\sigma_x = C_4^3.$$

This is called the conjugation of C_4 with σ_x. The fact that this conjugation results in C_4^3 defines C_4 and C_4^3 as being in the same class. When C_4 is conjugated with σ_y, σ_d, $\sigma_{d'}$, the result is C_4^3, but if it is conjugated with E, C_2, C_4, or C_4^3, we obtain a different result. For example, $C_2^{-1}C_4C_2$ means that starting from cross 1 we go to cross 2, then to cross 8, and finally to cross 5, since C_2^{-1} is, like C_2, the operation of a diad axis. But going from cross 1 to cross 5 is the same as C_4. The elements E, C_4, and C_4^3 give the same result. Thus the operation of conjugation of each of the elements of symmetry in turn with C_4 results either in C_4 itself or in C_4^3. For this reason the two elements of symmetry are said to be in the same class.

Similarly if σ_x is conjugated with C_4, we obtain $C_4^{-1}\sigma_xC_4$, which means the sequence of crosses in Fig. 12.1 1–5 (C_4), 5–4 (σ_x), 4–3 (C_4^{-1}). Now 1–3 is the same as σ_y, so that

$$C_4^{-1}\sigma_xC_4 = \sigma_y.$$

Thus we may show that conjugating σ_x with all the elements in turn yields σ_x and σ_y, so that these belong to the same class. Proceeding in the same way, we may show that σ_d and $\sigma_{d'}$ belong to the same class. C_2 and E when conjugated with the other elements of symmetry yield only themselves. For example $C_4^{-1}C_2C_4$ corresponds to the sequence of crosses in Fig. 12.1 1–5–8–2, which is the same as 1–2, namely C_2. It is clear that E must always be in a class by itself, since there is no other element like it, and, for example, $(C_4^{-1})EC_4 =$ 1–5–5–1 $= E$.

It is perhaps worth showing that C_4 and C_4^3 are not in the same class in point group 4 (C_4). In this point group the elements of symmetry are E, C_4, C_2, and C_4^3. If we conjugate C_4 with the other elements, we obtain

$$C_2^{-1}C_4C_2 = 1\text{–}2\text{–}8\text{–}5 = C_4,$$
$$(C_4^3)^{-1}C_4C_4^3 = 1\text{–}8\text{–}1\text{–}5 = C_4,$$
$$C_4^{-1}C_4C_4 = 1\text{–}5\text{–}2\text{–}5 = C_4.$$

Thus C_4 is in a class by itself. It may be shown in the same way that C_4^3 is also in a class by itself.

12.3. Multiplication tables for symmetry elements

In § 12.2 the operations of certain symmetry elements in a given point group were carried out one after the other. This process is called 'multiplication' of the symmetry elements. It is often necessary to perform such multiplications and, to assist in this, a so-called multiplication table is constructed. This table for point group 4*mm* is illustrated below (Table 12.1).

TABLE 12.1

Multiplication table for point group 4*mm* (C_{4v})

E	C_4	C_4^3	C_2	σ_x	σ_y	σ_d	$\sigma_{d'}$
C_4^3	E	C_2	C_4	σ_d	$\sigma_{d'}$	σ_y	σ_x
C_4	C_2	E	C_4^3	$\sigma_{d'}$	σ_d	σ_x	σ_y
C_2	C_4^3	C_4	E	σ_y	σ_x	$\sigma_{d'}$	σ_d
σ_x	σ_d	$\sigma_{d'}$	σ_y	E	C_2	C_4	C_4^3
σ_y	$\sigma_{d'}$	σ_d	σ_x	C_2	E	C_4^3	C_4
σ_d	σ_y	σ_x	$\sigma_{d'}$	C_4^3	C_4	E	C_2
$\sigma_{d'}$	σ_x	σ_y	σ_d	C_4	C_4^3	C_2	E

All the eight symmetry elements in this point group are set out along the top row. In the left-hand vertical column, the same symmetry elements are arranged so that the product of the element in the left-hand column with the element along the top row, which corresponds to a position on the diagonal (running from top left to bottom right) of the multiplication table, is E. This is not a necessary condition, but it is useful in subsequent work (see Chapter 13), and will be adopted throughout in this book. Thus, for example, C_4^3 must be multiplied by C_4 to obtain E, and so the second row begins with C_4^3. The first symmetry operation is in the left-hand column and the second operation is along the top row. This way of reading the multiplication tables must be observed in all cases because $AB \neq BA$ in general.

Most multiplication tables can be easily constructed from a stereogram of the equivalent general directions, e.g. Fig. 12.2. In

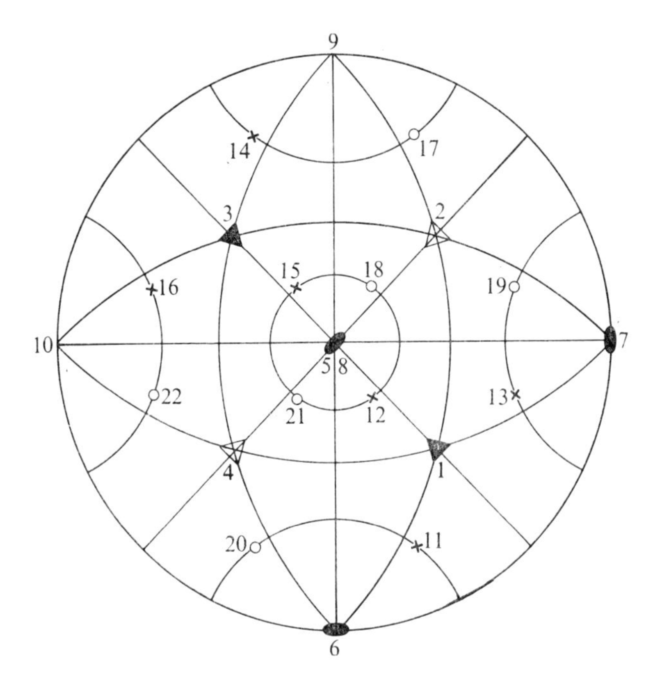

FIG. 12.2. Stereogram of equivalent general positions in point group 23.

the cubic system, the four triad axes do not lie in the same convenient planes as the axes of other crystal systems, and it is necessary to construct a table (Table 12.2) showing the operation of each element

TABLE 12.2

Movements of poles on cubic stereogram Fig. 12.2†

E	11	12	13	14	15	16	17	18	19	20	21	22
C_{3a}	13	11	12	22	20	21	16	14	15	19	17	18
C_{3a}^2	12	13	11	18	19	17	21	22	20	15	16	14
C_{3b}	19	17	18	16	14	15	22	20	21	13	11	12
C_{3b}^2	21	22	20	15	16	14	12	13	11	18	19	17
C_{3c}	22	20	21	13	11	12	19	17	18	16	14	15
C_{3c}^2	15	16	14	21	22	20	18	19	17	12	13	11
C_{3d}	16	14	15	19	17	18	13	11	12	22	20	21
C_{3d}^2	18	19	17	12	13	11	15	16	14	21	22	20
C_{2x}	20	21	22	17	18	19	14	15	16	11	12	13
C_{2y}	17	18	19	20	21	22	11	12	13	14	15	16
C_{2z}	14	15	16	11	12	13	20	21	22	17	18	19

† In Fig. 12.2 the axes are referred to the numbers as follows:
$C_{3a} = 1,\quad C_{3b} = 3,\quad C_{3c} = 2,\quad C_{3d} = 4,\quad C_{2x} = 6,\quad C_{2y} = 7,\quad C_{2z} = 5.$

of symmetry on each equivalent pole of the stereogram (see § 1.6). When this is done it is easy to draw up the multiplication table 12.3.

TABLE 12.3

Multiplication table for point group 23 (T)

E	C_{3a}^2	C_{3a}	C_{3b}^2	C_{3b}	C_{3c}^2	C_{3c}	C_{3d}^2	C_{3d}	C_{2x}	C_{2y}	C_{2z}
C_{3a}	E	C_{3a}^2	C_{2x}	C_{3d}^2	C_{2z}	C_{3b}^2	C_{2y}	C_{3c}^2	C_{3c}	C_{3b}	C_{3d}
C_{3a}^2	C_{3a}	E	C_{3c}	C_{2y}	C_{3d}	C_{2x}	C_{3b}	C_{2z}	C_{3b}^2	C_{3d}^2	C_{3c}^2
C_{3b}	C_{2x}	C_{3c}^2	E	C_{3b}^2	C_{2y}	C_{3d}^2	C_{2z}	C_{3a}^2	C_{3d}	C_{3a}	C_{3c}
C_{3b}^2	C_{3d}	C_{2y}	C_{3b}	E	C_{3a}	C_{2z}	C_{3c}	C_{2x}	C_{3a}^2	C_{3c}^2	C_{3d}^2
C_{3c}	C_{2z}	C_{3d}^2	C_{2y}	C_{3a}^2	E	C_{3c}^2	C_{2x}	C_{3b}^2	C_{3a}	C_{3d}	C_{3b}
C_{3c}^2	C_{3b}	C_{2x}	C_{3d}	C_{2z}	C_{3c}	E	C_{3a}	C_{2y}	C_{3d}^2	C_{3b}^2	C_{3a}^2
C_{3d}	C_{2y}	C_{3b}^2	C_{2z}	C_{3c}^2	C_{2x}	C_{3a}^2	E	C_{3d}^2	C_{3b}	C_{3c}	C_{3a}
C_{3d}^2	C_{3c}	C_{2z}	C_{3a}	C_{2x}	C_{3b}	C_{2y}	C_{3d}	E	C_{3c}^2	C_{3a}^2	C_{3b}^2
C_{2x}	C_{3c}^2	C_{3b}	C_{3d}^2	C_{3a}	C_{3a}^2	C_{3d}	C_{3b}^2	C_{3c}	E	C_{2z}	C_{2y}
C_{2y}	C_{3b}^2	C_{3d}	C_{3a}^2	C_{3c}	C_{3d}^2	C_{3b}	C_{3c}^2	C_{3a}	C_{2z}	E	C_{2x}
C_{2z}	C_{3d}^2	C_{3c}	C_{3c}^2	C_{3d}	C_{3b}^2	C_{3a}	C_{3a}^2	C_{3b}	C_{2y}	C_{2x}	E

(4, p. 477; 10, p. 72)

The four triad axes are equivalent, so that if a positive (anticlockwise) rotation for one of them is determined in a particular way, a positive rotation for the others must be determined in the same way. Thus, on a stereogram of a cubic crystal having the four triad axes represented by points at the same distance from the centre, a positive rotation appears anticlockwise for the two triad axes inclined upwards (triangles 1 and 3, Fig. 12.2), while it appears clockwise for the two triad axes inclined downwards (triangles 2 and 4). In Table 12.2, the twelve equivalent directions are marked 11–22 and are placed along the top row. The twelve symmetry elements, E–C_{2z}, are placed in the left-hand column. The table gives the corresponding directions before and after the operation of a given symmetry element. Thus direction 16 is converted to direction 14 by the axis C_{3b}^2.

The multiplication table 12.3 is drawn up so that the diagonal element is E. The multiplication of C_{3c} by C_{3d} is obtained from the above table as follows. (Note that C_{3c} is taken first.)

C_{3c} changes point 11 to point 22.

C_{3d} changes point 22 to point 21.

The product of C_{3c} and C_{3d} thus changes point 11 to point 21, which is the operation C_{3b}^2. It does not matter which point is chosen to start with, the product of C_{3c} and C_{3d} is always C_{3b}^2. Working in this way, we can construct the multiplication table.

Q.12.1. Write down the symmetry elements present in the point groups $3m(C_{3v})$, 422 (D_4), and 23 (T).

Q.12.2. Draw stereograms showing the distribution of the equivalent general directions in each of the point groups $3m$ (C_{3v}), 422 (D_4), and 23 (T). Mark in the symmetry elements.

Q.12.3. Construct the multiplication tables for the point groups $3m$ (C_{3v}) and 422 (D_4).

Q.12.4. Determine, by conjugating each element of symmetry in turn with each of the other elements, how many classes are present in each of the point groups $3m$ (C_{3v}), 422 (D_4), and 23 (T).

Answers

Q.12.1. Point group $3m$: $E, C_3, C_3^2, \sigma_{v_1}, \sigma_{v_2}, \sigma_{v_3}$.

Point group 422: $E, C_4, C_2, C_4^3, C_{2a}, C_{2b}, C'_{2c}, C'_{2d}$.

Point group 23: $E, C_{3a}, C_{3a}^2, C_{3b}, C_{3b}^2, C_{3c}, C_{3c}^2, C_{3d}, C_{3d}^2, C_{2x}, C_{2y}, C_{2z}$

Q.12.2. See Appendix 1.

Q.12.3.

TABLE 12.4

Multiplication table for point group 3m (C_{3v})

E	C_3	C_3^2	σ_{v1}	σ_{v2}	σ_{v3}
C_3^2	E	C_3	σ_{v3}	σ_{v1}	σ_{v2}
C_3	C_3^2	E	σ_{v2}	σ_{v3}	σ_{v1}
σ_{v1}	σ_{v3}	σ_{v2}	E	C_3^2	C_3
σ_{v2}	σ_{v1}	σ_{v3}	C_3	E	C_3^2
σ_{v3}	σ_{v2}	σ_{v1}	C_3^2	C_3	E

TABLE 12.5

Multiplication table for point group 422 (D_4)

E	C_4	C_2	C_4^3	C_{2a}†	C_{2b}	C'_{2c}	C'_{2d}
C_4^3	E	C_4	C_2	C'_{2c}	C'_{2d}	C_{2b}	C_{2a}
C_2	C_4^3	E	C_4	C_{2b}	C_{2a}	C'_{2d}	C'_{2c}
C_4	C_2	C_4^3	E	C'_{2d}	C'_{2c}	C_{2a}	C_{2b}
C_{2a}	C'_{2c}	C_{2b}	C'_{2d}	E	C_2	C_4	C_4^3
C_{2b}	C'_{2d}	C_{2a}	C'_{2c}	C_2	E	C_4^3	C_4
C'_{2c}	C_{2b}	C'_{2d}	C_{2a}	C_4^3	C_4	E	C_2
C'_{2d}	C_{2a}	C'_{2c}	C_{2b}	C_4	C_4^3	C_2	E

† $C_{2a} \parallel [100]$, $C_{2b} \parallel [010]$, $C'_{2c} \parallel [110]$, $C'_{2d} \parallel [1\bar{1}0]$.

Q.12.4.

Point group $3m$ $\sigma_{v1}^{-1} = \sigma_{v1}$, etc. $(C_3^2)^{-1} = C_3$, etc.

$$(C_3^2)^{-1}C_3C_3^2 = C_3C_3C_3^2 = C_3,$$
$$(\sigma_{v1})^{-1}C_3\sigma_{v1} = \sigma_{v1}C_3\sigma_{v1} = C_3^2,$$
$$(\sigma_{v2})^{-1}C_3\sigma_{v2} = \sigma_{v2}C_3\sigma_{v2} = C_3^2,$$
$$(\sigma_{v3})^{-1}C_3\sigma_{v3} = \sigma_{v3}C_3\sigma_{v3} = C_3^2.$$

Hence C_3 and C_3^2 are in the same class.

$$C_3^{-1}\sigma_{v1}C_3 = C_3^2\sigma_{v1}C_3 = \sigma_{v3},$$
$$(C_3^2)^{-1}\sigma_{v1}C_3^2 = C_3\sigma_{v1}C_3^2 = \sigma_{v2},$$
$$\sigma_{v2}^{-1}\sigma_{v1}\sigma_{v2} = \sigma_{v2}\sigma_{v1}\sigma_{v2} = \sigma_{v3},$$
$$\sigma_{v3}^{-1}\sigma_{v1}\sigma_{v3} = \sigma_{v3}\sigma_{v1}\sigma_{v3} = \sigma_{v2}.$$

Hence σ_{v1}, σ_{v2}, and σ_{v3} are in the same class, making altogether three classes.

Point group 422

$$C_2^{-1}C_4C_2 = C_2C_4C_2 = C_4,$$
$$(C_4^3)^{-1}C_4C_4^3 = C_4C_4C_4^3 = C_4,$$
$$(C_{2a})^{-1}C_4C_{2a} = C_{2a}C_4C_{2a} = C_4^3,$$
$$(C_{2b})^{-1}C_4C_{2b} = C_{2b}C_4C_{2b} = C_4^3.$$

Hence C_4 and C_4^3 belong to the same class.

$$C_4^{-1}C_{2a}C_4 = C_4^3C_{2a}C_4 = C_{2b},$$
$$(C'_{2c})^{-1}C_{2a}C'_{2c} = C'_{2c}C_{2a}C'_{2c} = C_{2b}.$$

Hence C_{2a} and C_{2b} are in the same class.

$$C_4^{-1}C'_{2c}C_4 = C_4^3C'_{2c}C_4 = C'_{2d},$$
$$C_2^{-1}C'_{2c}C_2 = C_2C'_{2c}C_2 = C'_{2c}.$$

Hence C'_{2c} and C'_{2d} are in the same class. E and C_2 are in classes by themselves, making five altogether.

Point Group 23 (refer to Table 12.3)

(1) C_{3a}

$$\left.\begin{aligned}
(C_{3a}^2)^{-1}C_{3a}C_{3a}^2 &= C_{3a}\\
(C_{3b})^{-1}C_{3a}C_{3b} &= C_{3d}\\
(C_{3b}^2)^{-1}C_{3a}C_{3b}^2 &= C_{3c}\\
(C_{3c})^{-1}C_{3a}C_{3c} &= C_{3b}\\
(C_{3c}^2)^{-1}C_{3a}C_{3c}^2 &= C_{3d}\\
(C_{3d})^{-1}C_{3a}C_{3d} &= C_{3c}\\
(C_{3d}^2)^{-1}C_{3a}C_{3d}^2 &= C_{3b}\\
(C_{2x})^{-1}C_{3a}C_{2x} &= C_{3d}\\
(C_{2y})^{-1}C_{3a}C_{2y} &= C_{3c}\\
(C_{2z})^{-1}C_{3a}C_{2z} &= C_{3b}
\end{aligned}\right\} C_{3a}, C_{3b}, C_{3c}, C_{3d} \text{ form one class.}$$

(2) C_{3a}^2

$$\left.\begin{aligned}
(C_{3a})^{-1}C_{3a}^2C_{3a} &= C_{3a}^2\\
(C_{3b})^{-1}C_{3a}^2C_{3b} &= C_{3d}^2\\
&\text{etc.}
\end{aligned}\right\} C_{3a}^2, C_{3b}^2, C_{3c}^2, C_{3d}^2 \text{ form one class.}$$

(3) C_{2x}

$$\left.\begin{aligned}
(C_{3a})^{-1}C_{2x}C_{3a} &= C_{2z}\\
(C_{3a}^2)^{-1}C_{2x}C_{3a}^2 &= C_{2y}
\end{aligned}\right\} C_{2x}, C_{2y}, C_{2z} \text{ form one class.}$$

Hence the classes are E, $4C_3$, $4C_3^2$, $3C_2$.

13

Representation of symmetry elements and classes by matrices

13.1. Regular representations

In the last chapter, the operation of a symmetry element was represented on a stereogram by the relative position of two or more crosses or circles. This is usually a satisfactory method of representation, but for certain purposes it is better to use a matrix representation. We shall again apply the method to the symmetry elements of point group 4*mm*. A matrix for each of the eight symmetry elements, E, C_4, C_4^3, C_2, σ_x, σ_y, σ_d, $\sigma_{d'}$, is arranged as follows. Starting from the multiplication table in which the diagonal is composed only of Es, we make an 8×8 matrix for each symmetry element by inserting '1' where it occurs and a dot or zero everywhere else. For convenience the multiplication table is repeated (Table 13.1), and beside it is given the matrix for C_4.

TABLE 13.1

Multiplication table for 4*mm* (C_{4v})

E	C_4	C_4^3	C_2	σ_x	σ_y	σ_d	$\sigma_{d'}$
C_4^3	E	C_2	C_4	σ_d	$\sigma_{d'}$	σ_y	σ_x
C_4	C_2	E	C_4^3	$\sigma_{d'}$	σ_d	σ_x	σ_y
C_2	C_4^3	C_4	E	σ_y	σ_x	$\sigma_{d'}$	σ_d
σ_x	σ_d	$\sigma_{d'}$	σ_y	E	C_2	C_4	C_4^3
σ_y	$\sigma_{d'}$	σ_d	σ_x	C_2	E	C_4^3	C_4
σ_d	σ_y	σ_x	$\sigma_{d'}$	C_4^3	C_4	E	C_2
$\sigma_{d'}$	σ_x	σ_y	σ_d	C_4	C_4^3	C_2	E

C_4

·	1	·	·	·	·	·	·
·	·	·	1	·	·	·	·
1	·	·	·	·	·	·	·
·	·	1	·	·	·	·	·
·	·	·	·	·	·	1	·
·	·	·	·	·	·	·	1
·	·	·	·	·	1	·	·
·	·	·	·	1	·	·	·

The corresponding matrix for C_4^3, obtained by putting a '1' wherever it occurs in the multiplication table and a dot everywhere else, is as follows.

$$C_4^3$$

$$\begin{array}{cccccccc}
\cdot & \cdot & 1 & \cdot & \cdot & \cdot & \cdot & \cdot \\
1 & \cdot & \cdot & \cdot & \cdot & \cdot & \cdot & \cdot \\
\cdot & \cdot & \cdot & 1 & \cdot & \cdot & \cdot & \cdot \\
\cdot & 1 & \cdot & \cdot & \cdot & \cdot & \cdot & \cdot \\
\cdot & \cdot & \cdot & \cdot & \cdot & \cdot & \cdot & 1 \\
\cdot & \cdot & \cdot & \cdot & \cdot & \cdot & 1 & \cdot \\
\cdot & \cdot & \cdot & \cdot & 1 & \cdot & \cdot & \cdot \\
\cdot & \cdot & \cdot & \cdot & \cdot & 1 & \cdot & \cdot
\end{array}.$$

Such arrangements of 1s and dots are called 'regular representations'. For a given point group there are many multiplication tables that can be drawn up with a diagonal consisting only of *E*s. In what follows, it is assumed that whichever order of the symmetry elements along the top line of the multiplication table is chosen is adhered to throughout the discussions of a given point group.

On inspection of the 'regular' matrices for C_4 and C_4^3, it will be seen that the columns of C_4^3 correspond in order with the rows of C_4. Therefore if the matrices are multiplied together according to the usual rules for matrix multiplication, i.e. multiplication of rows of the first matrix with columns of the second matrix, then the resulting matrix is 8×8 and has 1s along the diagonal only. Such a matrix is called a unit matrix, and, because the multiplication of C_4 and C_4^3 gives a unit matrix, C_4^3 is said to be the inverse matrix of C_4.

The matrix for C_2 is as follows, as may be seen by consulting the multiplication table.

$$C_2$$

$$\begin{array}{cccccccc}
\cdot & \cdot & \cdot & 1 & \cdot & \cdot & \cdot & \cdot \\
\cdot & \cdot & 1 & \cdot & \cdot & \cdot & \cdot & \cdot \\
\cdot & 1 & \cdot & \cdot & \cdot & \cdot & \cdot & \cdot \\
1 & \cdot & \cdot & \cdot & \cdot & \cdot & \cdot & \cdot \\
\cdot & \cdot & \cdot & \cdot & \cdot & 1 & \cdot & \cdot \\
\cdot & \cdot & \cdot & \cdot & 1 & \cdot & \cdot & \cdot \\
\cdot & \cdot & \cdot & \cdot & \cdot & \cdot & \cdot & 1 \\
\cdot & \cdot & \cdot & \cdot & \cdot & \cdot & 1 & \cdot
\end{array}.$$

Interchange of rows and columns in order would not alter this matrix, i.e. the inverse of the C_2 matrix is the C_2 matrix itself. The same result may be found for the pairs σ_x, σ_y and σ_d, $\sigma_{d'}$ as was found above for C_4 and C_4^3.

Q.13.1. Write down the regular representations for C_3 and σ_{v_1} from the multiplication table for point group $3m$ (Table 12.4).

13.2. Conjugation using matrices

In the last chapter, conjugation of symmetry elements was discussed in terms of stereograms of the general positions of points representing a given point group. The regular representations may be used for the same purpose. Thus in point group $4mm$ we have for the conjugation of C_4 with C_4^3, i.e. for $(C_4^3)^{-1}C_4C_4^3$, the following matrices.

$$(C_4^3)^{-1}C_4 = C_4C_4 = \begin{array}{cccccccc} \cdot & \cdot & \cdot & 1 & \cdot & \cdot & \cdot & \cdot \\ \cdot & \cdot & 1 & \cdot & \cdot & \cdot & \cdot & \cdot \\ \cdot & 1 & \cdot & \cdot & \cdot & \cdot & \cdot & \cdot \\ 1 & \cdot & \cdot & \cdot & \cdot & \cdot & \cdot & \cdot \\ \cdot & \cdot & \cdot & \cdot & \cdot & 1 & \cdot & \cdot \\ \cdot & \cdot & \cdot & \cdot & 1 & \cdot & \cdot & \cdot \\ \cdot & \cdot & \cdot & \cdot & \cdot & \cdot & \cdot & 1 \\ \cdot & \cdot & \cdot & \cdot & \cdot & \cdot & 1 & \cdot \end{array},$$

$$(C_4^3)^{-1}C_4C_4^3 = C_4C_4C_4^3 = \begin{array}{cccccccc} \cdot & 1 & \cdot & \cdot & \cdot & \cdot & \cdot & \cdot \\ \cdot & \cdot & \cdot & 1 & \cdot & \cdot & \cdot & \cdot \\ 1 & \cdot & \cdot & \cdot & \cdot & \cdot & \cdot & \cdot \\ \cdot & \cdot & 1 & \cdot & \cdot & \cdot & \cdot & \cdot \\ \cdot & \cdot & \cdot & \cdot & \cdot & \cdot & 1 & \cdot \\ \cdot & \cdot & \cdot & \cdot & \cdot & \cdot & \cdot & 1 \\ \cdot & \cdot & \cdot & \cdot & \cdot & 1 & \cdot & \cdot \\ \cdot & \cdot & \cdot & \cdot & 1 & \cdot & \cdot & \cdot \end{array} = C_4.$$

Similarly, if C_4 is conjugated with σ_x we obtain

$$\sigma_x^{-1}C_4 = \sigma_x C_4 = \begin{array}{cccccccc} \cdot & \cdot & \cdot & \cdot & \cdot & \cdot & 1 & \cdot \\ \cdot & \cdot & \cdot & \cdot & 1 & \cdot & \cdot & \cdot \\ \cdot & \cdot & \cdot & \cdot & \cdot & 1 & \cdot & \cdot \\ \cdot & \cdot & \cdot & \cdot & \cdot & \cdot & \cdot & 1 \\ \cdot & 1 & \cdot & \cdot & \cdot & \cdot & \cdot & \cdot \\ \cdot & \cdot & 1 & \cdot & \cdot & \cdot & \cdot & \cdot \\ 1 & \cdot & \cdot & \cdot & \cdot & \cdot & \cdot & \cdot \\ \cdot & \cdot & \cdot & 1 & \cdot & \cdot & \cdot & \cdot \end{array},$$

$$\sigma_x^{-1}C_4\sigma_x = \sigma_x C_4 \sigma_x = \begin{array}{cccccccc} \cdot & \cdot & 1 & \cdot & \cdot & \cdot & \cdot & \cdot \\ 1 & \cdot & \cdot & \cdot & \cdot & \cdot & \cdot & \cdot \\ \cdot & \cdot & \cdot & 1 & \cdot & \cdot & \cdot & \cdot \\ \cdot & 1 & \cdot & \cdot & \cdot & \cdot & \cdot & \cdot \\ \cdot & \cdot & \cdot & \cdot & \cdot & \cdot & \cdot & 1 \\ \cdot & \cdot & \cdot & \cdot & \cdot & \cdot & 1 & \cdot \\ \cdot & \cdot & \cdot & \cdot & 1 & \cdot & \cdot & \cdot \\ \cdot & \cdot & \cdot & \cdot & \cdot & 1 & \cdot & \cdot \end{array} = C_4^3.$$

Thus C_4 and C_4^3 are in the same class, as we have seen already in Chapter 12.

13.3. Representation of classes of symmetry elements by matrices

The matrix representation affords a method of representing the class of symmetry elements containing two or more elements. This representation is achieved simply by addition of the matrices. Thus the class containing C_4 and C_4^3 in point group $4mm$ is represented by

$$\begin{array}{cccccccc} \cdot & 1 & 1 & \cdot & \cdot & \cdot & \cdot & \cdot \\ 1 & \cdot & \cdot & 1 & \cdot & \cdot & \cdot & \cdot \\ 1 & \cdot & \cdot & 1 & \cdot & \cdot & \cdot & \cdot \\ \cdot & 1 & 1 & \cdot & \cdot & \cdot & \cdot & \cdot \\ \cdot & \cdot & \cdot & \cdot & \cdot & \cdot & 1 & 1 \\ \cdot & \cdot & \cdot & \cdot & \cdot & \cdot & 1 & 1 \\ \cdot & \cdot & \cdot & \cdot & 1 & 1 & \cdot & \cdot \\ \cdot & \cdot & \cdot & \cdot & 1 & 1 & \cdot & \cdot \end{array}.$$

The addition of the regular matrices for the symmetry elements in the same class always results in a matrix that is diagonally symmetrical. Those symmetry elements that occur in classes by themselves, e.g. C_2 in $4mm$, have regular matrices that are diagonally symmetrical. Thus there is a universal rule that the matrices, and not only the regular matrices, representing classes of symmetry elements are diagonally symmetrical.

Q.13.2. In point group $3m$, σ_{v_1}, σ_{v_2}, and σ_{v_3} are in the same class. Write down the matrix representation for this class, derived from the regular representations of the individual symmetry elements.

13.4. The multiplication of classes

The matrix representations of classes of symmetry elements can be multiplied together. We shall denote the classes in point group $4mm$ by $\mathscr{C}_1$–$\mathscr{C}_5$, where

$$\mathscr{C}_1 = E, \quad \mathscr{C}_2 = C_4 + C_4^3, \quad \mathscr{C}_3 = C_2, \quad \mathscr{C}_4 = \sigma_x + \sigma_y,$$
$$\mathscr{C}_5 = \sigma_d + \sigma_{d'}.$$

The product of $\mathscr{C}_2$ and $\mathscr{C}_4$ (see Table 13.1 for $\mathscr{C}_4$) may be represented by

$$\begin{array}{cccccccc}
\cdot & 1 & 1 & \cdot & \cdot & \cdot & \cdot & \cdot \\
1 & \cdot & \cdot & 1 & \cdot & \cdot & \cdot & \cdot \\
1 & \cdot & \cdot & 1 & \cdot & \cdot & \cdot & \cdot \\
\cdot & 1 & 1 & \cdot & \cdot & \cdot & \cdot & \cdot \\
\cdot & \cdot & \cdot & \cdot & \cdot & \cdot & 1 & 1 \\
\cdot & \cdot & \cdot & \cdot & \cdot & \cdot & 1 & 1 \\
\cdot & \cdot & \cdot & \cdot & 1 & 1 & \cdot & \cdot \\
\cdot & \cdot & \cdot & \cdot & 1 & 1 & \cdot & \cdot
\end{array}
\quad \times \quad
\begin{array}{cccccccc}
\cdot & \cdot & \cdot & \cdot & 1 & 1 & \cdot & \cdot \\
\cdot & \cdot & \cdot & \cdot & \cdot & \cdot & 1 & 1 \\
\cdot & \cdot & \cdot & \cdot & \cdot & \cdot & 1 & 1 \\
\cdot & \cdot & \cdot & \cdot & 1 & 1 & \cdot & \cdot \\
1 & \cdot & \cdot & 1 & \cdot & \cdot & \cdot & \cdot \\
1 & \cdot & \cdot & 1 & \cdot & \cdot & \cdot & \cdot \\
\cdot & 1 & 1 & \cdot & \cdot & \cdot & \cdot & \cdot \\
\cdot & 1 & 1 & \cdot & \cdot & \cdot & \cdot & \cdot
\end{array}$$

which, when rows of $\mathscr{C}_2$ are multiplied by columns of $\mathscr{C}_4$, gives

$$\begin{array}{cccccccc}
\cdot & \cdot & \cdot & \cdot & \cdot & \cdot & 2 & 2 \\
\cdot & \cdot & \cdot & \cdot & 2 & 2 & \cdot & \cdot \\
\cdot & \cdot & \cdot & \cdot & 2 & 2 & \cdot & \cdot \\
\cdot & \cdot & \cdot & \cdot & \cdot & \cdot & 2 & 2 \\
\cdot & 2 & 2 & \cdot & \cdot & \cdot & \cdot & \cdot \\
\cdot & 2 & 2 & \cdot & \cdot & \cdot & \cdot & \cdot \\
2 & \cdot & \cdot & 2 & \cdot & \cdot & \cdot & \cdot \\
2 & \cdot & \cdot & 2 & \cdot & \cdot & \cdot & \cdot
\end{array}$$

From the multiplication table (13.1) it can be seen that the matrix representation of $\mathscr{C}_5 = \sigma_d + \sigma_{d'}$ is given by

$$\begin{array}{cccccccc}
\cdot & \cdot & \cdot & \cdot & \cdot & \cdot & 1 & 1 \\
\cdot & \cdot & \cdot & \cdot & 1 & 1 & \cdot & \cdot \\
\cdot & \cdot & \cdot & \cdot & 1 & 1 & \cdot & \cdot \\
\cdot & \cdot & \cdot & \cdot & \cdot & \cdot & 1 & 1 \\
\cdot & 1 & 1 & \cdot & \cdot & \cdot & \cdot & \cdot \\
\cdot & 1 & 1 & \cdot & \cdot & \cdot & \cdot & \cdot \\
1 & \cdot & \cdot & 1 & \cdot & \cdot & \cdot & \cdot \\
1 & \cdot & \cdot & 1 & \cdot & \cdot & \cdot & \cdot
\end{array} .$$

Thus

$$\mathscr{C}_2\mathscr{C}_4 = 2\mathscr{C}_5.$$

This result can also be obtained without the use of the matrices. We can write

$$\begin{aligned}
\mathscr{C}_2\mathscr{C}_4 &= (C_4 + C_4^3)(\sigma_x + \sigma_y) \\
&= C_4\sigma_x + C_4\sigma_y + C_4^3\sigma_x + C_4^3\sigma_y,
\end{aligned}$$

and from the multiplication table we obtain

$$\mathscr{C}_2\mathscr{C}_4 = \sigma_d + \sigma_{d'} + \sigma_{d'} + \sigma_d = 2\mathscr{C}_5$$

The factor 2 is called a *class multiplication coefficient*, c_{245}. (6, p. 18; 10, p. 75).

Q.13.3. In point group $3m$ denote the classes E, $C_3 + C_3^2$ and $\sigma_{v_1} + \sigma_{v_2} + \sigma_{v_3}$ by $\mathscr{C}_1$, $\mathscr{C}_2$ and $\mathscr{C}_3$ respectively.

From the multiplication table (Table 12.4) write down the matrix representation for classes $\mathscr{C}_2$ and $\mathscr{C}_3$ and find the product $\mathscr{C}_2\mathscr{C}_3$.

Find the value of $\mathscr{C}_2\mathscr{C}_3$ by taking the products of the individual symmetry elements and using the multiplication table to find the values of these products.

Q.13.4. Find the class multiplication coefficients for $\mathscr{C}_4\mathscr{C}_5$ and $\mathscr{C}_3\mathscr{C}_4$ in point group $4mm$ using the multiplication table in §13.1.

Q.13.5. In the point group 23 there are the classes $E, 4C_3, 4C_3^2, 3C_2$, which are denoted $\mathscr{C}_1, \mathscr{C}_2, \mathscr{C}_3, \mathscr{C}_4$ respectively. Find the class multiplication coefficients for $\mathscr{C}_2\mathscr{C}_3$ and $\mathscr{C}_2\mathscr{C}_4$. (The multiplication table is given in §12.3.)

Answers

Q.13.1.

TABLE 13.2

Multiplication table for point group $3m$ (C_{3v})

E	C_3	C_3^2	σ_{v1}	σ_{v2}	σ_{v3}
C_3^2	E	C_3	σ_{v3}	σ_{v1}	σ_{v2}
C_3	C_3^2	E	σ_{v2}	σ_{v3}	σ_{v1}
σ_{v1}	σ_{v3}	σ_{v2}	E	C_3^2	C_3
σ_{v2}	σ_{v1}	σ_{v3}	C_3	E	C_3^2
σ_{v3}	σ_{v2}	σ_{v1}	C_3^2	C_3	E

The regular representation of C_3 and σ_{v_1} are

$$\overset{C_3}{\begin{array}{cccccc} . & 1 & . & . & . & . \\ . & . & 1 & . & . & . \\ 1 & . & . & . & . & . \\ . & . & . & . & . & 1 \\ . & . & . & 1 & . & . \\ . & . & . & . & 1 & . \end{array}} \quad \text{and} \quad \overset{\sigma_{v_1}}{\begin{array}{cccccc} . & . & . & 1 & . & . \\ . & . & . & . & 1 & . \\ . & . & . & . & . & 1 \\ 1 & . & . & . & . & . \\ . & 1 & . & . & . & . \\ . & . & 1 & . & . & . \end{array}}.$$

Q.13.2. Class $(\sigma_{v_1}+\sigma_{v_2}+\sigma_{v_3})$:

$$\begin{array}{cccccc} . & . & . & 1 & 1 & 1 \\ . & . & . & 1 & 1 & 1 \\ . & . & . & 1 & 1 & 1 \\ 1 & 1 & 1 & . & . & . \\ 1 & 1 & 1 & . & . & . \\ 1 & 1 & 1 & . & . & . \end{array}.$$

Q.13.3. $C_3+C_3^2 =$

$$\begin{array}{cccccc} . & 1 & 1 & . & . & . \\ 1 & . & 1 & . & . & . \\ 1 & 1 & . & . & . & . \\ . & . & . & . & 1 & 1 \\ . & . & . & 1 & . & 1 \\ . & . & . & 1 & 1 & . \end{array} = \mathscr{C}_2,$$

$$\sigma_{v_1}+\sigma_{v_2}+\sigma_{v_3} = \begin{matrix} \cdot & \cdot & \cdot & 1 & 1 & 1 \\ \cdot & \cdot & \cdot & 1 & 1 & 1 \\ \cdot & \cdot & \cdot & 1 & 1 & 1 \\ 1 & 1 & 1 & \cdot & \cdot & \cdot \\ 1 & 1 & 1 & \cdot & \cdot & \cdot \\ 1 & 1 & 1 & \cdot & \cdot & \cdot \end{matrix} = \mathscr{C}_3,$$

$$\mathscr{C}_2\mathscr{C}_3 = \begin{matrix} \cdot & \cdot & \cdot & 2 & 2 & 2 \\ \cdot & \cdot & \cdot & 2 & 2 & 2 \\ \cdot & \cdot & \cdot & 2 & 2 & 2 \\ 2 & 2 & 2 & \cdot & \cdot & \cdot \\ 2 & 2 & 2 & \cdot & \cdot & \cdot \\ 2 & 2 & 2 & \cdot & \cdot & \cdot \end{matrix} = 2\mathscr{C}_3.$$

Or, we may write

$$\begin{aligned} \mathscr{C}_2\mathscr{C}_3 &= (C_3+C_3^2)(\sigma_{v_1}+\sigma_{v_2}+\sigma_{v_3}) \\ &= C_3\sigma_{v_1}+C_3\sigma_{v_2}+C_3\sigma_{v_3}+C_3^2\sigma_{v_1}+C_3^2\sigma_{v_2}+C_3^2\sigma_{v_3} \\ &= \sigma_{v_3}+\sigma_{v_1}+\sigma_{v_2}+\sigma_{v_2}+\sigma_{v_3}+\sigma_{v_1} \\ &= 2\mathscr{C}_3. \end{aligned}$$

Q.13.4. $$\begin{aligned} \mathscr{C}_4\mathscr{C}_5 &= (\sigma_x+\sigma_y)(\sigma_d+\sigma_{d'}) \\ &= \sigma_x\sigma_d+\sigma_x\sigma_{d'}+\sigma_y\sigma_d+\sigma_y\sigma_{d'} \\ &= C_4^3+C_4+C_4+C_4^3 \\ &= 2\mathscr{C}_2. \end{aligned}$$

Hence $c_{452} = 2.$

$$\begin{aligned} \mathscr{C}_3\mathscr{C}_4 &= C_2(\sigma_x+\sigma_y) \\ &= C_2\sigma_x+C_2\sigma_y \\ &= \sigma_y+\sigma_x \\ &= \mathscr{C}_4. \end{aligned}$$

Hence $c_{344} = 1.$

Q.13.5. $\mathscr{C}_2\mathscr{C}_3 = (C_{3a}+C_{3b}+C_{3c}+C_{3d})(C_{3a}^2+C_{3b}^2+C_{3c}^2+C_{3d}^2)$

$= C_{3a}C_{3a}^2+C_{3a}C_{3b}^2+C_{3a}C_{3c}^2+C_{3a}C_{3d}^2$ etc.

$= E+C_{2y}+C_{2x}+C_{2z}+C_{2y}+E+C_{2z}+C_{2x}+$

$+C_{2x}+C_{2z}+E+C_{2y}+C_{2z}+C_{2x}+C_{2y}+E$

$= 4\mathscr{C}_1+4\mathscr{C}_4$

Hence, $c_{231} = 4$ and $c_{234} = 4$

$\mathscr{C}_2\mathscr{C}_4 = (C_{3a}+C_{3b}+C_{3c}+C_{3d})(C_{2x}+C_{2y}+C_{2z})$

$= C_{3b}+C_{3a}+C_{3d}+C_{3c}+C_{3d}+C_{3c}+$

$+C_{3b}+C_{3a}+C_{3c}+C_{3d}+C_{3a}+C_{3b}$

$= 3(C_{3a}+C_{3b}+C_{3c}+C_{3d})$

$= 3\mathscr{C}_2.$

Hence, $c_{242} = 3$

14

General principles of the construction of character tables for the thirty-two point groups

14.1. Symbols associated with axes of rotation of higher order than twofold

IN Chapter 11 the character tables for the point groups involving only diad axes, planes, and centres of symmetry were described. These eight character tables contain only $+1$ and -1. This is because a coordinate can only be reversed in sign by a diad axis or plane or centre of symmetry. In point groups containing triad, tetrad, or hexad axes, this need not be true. Special symbols are used to represent rotation through the angle appropriate for the axis which is present. The Argand diagram is used, in which a rotating vector of unit length makes an angle ϕ with the horizontal axis (Fig. 14.1). Along the axis OA is plotted $\cos\phi$; along the per-

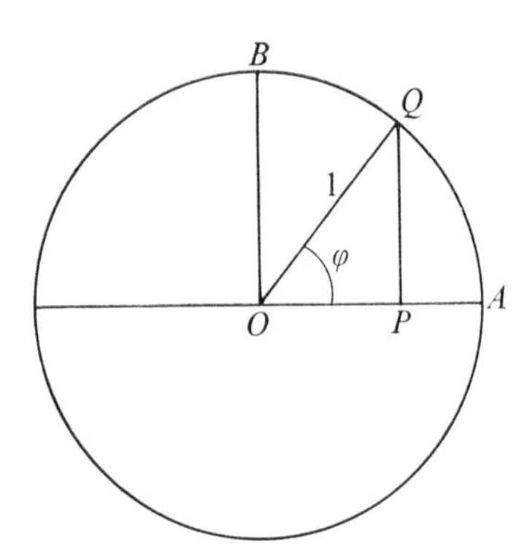

FIG. 14.1. The Argand diagram representing the rotation of unit vector OQ. Real distances are measured along axis OA and imaginary distances along axis OB.

pendicular axis OB, sin ϕ is plotted, and coordinates along this axis are multiplied by i $(=\sqrt{-1})$. The symbol used to define the rotation of OQ is ε where

$$\varepsilon = e^{i\phi}.$$

In the Argand diagram, since $OP = \cos\phi$ and $PQ = \sin\phi$, the length $OQ = 1$. Further, we have

$$e^{i\phi} = \cos\phi + i\sin\phi.$$

To find the real length of ε we multiply $e^{i\phi}$ by its complex conjugate and take the square root of the result. Writing ε^* for $e^{-i\phi}$ we have

$$\varepsilon\varepsilon^* = e^{i\phi}e^{-i\phi} = e^0 = 1.$$

The real length of ε is thus unity, i.e. it is equal to OQ.

Tetrad axis

This symbolism is applied to the tetrad, triad, and hexad axes in the following way. For a tetrad axis $\phi = 2\pi/4$ and

$$\varepsilon = \exp\frac{i2\pi}{4} = \cos\frac{2\pi}{4} + i\sin\frac{2\pi}{4} = i.$$

A second rotation of $2\pi/4$ corresponds to ε^2 and is equal to

$$\varepsilon^2 = \exp\frac{i2\pi}{2} = \cos\frac{2\pi}{2} + i\sin\frac{2\pi}{2} = -1.$$

A third rotation corresponds to

$$\varepsilon^3 = \exp i\frac{3}{4}\times 2\pi = \cos\frac{3}{4}\times 2\pi + i\sin\frac{3}{4}\times 2\pi = -i.$$

It is usual to write ε^* instead of ε^3, since

$$\varepsilon^* = \exp\left(-i\frac{2\pi}{4}\right) = \cos\frac{2\pi}{4} - i\sin\frac{2\pi}{4} = -i = \varepsilon^3.$$

Triad axis

For a triad axis $\varepsilon = \exp i2\pi/3$ and the values of ε for one and two rotations of $2\pi/3$ are

$$\varepsilon = \exp i2\pi/3 = \cos 2\pi/3 + i \sin 2\pi/3 = -\frac{1}{2} + i\frac{\sqrt{3}}{2},$$

$$\varepsilon^2 = \exp i\frac{2}{3}\times 2\pi = \cos\frac{2}{3}\times 2\pi + i \sin\frac{2}{3}\times 2\pi = -\frac{1}{2} - i\frac{\sqrt{3}}{2}.$$

It is usual to write ε^* instead of ε^2 because it has the same value.

Hexad axis

For a hexad axis $\varepsilon = \exp i2\pi/6$, and the values of ε for successive rotations of $2\pi/6$ are

$$\varepsilon = \exp i2\pi/6 = \cos 2\pi/6 + i \sin 2\pi/6 = \frac{1}{2} + i\frac{\sqrt{3}}{2},$$

$$\varepsilon^2 = \exp i\, 2\pi/3 = \cos 2\pi/3 + i \sin 2\pi/3 = -\frac{1}{2} + i\frac{\sqrt{3}}{2},$$

$$\varepsilon^3 = \exp i\, 2\pi/2 = \cos 2\pi/2 + i \sin 2\pi/2 = -1,$$

$$\varepsilon^4 = \varepsilon^3 \times \varepsilon = -\frac{1}{2} - i\frac{\sqrt{3}}{2} = \varepsilon^{*2},$$

$$\varepsilon^5 = \varepsilon^3 \times \varepsilon^2 = \frac{1}{2} - i\frac{\sqrt{3}}{2} = \varepsilon^*.$$

14.2. The identity operator E

In the character tables involving no axes of higher order of symmetry than two, namely 1, $\bar{1}$, 2, *m*, 2/*m*, 222, 2*mm*, *mmm*, the identity operator E has been put equal to unity. In point groups having axes of higher symmetry than two, and having more than one element of symmetry in at least one class, the numerical value of E may be 2 or 3. In the eight point groups mentioned above the coordinates are not 'mixed' by any of the symmetry elements. Consider for example the coordinates of equivalent points in point group 222. These are as follows.

$$x, y, z;\ x, \bar{y}, \bar{z},\ \bar{x}, y, \bar{z};\ \bar{x}, \bar{y}, z.$$

Now compare these with the corresponding coordinates in point group 4*mm*, which are

$$x, y, z;\ y, x, z;\ \bar{y}, x, z;\ \bar{x}, y, z;\ \bar{x}, \bar{y}, z;\ \bar{y}, \bar{x}, z;\ y, \bar{x}, z;\ x, \bar{y}, z.$$

Here there are pairs such as x, y, z and y, x, z where the coordinates of points along the two horizontal axes are not independently variable. Whereas in point group 222, x is always in the left-hand position of the three coordinates, in point group $4mm$, x may be either in the left-hand or in the middle position. The matrices that provide for the transformations corresponding to the points in 222 are

$$\begin{matrix} 1 & 0 & 0 \\ 0 & -1 & 0 \\ 0 & 0 & -1, \end{matrix} \qquad \begin{matrix} -1 & 0 & 0 \\ 0 & 1 & 0 \\ 0 & 0 & -1, \end{matrix} \qquad \begin{matrix} -1 & 0 & 0 \\ 0 & -1 & 0 \\ 0 & 0 & 1, \end{matrix}$$

and no non-zero terms occur off the diagonal in any of the matrices. In $4mm$ the matrices are such as

$$\begin{matrix} 0 & 1 & 0 \\ 1 & 0 & 0 \\ 0 & 0 & 1 \end{matrix}$$

for y, x, z, and there are two 1s in off-diagonal positions. The same is true of three other matrices for the point group $4mm$. These matrices are of the form

$$\begin{matrix} F & G & . \\ H & I & . \\ . & . & J, \end{matrix}$$

where F, G, H, I can be 1 or 0 and J is ± 1. The matrix $FGHI$ is described as two-dimensional, and for the identity operator it must be

$$\begin{matrix} 1 & 0 \\ 0 & 1. \end{matrix}$$

The character of this matrix (sum of diagonal terms) is 2. Thus where the two coordinates x and y are involved together in a representation, the value of E is 2. In some point groups having an E representation, the individual representations are one-dimensional but the symmetry elements link two such representations (see $4, \bar{4}, 4/m, 3, \bar{3}, 6, \bar{6}, 6/m, 23, m3$, Appendix 1).

In the cubic point groups, the triad axis [111] requires that the coordinates x, y, z be associated with y, z, x and z, x, y. The transformation matrices for these points are

$$\begin{matrix} 0 & 1 & 0 \\ 0 & 0 & 1 \\ 1 & 0 & 0, \end{matrix} \qquad \begin{matrix} 0 & 0 & 1 \\ 1 & 0 & 0 \\ 0 & 1 & 0. \end{matrix}$$

The general form of such matrices is

$$\begin{matrix} A & B & C \\ D & E & F \\ G & H & K, \end{matrix}$$

where each letter can be $+1$, 0, or -1. Such a matrix is called three-dimensional, and for it the identity operator is

$$\begin{matrix} 1 & 0 & 0 \\ 0 & 1 & 0 \\ 0 & 0 & 1. \end{matrix}$$

This has the character 3. However, not all representations in cubic point groups are three-dimensional as is shown in § 14.3.

14.3. Class multiplication coefficients and their use in determining irreducible representations

Within a given point group, the product of two classes of symmetry elements is always equal to one class or the sum of several classes in the same point group. We have seen, in § 13.4, that, in point group $4mm$, $\mathscr{C}_2\mathscr{C}_4 = 2\mathscr{C}_5$. In general we can write

$$\mathscr{C}_i\mathscr{C}_j = \sum_s c_{ij\,s}\mathscr{C}_s. \qquad (14.1)$$

In the above example, $c_{245} = 2$. The terms $c_{ij\,s}$ are called class multiplication coefficients, and they play a very important part in determining the character tables of the point groups.

In the eight point groups involving only diad axes, planes, and centres of symmetry, the coordinates x, y, z are not linked together, and the matrices representing the coordinates of equivalent positions are all 1×1 (one-dimensional). The characters of such matrices can be only $+1$ or -1. In other point groups there may be 2×2 (two-dimensional) matrices that represent the coordinates of equivalent points, and the characters may be $+2$, -2, $+1$, -1, and 0. Some of the matrices that have these characters are the following.

$$\begin{matrix} 1 & 0 \\ 0 & 1, \end{matrix} \qquad \begin{matrix} -1 & 0 \\ 0 & -1, \end{matrix} \qquad \begin{matrix} \frac{1}{2} & 0 \\ 0 & \frac{1}{2}, \end{matrix} \qquad \begin{matrix} -\frac{1}{2} & 0 \\ 0 & -\frac{1}{2}, \end{matrix} \qquad \begin{matrix} -1 & 0 \\ 0 & +1. \end{matrix}$$

The object of our present study is to determine the value, within a given point group, of the character of the matrix corresponding to a given class of symmetry and a given representation. The representations with which we shall be concerned here are all known as 'irreducible'. In Chapters 15–19 there will be many examples of 'reducible' representations that can be resolved into smaller and simpler irreducible representations. The sum of any two or more irreducible representations gives a reducible representation.

Equation (14.1) gives the relation between the products of the classes and the class multiplication coefficients. A similar relation can be shown to exist between the characters of the matrices corresponding to the classes. The symbol h, with suitable subscript, stands for the number of symmetry elements in a class. For example in $4mm$ the class made up of C_4 and C_4^3 has two elements and $h = 2$. The symbol χ, also with suitable subscript, stands for the character of the matrix representing a given class. χ_E is the character of the identity operator, which, as shown in § 14.2, must be 1 or 2 or 3. The equation containing these symbols, which is derived from equation 14.1 (see Appendix 5), is as follows.

$$h_i h_j \chi_i \chi_j = \chi_E \sum_s c_{ij\,s} h_s \chi_s. \tag{14.2}$$

A quantity x_i is introduced by the equation

$$x_i \chi_E = h_i \chi_i. \tag{14.3}$$

Substituting (14.3) in (14.2), we obtain

$$x_i x_j (\chi_E)^2 = \chi_E \sum_s c_{ij\,s} x_s \chi_E,$$

or

$$x_i x_j = \sum_s c_{ij\,s} x_s.$$

We may write

$$x_i = \sum_s x_s \delta_{is}, \text{ where } \delta_{is} = 1 \text{ when } s = i$$
$$= 0 \text{ when } s \neq i$$

and hence

$$\sum_s x_s x_j \delta_{is} = \sum_s c_{ijs} x_s,$$
$$\sum_s (c_{ijs} - x_j \delta_{is}) x_s = 0. \tag{14.4}$$

Equation (14.4) when written out in full gives a determinantal equation, the solution of which provides values of x_i from which the characters χ_i can be found.

14.4. Application of the general principles to point group 3 (C_3)

In point group 3 there are three elements of symmetry, namely E, C_3, C_3^2, and they each form a separate class. These classes are denoted $\mathscr{C}_1, \mathscr{C}_2, \mathscr{C}_3$ respectively, and the multiplication of the classes is given by Table 14.1.

TABLE 14.1

Multiplication table for $\mathscr{C}_i \mathscr{C}_j$ in point group 3

i \ j	1	2	3
1	$\mathscr{C}_1$	$\mathscr{C}_2$	$\mathscr{C}_3$
3	$\mathscr{C}_3$	$\mathscr{C}_1$	$\mathscr{C}_2$
2	$\mathscr{C}_2$	$\mathscr{C}_3$	$\mathscr{C}_1$

It may be noted that, from the definition of the classes, $\mathscr{C}_i \times \mathscr{C}_j$ must be equal to $\mathscr{C}_j \times \mathscr{C}_i$, since each class is represented by a diagonally symmetrical matrix.

When $j = 1$ we have from this table

$$\mathscr{C}_1 \mathscr{C}_1 = \mathscr{C}_1 \text{ and } c_{111} = 1,$$
$$\mathscr{C}_2 \mathscr{C}_1 = \mathscr{C}_2 \text{ and } c_{212} = 1,$$
$$\mathscr{C}_3 \mathscr{C}_1 = \mathscr{C}_3 \text{ and } c_{313} = 1,$$

and all other class multiplication coefficients c_{i1s} are zero. Inserting these c_{i1s} values into equation (14.4), we obtain the determinantal equation,

$$\begin{array}{c|ccc|l} i \backslash s & 1 & 2 & 3 & \\ 1 & (1-x_1) & 0 & 0 & \\ 2 & 0 & (1-x_1) & 0 & = 0. \\ 3 & 0 & 0 & (1-x_1) & \end{array}$$

This is the cubic equation

$$(1-x_1)^3 = 0.$$

The three roots of this equation are 1, 1, 1. Because $\mathscr{C}_1 = E$ in every point group, there must always be an expression $(1-x_1)^n = 0$, leading to a solution consisting only of $+1$s. (In subsequent discussion this solution is omitted.)

We now put $j = 2$. From Table 14.1 we obtain

$$\mathscr{C}_1\mathscr{C}_2 = \mathscr{C}_2 \text{ and } c_{122} = 1,$$

$$\mathscr{C}_2\mathscr{C}_2 = \mathscr{C}_3 \text{ and } c_{223} = 1,$$

$$\mathscr{C}_3\mathscr{C}_2 = \mathscr{C}_1 \text{ and } c_{321} = 1.$$

The corresponding determinantal equation is, for class $\mathscr{C}_2$,

$$\begin{vmatrix} -x_2 & 1 & 0 \\ 0 & -x_2 & 1 \\ 1 & 0 & -x_2 \end{vmatrix} = 0,$$

which gives $-x_2^3+1 = 0$. The three roots are therefore the cube roots of $+1$, which in § 14.1 were shown to be equal to 1, ε, and ε^2 where $\varepsilon = \exp \mathrm{i}2\pi/3$.

Put $j = 3$ for class $\mathscr{C}_3$. Then

$$\mathscr{C}_1\mathscr{C}_3 = \mathscr{C}_3 \text{ and } c_{133} = 1,$$

$$\mathscr{C}_2\mathscr{C}_3 = \mathscr{C}_1 \text{ and } c_{231} = 1,$$

$$\mathscr{C}_3\mathscr{C}_3 = \mathscr{C}_2 \text{ and } c_{332} = 1.$$

The corresponding equation is

$$\begin{vmatrix} -x_3 & 0 & 1 \\ 1 & -x_3 & 0 \\ 0 & 1 & -x_3 \end{vmatrix} = 0.$$

Hence,

$$-x_3^3+1 = 0,$$

$$x_3 = 1, \varepsilon, \varepsilon^2.$$

To sum up these results, we can write x_i for the three classes:

$\mathscr{C}_1(E)$	$\mathscr{C}_2(C_3)$	$\mathscr{C}_3(C_3^2)$
1	1	1
1	ε	ε
1	ε^2	ε^2 .

From these values of x_i it is necessary to derive the corresponding characters χ_i from the equation (14.3). A result of group theory is that the sum of the squares of χ_E for all the representations is equal to the number of elements of symmetry, g, in the point group (see Appendix 6). In point group 3, $g = 3$ and there are three representations corresponding to x_1, x_2, x_3. Only if χ_E is unity for each representation can this rule be satisfied. Inserting this value of χ_E and $h_i = 1$ in equation (14.3), we obtain

$$\chi_i = x_i.$$

The character table can therefore be drawn up as in Table 14.2, using the values of x_1, x_2, x_3 already found.

TABLE 14.2

Character table for point group 3 (C_3)

E	C_3	C_3^2	
1	1	1	z, R_z
1	ε	ε^2	x, y, R_x, R_y
1	ε^2	ε	

The order of ε and ε^2 in the second and third rows of Table 14.2 has been chosen to satisfy another rule of group theory. This requires that the sum of the products of the terms in any horizontal row (representation) of the character table by the number of symmetry elements in the corresponding class shall be zero (see Appendix 6). In this case, there is only one symmetry element in each class, so multiplication of the second row by h_i gives a zero sum, namely

$$1\times 1+1\times\varepsilon+1\times\varepsilon^2 = 1+\cos 2\pi/3+\mathrm{i}\sin 2\pi/3+\cos\frac{2}{3}2\pi+\mathrm{i}\sin\frac{2}{3}2\pi$$

$$= 1-\frac{1}{2}+\mathrm{i}\frac{\sqrt{3}}{2}-\frac{1}{2}-\mathrm{i}\frac{\sqrt{3}}{2} = 0.$$

Thus the ε and ε^2 terms are arranged to occur in the same horizontal rows.

In the fourth column of Table 14.2, coordinates x, y, z and small rotations R_x, R_y, R_z are associated with the three representations. The coordinates of general equivalent points can be written

$$x, y, z; \quad -\frac{x}{2}+\frac{\sqrt{3}}{2}y, -\frac{\sqrt{3}}{2}x-\frac{1}{2}y, z; \quad -\frac{x}{2}-\frac{\sqrt{3}}{2}y, \frac{\sqrt{3}}{2}x-\frac{1}{2}y, z.$$

The coordinate z is the same for all three elements of symmetry and it therefore corresponds to the first representation, namely, $+1$, $+1$, $+1$. The coordinates x and y are 'mixed' in the sense discussed in § 14.2. The representation is therefore two-dimensional, but in a special way. The identity operator has two characters of $+1$ rather than one character of 2, and so the two representations are bracketed together. The small rotation R_z corresponds to a small rotation about the triad axis and is clearly consistent with all three elements of symmetry. R_z therefore also belongs to the first representation. Fig. 14.2 shows the three general equivalent directions 1, 2, 3 and the direction of the X axis in its original position and after the operation of C_3 and C_3^2. The tangents to the small circles indicate the rotation R_x associated with the representation 1, ε, ε^2. The arrows satisfy the symmetries of C_3 and C_3^2 for both representations. Since there is no necessary orientation of the axes x and y in the plane perpendicular to z, whatever applies to R_x also applies to R_y. Thus x, y, R_x, and R_y all apply to the second and third representations when bracketed together.

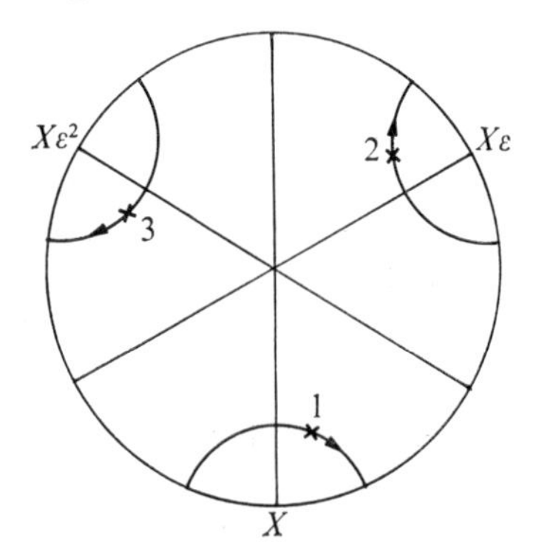

FIG. 14.2. Diagram showing three general equivalent directions 1, 2, 3 and the arrows associated with the small rotation R_x, in point group 3 (C_3).

14.5. Point group 32 (D_3)

In point group 32 there are the following elements of symmetry,

$$E, C_3, C_3^2, C_{2x}, C_{2y}, C_{2u}.$$

These can be shown by conjugation to belong to three classes, namely

$$\mathscr{C}_1 = E, \quad \mathscr{C}_2 = C_3+C_3^2, \quad \mathscr{C}_3 = C_{2x}+C_{2y}+C_{2u}.$$

The multiplication table for the elements of symmetry is as shown in Table 14.3.

TABLE 14.3

Multiplication table for elements of symmetry in point group 32 (D_3)

E	C_3	C_3^2	C_{2x}	C_{2y}	C_{2u}
C_3^2	E	C_3	C_{2u}	C_{2x}	C_{2y}
C_3	C_3^2	E	C_{2y}	C_{2u}	C_{2x}
C_{2x}	C_{2u}	C_{2y}	E	C_3^2	C_3
C_{2y}	C_{2x}	C_{2u}	C_3	E	C_3^2
C_{2u}	C_{2y}	C_{2x}	C_3^2	C_3	E

From this the multiplication table for the classes can be drawn up as shown in Table 14.4.

TABLE 14.4

Multiplication table giving $\mathscr{C}_i\mathscr{C}_j$ for classes of symmetry in point group 32 (D_3)

i \ j	1	2	3
1	$\mathscr{C}_1$	$\mathscr{C}_2$	$\mathscr{C}_3$
2	$\mathscr{C}_2$	$2\mathscr{C}_1+\mathscr{C}_2$	$2\mathscr{C}_3$
3	$\mathscr{C}_3$	$2\mathscr{C}_3$	$3\mathscr{C}_1+3\mathscr{C}_2$

From this table the class multiplication coefficients c_{ijs} can be read off. As stated earlier, the solution of equation (14.4) when $j = 1$ will be omitted because all the roots of the determinantal equation must be unity. Putting $j = 2$, we get

$$\mathscr{C}_1\mathscr{C}_2 = \mathscr{C}_2 \quad \text{and } c_{122} = 1,$$
$$\mathscr{C}_2\mathscr{C}_2 = 2\mathscr{C}_1+\mathscr{C}_2 \text{ and } c_{221} = 2,\ c_{222} = 1,$$
$$\mathscr{C}_3\mathscr{C}_2 = 2\mathscr{C}_3 \quad \text{and } c_{323} = 2.$$

When $j = 3$,

$$\mathscr{C}_1\mathscr{C}_3 = \mathscr{C}_3 \quad \text{and } c_{133} = 1,$$
$$\mathscr{C}_2\mathscr{C}_3 = 2\mathscr{C}_3 \quad \text{and } c_{233} = 2,$$
$$\mathscr{C}_3\mathscr{C}_3 = 3\mathscr{C}_1+3\mathscr{C}_2 \text{ and } c_{331} = 3,\ c_{332} = 3.$$

The determinantal equations are as follows.
When $j = 2$ (class $\mathscr{C}_2$),

$$\begin{vmatrix} -x_2 & 1 & \cdot \\ 2 & (1-x_2) & \cdot \\ \cdot & \cdot & (2-x_2) \end{vmatrix} = 0.$$

This may be reduced by a process known as triangulation. The value of a determinant remains unchanged when any row is multiplied by a given factor and added, term by corresponding term, to any other row. Thus the determinant is written in the form

$$\begin{vmatrix} -x_2 & 1 & \cdot \\ \cdot & \left(\dfrac{2}{x_2}+1-x_2\right) & \cdot \\ \cdot & \cdot & (2-x_2) \end{vmatrix} = 0.$$

We have obtained the second row by multiplying the first row by $2/x_2$ and adding this to the second row. The value of this determinant is simply the product of the diagonal terms, namely

$$x_2\frac{(2+x_2-x_2^2)}{x_2}(2-x_2) = (2-x_2)^2(1+x_2) = 0.$$

Finally,

$$x_2 = 2, 2, -1.$$

When $j = 3$ (class $\mathscr{C}_3$),

$$\begin{vmatrix} -x_3 & \cdot & 1 \\ \cdot & -x_3 & 2 \\ 3 & 3 & -x_3 \end{vmatrix} = 0,$$

or

$$x_3(x_3^2-9) = 0, \text{ and } x_3 = 3, -3, 0.$$

The rule for the sum of the squares of the characters for the E column of the character table is now applied. There are six elements of symmetry, and it is clear that we can write

$$1^2+1^2+2^2 = 6$$

and deduce that χ_E is 1 for two of the representations and 2 for the third. The values of h in the three classes are $h_1 = 1, h_2 = 2, h_3 = 3$. These values of χ_E and h_i are put into equation (14.3), namely

$$\chi_i = \frac{\chi_E}{h_i} x_i.$$

For the first column of the character table we have $h_1 = 1$ and $\chi_E = 1, 1, 2$. For the second column we have $h_2 = 2$ and $x_2 = 2, 2, -1$. The value of χ_E is 1, 1, 2, and, inserting these values into equation (14.3), we obtain the characters 1, 1, -1 in the three horizontal rows (representations). In the third column, $h_3 = 3$ and $x_3 = 3, -3, 0$, and this therefore leads to the characters 1, -1, 0 in the first, second, and third representations. Since χ_E is 1 in the first two representations and the characters under C_3 are $+1$, they are designated A_1, A_2, and since it is 2 in the third representation, i.e. it is two-dimensional, it has the designation E (which has nothing to do with E at the head of the first column). Thus the character table is as Table 14.5.

TABLE 14.5

Character table for point group 32 (D_3)

	E	$2C_3$	$3C_2$		
A_1	1	1	1		x^2+y^2, z^2
A_2	1	1	-1	z, R_z	
E	2	-1	0	(x, y) (R_x, R_y)	

The coordinates of equivalent points are

$$x, y, z;\quad -\frac{x}{2}+\frac{\sqrt{3}}{2}y,\ -\frac{\sqrt{3}}{2}x-\frac{1}{2}y,\ z;\quad -\frac{x}{2}-\frac{\sqrt{3}}{2}y,\ \frac{\sqrt{3}}{2}x-\frac{1}{2}y,\ z;$$

$$x, \bar{y}, \bar{z};\quad -\frac{x}{2}-\frac{\sqrt{3}}{2}y,\ -\frac{\sqrt{3}}{2}x+\frac{1}{2}y,\ \bar{z};\quad -\frac{x}{2}+\frac{\sqrt{3}}{2}y,\ \frac{\sqrt{3}}{2}x+\frac{1}{2}y,\ \bar{z}.$$

The first three sets of x, y, z values correspond to C_3 or C_3^2, and z is unchanging, so a $+1$ is appropriate in the second column for this coordinate. The second three sets of x, y, z values correspond to the transformations due to C_2, and here z is negative, so a -1 is appropriate in the third column. Thus the z coordinate corresponds to the second representation. Similarly, R_z conforms to C_3 but contravenes C_2, and it also applies, therefore, to the second representation. In the coordinates given above, x and y are 'mixed' and therefore correspond to the third two-dimensional representation. The same applies to R_x and R_y, which also belong to this representation. It should be noted that the characters 2, -1, 0 in the third representation correspond to the following two-dimensional matrices,

$$\begin{matrix} 1 & 0 \\ 0 & 1, \end{matrix} \qquad \begin{matrix} \frac{-1}{2} & \frac{\sqrt{3}}{2} \\ \frac{-\sqrt{3}}{2} & \frac{-1}{2}, \end{matrix} \qquad \begin{matrix} 1 & 0 \\ 0 & -1, \end{matrix}$$

which refer respectively to the identity operator, a triad axis C_3, C_3^2 normal to X and Y, and a diad axis parallel to X. The right-hand column of Table 14.5 carries x^2+y^2 and z^2. These expressions belong to the first representation, because x^2+y^2 and z^2 are clearly of the same value for each of the equivalent general positions. Some other functions of the coordinates may be placed in this column (see Appendix 1).

14.6. Point group 4 (C_4)

TABLE 14.6

Multiplication table for the symmetry elements

E	C_4	C_2	C_4^3
C_4^3	E	C_4	C_2
C_2	C_4^3	E	C_4
C_4	C_2	C_4^3	E

TABLE 14.7

Multiplication table giving $\mathscr{C}_i\mathscr{C}_j$ for the classes of symmetry

$\mathscr{C}_1 = E, \mathscr{C}_2 = C_4, \mathscr{C}_3 = C_2, \mathscr{C}_4 = C_4^3$

i \ j	1	2	3	4
1	$\mathscr{C}_1$	$\mathscr{C}_2$	$\mathscr{C}_3$	$\mathscr{C}_4$
2	$\mathscr{C}_2$	$\mathscr{C}_3$	$\mathscr{C}_4$	$\mathscr{C}_1$
3	$\mathscr{C}_3$	$\mathscr{C}_4$	$\mathscr{C}_1$	$\mathscr{C}_2$
4	$\mathscr{C}_4$	$\mathscr{C}_1$	$\mathscr{C}_2$	$\mathscr{C}_3$

The class multiplication coefficients, c_{ijs}, are therefore

$$\text{for } j = 2,\quad c_{122} = 1,\quad c_{223} = 1,\quad c_{324} = 1,\quad c_{421} = 1;$$

$$\text{for } j = 3,\quad c_{133} = 1,\quad c_{234} = 1,\quad c_{331} = 1,\quad c_{432} = 1;$$

$$\text{for } j = 4,\quad c_{144} = 1,\quad c_{241} = 1,\quad c_{342} = 1,\quad c_{443} = 1.$$

The corresponding determinantal equation for $j = 2$ is

$$\begin{vmatrix} -x_2 & 1 & \cdot & \cdot \\ \cdot & -x_2 & 1 & \cdot \\ \cdot & \cdot & -x_2 & 1 \\ 1 & \cdot & \cdot & -x_2 \end{vmatrix} = 0,$$

which on multiplying out gives

$$x_2^4 = 1, \text{ or } x_2 = 1, -1, \mathrm{i}, -\mathrm{i},$$

these being the four roots of unity (see § 14.1).

For $j = 3$, the determinantal equation is

$$\begin{vmatrix} -x_3 & \cdot & 1 & \cdot \\ \cdot & -x_3 & \cdot & 1 \\ 1 & \cdot & -x_3 & \cdot \\ \cdot & 1 & \cdot & -x_3 \end{vmatrix} = 0.$$

On evaluating this determinant, we have

$$x_3^4 - 2x_3^2 + 1 = 0,$$

or

$$(x_3 - 1)^2(x_3 + 1)^2 = 0.$$

The roots of x_3 are therefore $+1, +1, -1, -1$.

For $j = 4$, the determinantal equation is

$$\begin{vmatrix} -x_4 & \cdot & \cdot & 1 \\ 1 & -x_4 & \cdot & \cdot \\ \cdot & 1 & -x_4 & \cdot \\ \cdot & \cdot & 1 & -x_4 \end{vmatrix} = 0,$$

which again gives

$$x_4^4 - 1 = 0,$$

and the roots are $1, -1, \mathrm{i}, -\mathrm{i}$.

Since $g = 4$, the values of χ_E must be unity for each representation, since

$$1^2+1^2+1^2+1^2 = 4.$$

Thus the same relation, namely

$$\chi_i = x_i$$

holds in this point group as in point group 3. The character table is therefore drawn up as in Table 14.8.

TABLE 14.8

Character table for point group 4 (C_4)

	E	C_4	C_2	C_4^3		
A	1	1	1	1	z, R_z	x^2+y^2, z^2
B	1	−1	1	−1		$(x^2-y^2), xy$
E	1	i	−1	−i	(x, y) (R_x, R_y)	yz, xz
	1	−i	−1	i		

The arrangement of the χ_i within each column has been made to satisfy the rule that because each class has only one element of symmetry, the sum of the products in pairs of the terms in the first and any other row is zero. The second representation corresponds to x^2-y^2 or xy since the coordinates of the general equivalent positions are

$$x, y, z; \quad \bar{y}, x, z; \quad \bar{x}, \bar{y}, z; \quad y, \bar{x}, z.$$

The signs of the products of the first two coordinates are +1, −1, +1, −1 and agree with the characters in the second representation. The same is true of the difference between the squares of the first and second coordinates.

14.7. Point group 23 (T)

The elements of symmetry are grouped into four classes, namely $\mathscr{C}_1 = E$, $\mathscr{C}_2 = 4C_3$, $\mathscr{C}_3 = 4C_3^2$, $\mathscr{C}_4 = 3C_2$. The multiplication table for the elements of symmetry is given in Table 12.3, and from it the multiplication table for the classes can be derived (Table 14.9).

TABLE 14.9

Multiplication table for the classes of point group 23 (T)

i \ j	1	2	3	4
1	$\mathscr{C}_1$	$\mathscr{C}_2$	$\mathscr{C}_3$	$\mathscr{C}_4$
2	$\mathscr{C}_2$	$4\mathscr{C}_3$	$4\mathscr{C}_1+4\mathscr{C}_4$	$3\mathscr{C}_2$
3	$\mathscr{C}_3$	$4\mathscr{C}_1+4\mathscr{C}_4$	$4\mathscr{C}_2$	$3\mathscr{C}_3$
4	$\mathscr{C}_4$	$3\mathscr{C}_2$	$3\mathscr{C}_3$	$3\mathscr{C}_1+2\mathscr{C}_4$

The determinantal equations can be written down as follows.

For $j = 2$ (class $4C_3$),

$$\begin{vmatrix} -x_2 & 1 & \cdot & \cdot \\ \cdot & -x_2 & 4 & \cdot \\ 4 & \cdot & -x_2 & 4 \\ \cdot & 3 & \cdot & -x_2 \end{vmatrix} = 0,$$

which on multiplication gives

$$x_2(x_2^3-64) = 0.$$

The roots are 0, $\sqrt[3]{64}$, or

$$0, 4, 4\varepsilon, 4\varepsilon^*, \text{ where } \varepsilon = \exp \mathrm{i}\, 2\pi/3.$$

For $j = 3$ (class $4C_3^2$),

$$\begin{vmatrix} -x_3 & \cdot & 1 & \cdot \\ 4 & -x_3 & \cdot & 4 \\ \cdot & 4 & -x_3 & \cdot \\ \cdot & \cdot & 3 & -x_3 \end{vmatrix} = 0,$$

which leads to the same result as for $j = 2$.

For $j = 4$ (class $3C_2$),

$$\begin{vmatrix} -x_4 & \cdot & \cdot & 1 \\ \cdot & (3-x_4) & \cdot & \cdot \\ \cdot & \cdot & (3-x_4) & \cdot \\ 3 & \cdot & \cdot & (2-x_4) \end{vmatrix} = 0,$$

which leads to

$$(3-x_4)^3(1+x_4) = 0,$$

giving the roots 3, 3, 3, −1.

Summarizing these results, we have for the values of x_i

E	$4C_3$	$4C_3^2$	$3C_2$
1	4	4	3
1	4ε	$4\varepsilon^*$	3
1	$4\varepsilon^*$	4ε	3
1	0	0	−1.

There are twelve elements of symmetry, and the values of χ_E can be deduced as 1, 1, 1, 3, since

$$1^2+1^2+1^2+3^2 = 12.$$

Using the expression

$$\chi_i = \chi_E \frac{x_i}{h_i},$$

we can write down the character table as shown in Table 14.10.

TABLE 14.10
Character table for point group 23 (*T*)

	E	$4C_3$	$4C_3^2$	$3C_2$	$\varepsilon = \exp 2\pi i/3$	
A	1	1	1	1		$x^2+y^2+z^2$
E	1	ε	ε^*	1		$x^2-y^2,\ 2z^2-x^2-y^2$
	1	ε^*	ε	1		
T	3	0	0	−1	$(x, y, z)\ (R_x, R_y, R_z)$	yz, zx, xy

The second representation is chosen so that the products of its characters by the values of h_i in the four classes add up to zero. Thus

$$\begin{aligned} 1\times 1+4\varepsilon+4\varepsilon^2+3\times 1 &= 4(1+\varepsilon+\varepsilon^2) \\ &= 0. \end{aligned}$$

With this choice for the second representation, the third follows automatically. The -1 character must occur in the fourth representation, so as to make the sum of the products of the characters in the first and fourth columns zero (see Rule 4, Appendix 6). Thus

$$1\times 1+1\times 1+1\times 1+3\times -1 = 0.$$

The second and third representations taken together form a two-dimensional representation as in point group 3. The fourth representation is designated T to show that it is three-dimensional. This also is clear from the coordinates of the general equivalent positions, which are as follows,

$$\begin{array}{llll} x, y, z; & \bar{x}, y, \bar{z}; & \bar{x}, \bar{y}, z; & x, \bar{y}, \bar{z}; \\ y, z, x; & \bar{y}, z, \bar{x}; & \bar{y}, \bar{z}, x; & y, \bar{z}, \bar{x}; \\ z, x, y; & \bar{z}, x, \bar{y}; & \bar{z}, \bar{x}, y; & z, \bar{x}, \bar{y}. \end{array}$$

It should be noted that the characters of the transformation matrices for the elements of symmetry correspond to the characters in this fourth representation. The matrices are as follows.

$$\text{For } E, \quad \begin{array}{ccc} 1 & \cdot & \cdot \\ \cdot & 1 & \cdot \\ \cdot & \cdot & 1, \end{array} \qquad \text{for } C_{3a}, \quad \begin{array}{ccc} \cdot & 1 & \cdot \\ \cdot & \cdot & 1 \\ 1 & \cdot & \cdot, \end{array}$$

$$\text{for } C_{3a}^2, \quad \begin{array}{ccc} \cdot & \cdot & 1 \\ 1 & \cdot & \cdot \\ \cdot & 1 & \cdot \end{array} \qquad \text{for } C_{2x}, \quad \begin{array}{ccc} 1 & \cdot & \cdot \\ \cdot & -1 & \cdot \\ \cdot & \cdot & -1 \end{array}$$

and the characters are 3, 0, 0, -1 respectively. R_x is a small rotation consistent with C_{2x} but inconsistent with C_{2y} and C_{2z}. Hence its character is -1 under $3C_2$. Similarly R_x is inconsistent with the four trigonal axes and its character under them is zero. The same considerations apply to R_y and R_z. In the right-hand column yz appears in the fourth representation. C_{3a} changes x, y, z into y, z, x into z, x, y, and the product of the second and third coordinates changes in going from one equivalent point to another. To this absence of any correlation, either positive or negative, corresponds the character zero. The axes C_{2x}, C_{2y}, C_{2z} change yz to $\bar{y}\bar{z}$, $y\bar{z}$, and $\bar{y}z$ respectively. The sign of the product is twice negative and once positive. The corresponding character is -1.

14.8. Point group $\bar{3}m$ (D_{3d})

Point group $\bar{3}m$ is derived from point group 32 by addition to the latter of a centre of symmetry. It was shown in § 11.2 that addition of a centre of symmetry to a point group changes the character table in such a way that three of the four quadrants are identical with the acentric table and the fourth has all the signs reversed. Thus we may write down the character table for $\bar{3}m$ as in Table 14.11.

TABLE 14.11

Character table for point group $\bar{3}m$ (D_{3d})

	E	$2C_3$	$3C_2$	I	$2S_6$	$3\sigma_d$		
A_{1g}	1	1	1	1	1	1		$x^2+y^2+z^2$
A_{2g}	1	1	-1	1	1	-1	R_z	
E_g	2	-1	0	2	-1	0	R_x, R_y	(x^2-y^2, xy) (xz, yz)
A_{1u}	1	1	1	-1	-1	-1		
A_{2u}	1	1	-1	-1	-1	1	z	
E_u	2	-1	0	-2	1	0	(x, y)	

The first two representations are one-dimensional and correspond to $+1$ under C_3, and are therefore designated A. The subscript 'g' (*gerade*) corresponds to the fact that the characters in columns E and I are of the same sign. The fourth and fifth representations are also one-dimensional and denoted A but with a subscript 'u' (*ungerade*). This corresponds to the sign of the characters in columns E and I being opposite to one another. The third and sixth representations are two-dimensional and the subscripts have the same meaning as for the As.

The assignation of z to the A_{2u} representation follows from the fact that the z coordinate is unchanged by E, C_3, C_3^2, and σ_d, but reversed by C_{2x}, C_{2y}, C_{2u}, I, S_6, S_6^5. The coordinates x and y are assigned to E_u because they are 'mixed' and therfore occur in a two-dimensional representation, and their signs are reversed by I. R_z occurs in the A_{2g} representation because it contravenes C_2 and σ_d, for both of which the character is -1. It does not contravene the other elements of symmetry. R_x and R_y both contravene C_3 and S_6 but do not contravene I. Thus they belong to representation E_g (3, p. 56; 6, p. 28; 10, p. 80).

14.9. Orthogonality relations (see also Appendix 6)

Use has been made of certain rules, which apply to the sums of the products of the characters, taken in pairs, that occur in the columns and rows of the character tables. These relations bear the name 'orthogonal' because they are a generalized form of the sum of the products of the direction cosines of two mutually perpendicular lines. The rules may be designated 'vertical' and 'horizontal', the former applying to vertical columns and the latter to horizontal rows. For real characters the rules are as follows.

(1) The sum of the products of the corresponding characters in any two vertical columns of a given character table is equal to zero, except for the sum of the products of the terms in the first column by themselves, i.e. the sum of the squares of the characters in the E column. This sum is equal to the total number of symmetry elements present.

(2) The sum of the products of the characters in any horizontal row of a given character table with the number of elements of symmetry in the corresponding class is zero, except for the top row for which the sum is equal to the total number of symmetry elements present.

These rules are needed to get the proper sequence of characters down a given vertical column. The roots of x give no indication where the corresponding character should occur in the column for a particular class, but it is always easy to see by inspection with the help of these two rules in which horizontal row a given character should be placed. The order in which the horizontal rows occur in the character table is arbitrary or conventional. Consultation of Appendix 1 will show the usual arrangements of A, B, E, T (2, p. 229; 12, p. 116). For the treatment of complex characters see Appendix 6.

Q.14.1. Find the character table for point group $4mm(C_{4v})$ using the multiplication table 12.1.

Q.14.2. Using the character table for point group 4 (C_4) given in §14.6, write down the character table for point group $4/m(C_{4h})$. Assign x, y, z and R_x, R_y, R_z to appropriate representations.

Answers

Q.14.1. There are five classes in this point group, namely, $\mathscr{C}_1 = E$, $\mathscr{C}_2 = C_4+C_4^3$, $\mathscr{C}_3 = C_2$, $\mathscr{C}_4 = \sigma_x+\sigma_y$, $\mathscr{C}_5 = \sigma_d+\sigma_{d'}$. The multiplication table for the classes can be drawn up as shown in Table 14.12.

TABLE 14.12

Multiplication table for the classes of point group 4mm (C_{4v})

i \ j	1	2	3	4	5
1	$\mathscr{C}_1$	$\mathscr{C}_2$	$\mathscr{C}_3$	$\mathscr{C}_4$	$\mathscr{C}_5$
2	$\mathscr{C}_2$	$2\mathscr{C}_1+2\mathscr{C}_3$	$\mathscr{C}_2$	$2\mathscr{C}_5$	$2\mathscr{C}_4$
3	$\mathscr{C}_3$	$\mathscr{C}_2$	$\mathscr{C}_1$	$\mathscr{C}_4$	$\mathscr{C}_5$
4	$\mathscr{C}_4$	$2\mathscr{C}_5$	$\mathscr{C}_4$	$2\mathscr{C}_1+2\mathscr{C}_3$	$2\mathscr{C}_2$
5	$\mathscr{C}_5$	$2\mathscr{C}_4$	$\mathscr{C}_5$	$2\mathscr{C}_2$	$2\mathscr{C}_1+2\mathscr{C}_3$

The determinantal equations can be written down as follows

For $j = 2$ (class $\mathscr{C}_2$), we have

$$\begin{vmatrix} -x_2 & 1 & \cdot & \cdot & \cdot \\ 2 & -x_2 & 2 & \cdot & \cdot \\ \cdot & 1 & -x_2 & \cdot & \cdot \\ \cdot & \cdot & \cdot & -x_2 & 2 \\ \cdot & \cdot & \cdot & 2 & -x_2 \end{vmatrix} = 0.$$

This reduces to

$$x_2(x_2^2-4)^2 = 0,$$

and the roots are 0, +2, −2, +2, −2.

For $j = 3$ (class $\mathscr{C}_3$), we have

$$\begin{vmatrix} -x_3 & \cdot & 1 & \cdot & \cdot \\ \cdot & 1-x_3 & \cdot & \cdot & \cdot \\ 1 & \cdot & -x_3 & \cdot & \cdot \\ \cdot & \cdot & \cdot & 1-x_3 & \cdot \\ \cdot & \cdot & \cdot & \cdot & 1-x_3 \end{vmatrix} = 0.$$

This reduces to

$$(1-x_3^2)(1-x_3)^3 = 0,$$

and the roots are 1, 1, 1, 1, −1.

For $j = 4$ (class $\mathscr{C}_4$), we have

$$\begin{vmatrix} -x_4 & . & . & 1 & . \\ . & -x_4 & . & . & 2 \\ . & . & -x_4 & 1 & . \\ 2 & . & 2 & -x_4 & . \\ . & 2 & . & . & -x_4 \end{vmatrix} = 0.$$

This reduces to

$$x_4(x_4^2-4)^2 = 0,$$

and the roots are 0, +2, −2, +2, −2.

For $j = 5$ (class $\mathscr{C}_5$), we have

$$\begin{vmatrix} -x_5 & . & . & . & 1 \\ . & -x_5 & . & 2 & . \\ . & . & -x_5 & . & 1 \\ . & 2 & . & -x_5 & . \\ 2 & . & 2 & . & -x_5 \end{vmatrix} = 0.$$

This reduces to

$$x_5(x_5^2-4)^2 = 0,$$

and the roots are 0, +2, −2, +2, −2.

There are eight elements of symmetry in the five classes, and the only values of χ_E that can satisfy the orthogonality rule are 1, 1, 1, 1, 2, since

$$1^2+1^2+1^2+1^2+2^2 = 8.$$

The equation (14.3) and the orthogonality rules enable the character table shown in Appendix 1 to be drawn up.

Q.14.2. See Appendix 1.

15

Space groups and wave-vector groups

15.1. Bravais lattices, point groups, and space groups

It is not the intention of this chapter to derive from first principles the 14 Bravais lattices, the 32 point groups, and the 230 space groups. That has already been done in numerous textbooks, which are referred to in the Bibliography given at the end of this book. The *International tables for X-ray crystallography* set out all of these groups in Volume 1 of the series. Only the features necessary for the group-theory approach will be considered here. The point groups we are concerned with are known as *finite groups*, since they are restricted to arrangements of symmetry elements about a given point. The Bravais lattices are known as *infinite groups*, because the unit cells are repeated indefinitely in three dimensions and completely fill space. The Bravais lattices are built up by the repetition of lattice vectors **a**, **b**, **c** which may be inclined to one another at arbitrary angles α, β, γ, or at special angles such as 90°, or 60°, or 120°. If two symmetry elements in a point group are denoted A, B then we have seen in §12.1 that, in general, the products AB and BA bear the relation

$$AB \neq BA.$$

A group having the property of equality, rather than non-equality, in the product of all the elements A and B is called Abelian. Thus many of the point groups are non-Abelian. This is in contrast with the Bravais lattices. A lattice displacement **a** followed by a lattice displacement **b** brings the lattice into the same position as if the displacements were carried out in the opposite order, i.e. **b** before **a**. Thus the Bravais lattices form infinite Abelian groups. (The formal requirements which all groups must satisfy are given in Appendix 4.)

Space groups are the infinite groups that result from a combination of Bravais lattices and the symmetry operations found in point groups and certain additional symmetry operations. A plane of symmetry in one of the thirty two point groups is a *reflection plane*, i.e. the symmetry operation corresponds to that which is produced by reflection in a mirror. A plane of symmetry in one of the 230 space groups may be a reflection plane but it may instead be a glide plane of symmetry (Fig. 15.1). The amount of the translation is

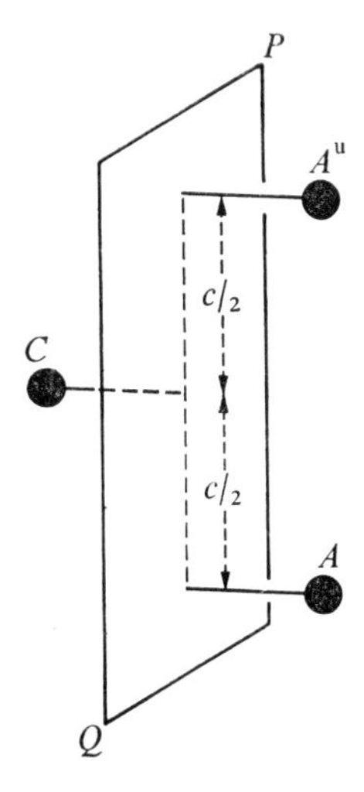

FIG. 15.1. Diagram illustrating the operation of a glide plane. Point A is reflected across the plane PQ and translated along the z axis a distance $c/2$. Point C is similarly reflected and translated to A^u, the point corresponding to A in the cell above it.

usually one-half of one of the translations **a** or **b** or **c**, but in certain space groups there are translations which involve translations such as $(\mathbf{a}+\mathbf{b})/2$. Similarly, the operations of diad, triad, tetrad, and hexad axes may be the same in a given space group as in a corresponding point group, but they need not be so. It is possible to form infinite groups with *screw axes*, which combine translation with rotation. The rotations are the same as in the point groups, and the translations are simple fractions of one or more of the basic translations. All of these axes, including those found in point groups, are illustrated in Fig. 15.2.

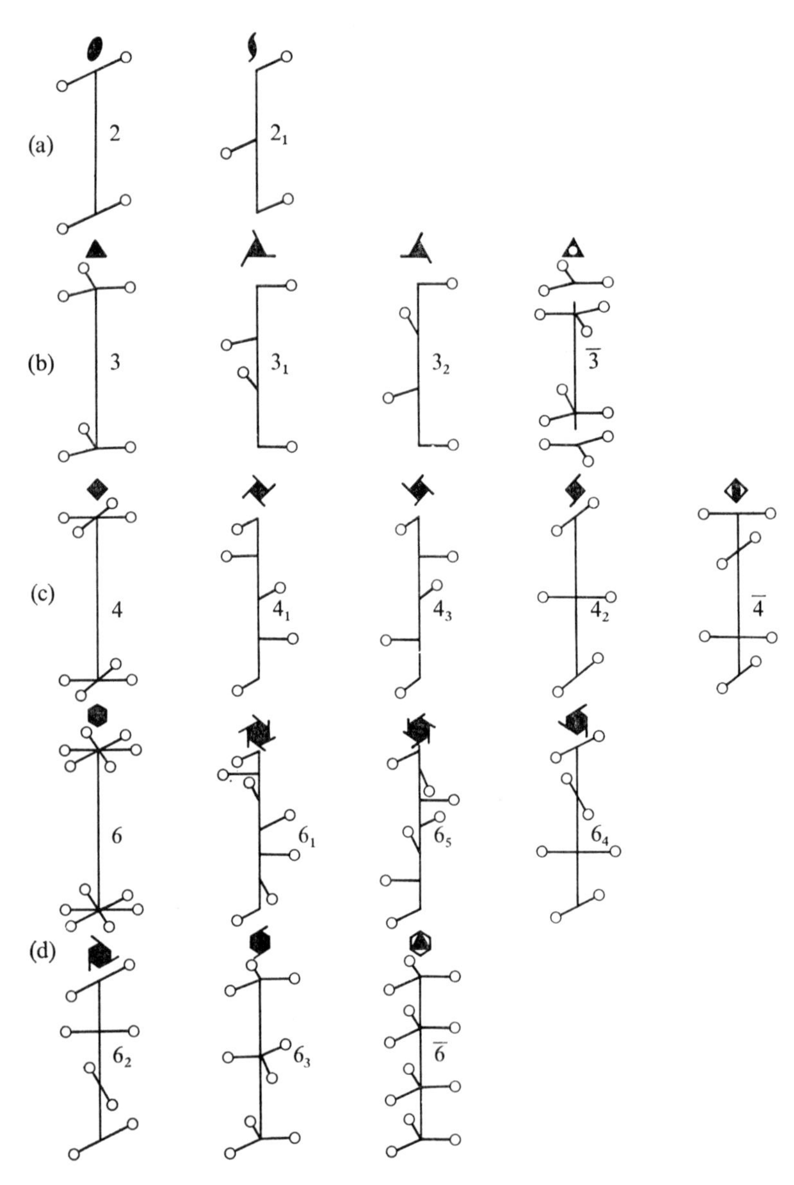

FIG. 15.2. Diagrams representing all the types of rotational, screw, and inversion–rotation axes possible in the 230 space groups.

Diad axes may be purely rotatory or screw. The international symbols for these axes are 2 and 2_1 respectively (Fig. 15.2(a)). Triad axes may be of four kinds, denoted respectively 3, 3_1 3_2, and $\bar{3}$ (Fig. 15.2(b)). The axes 3 and $\bar{3}$ we have encountered in the point groups, but 3_1 and 3_2 are found in the space groups. The 3_1 and 3_2 axes are like right-handed and left-handed screws respectively. There are five tetrad axes denoted 4, 4_1, 4_3, 4_2, and $\bar{4}$ respectively (Fig. 15.2(c)). Axes 4 and $\bar{4}$ occur in the point groups. Axes 4_1 and 4_3 have translations $\mathbf{c}/4$ and wind like right-handed and left-handed screws respectively. Axis 4_2 has a translation of $\mathbf{c}/2$ combined with a rotation of 90° for pairs of points at the same height along the c axis. There are seven hexad axes, denoted 6, 6_1, 6_5, 6_2, 6_4, 6_3, and $\bar{6}$ respectively (Fig. 15.2(d)). Axes 6 and $\bar{6}$ are present in the point groups. 6_1 and 6_5 have translations of single points through $c/6$, the sense of winding being right-handed and left-handed respectively. 6_2 and 6_4 have translations of pairs of points through $c/3$ and left-handed and right-handed sense of winding respectively. 6_3 has a translation for groups of three points through $\mathbf{c}/2$ with a rotation of $2\pi/2$. The way in which the elements of symmetry characteristic of space groups may be combined will now be illustrated by some examples.

Space group $P222_1$ (D_2^2)

The space group $P222_1$ (D_2^2) has the distribution of equivalent points and symmetry elements, in one unit cell, as shown in Fig. 15.3. As mentioned above, the space group is an infinite group and what is shown in Fig. 15.3(a) is the arrangement of equivalent

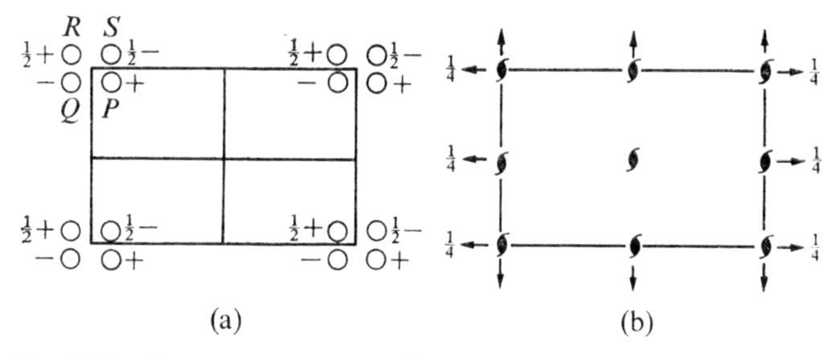

FIG. 15.3. Space group $P222_1$ (D_2^2). Diagrams representing
(a) the distribution of equivalent points,
(b) the distribution of symmetry elements.

points, and in Fig. 15.3(b) the symmetry elements, for only one unit cell. This is usually called a 'factor' or 'quotient' group of the space group. Fig. 15.3(a) strictly shows the equivalent points belonging to four unit cells. The contents of only one unit cell belong to the factor group. The coordinates of equivalent points, marked P, Q, R, S are as follows,

$$x, y, z; \quad x, \bar{y}, \bar{z}; \quad \bar{x}, \bar{y}, \tfrac{1}{2}+z; \quad \bar{x}, y, \tfrac{1}{2}-z.$$

Starting from any point, say P, we can obtain the other points by certain symmetry operations or successions of such operations. Thus point Q is derived from P by the operation of a diad axis parallel to the X axis at a level zero, i.e. at the base of the unit cell. In Fig. 15.3(a), this is expressed by writing the minus sign against Q to show that it is as much below the base of the unit cell as P is above it. The operation of the diad axis represented by the arrows pointing up and down the paper in Fig. 15.3(b) can be expressed by the equation

$$Q = C_{2x}P.$$

To reach R from P it is necessary to rotate 180° about the axis perpendicular to the paper and passing through the corner of the unit cell, and then to translate a distance $\mathbf{c}/2$ upwards from the paper. The screw axis at the corner of Fig. 15.3(b) expresses this operation, and it may also be expressed by the equation

$$R = T_{\mathbf{c}/2}C_{2z}P,$$

where $T_{c/2}$ stands for a translation of $\mathbf{c}/2$. Such equations, like the products of symmetry elements in Chapter 14 are read from right to left, and this equation expresses what has been stated above. Finally, a rotation about the diad axis at level 1/4 parallel to the y axis takes P to S. This may be expressed by the equation

$$S = T_{\mathbf{c}/2}C_{2y}P.$$

It is equally possible to go from R to S by operating with the axis C_{2x}, i.e.

$$S = C_{2x}R = C_{2x}T_{\mathbf{c}/2}C_{2z}P = T_{\mathbf{c}/2}C_{2y}P.$$

Q.15.1. Draw diagrams of the equivalent general positions and of the distribution of symmetry elements in space group $P2_12_12$ (D_2^3).

Space group Cmc (C_{2v}^{12})

An example involving glide planes, *Cmc* (C_{2v}^{12}), will now be discussed. The diagrams showing the distribution of equivalent points and symmetry elements are given in Figs 15.4(a), (b). The first symbol *C* indicates a *C*-face-centred lattice. In every unit cell there is therefore a lattice point obtained from the one at the origin by a translation of $\frac{1}{2}(\mathbf{a}+\mathbf{b})$. The environment of equivalent points about this *C*-face centring point must be the same as the environment of the origin. Thus, starting from the point *P* (Fig. 15.4(a)) we arrive with this translation of $\frac{1}{2}(\mathbf{a}+\mathbf{b})$ at point *R*, which has the same coordinates relative to the *C* face centre as *P* has relative to the origin at the corner of the cell.

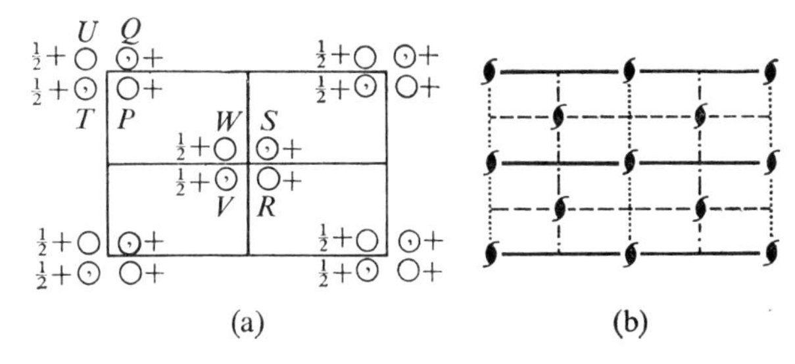

FIG. 15.4. Space group *Cmc* (C_{2v}^{12}). Diagrams representing
(a) the distribution of equivalent points,
(b) the distribution of symmetry elements.

The second symbol *m* implies a reflection plane perpendicular to the x axis. The first, second, and third symbols after the capital letter *C* refer to the x, y, and z axes respectively. (In this case the third symbol, 2_1, is omitted because it is not needed.) The *m* therefore takes *P* to *Q*, which is at the same height (+) above the base of the cell. The comma inserted into the open circle representing *Q* indicates that it bears a mirror-image relation to the point at *P*. Suppose *P* were a molecule of d-tartaric acid, then *Q* would be a molecule of l-tartaric acid. (It should not be assumed that this racemic compound occurs in this space group.) In detailed discussion of space groups, planes of symmetry perpendicular to the x, y and z axes are denoted m_1, m_2, and m_3 respectively.

The third symmetry element c stands for a glide plane normal to the y axis (since it occurs in the second place after the C), gliding in the direction of the z axis. The translation is $\mathbf{c}/2$, so that starting from point P, level $+$, we arrive at point T, level $\frac{1}{2}+$. The point T has a comma.

Thus, starting from P, we arrive at Q by the operation m_1, which may be written

$$Q = m_1 P.$$

Similarly, starting from P we arrive at T by the operation $T_{\mathbf{c}/2} m_2$ so we may write

$$T = T_{\mathbf{c}/2} m_2 P.$$

The point U is derived from point T by the plane m, so we may write

$$U = m_1 T = m_1 T_{\mathbf{c}/2} m_2 P.$$

Now,

$$m_1 m_2 = C_{2z}$$

as we can see in Fig. 15.4(a) by ignoring the $\frac{1}{2}$ at the points T and U. Thus we also have

$$U = T_{\mathbf{c}/2} C_{2z} P.$$

The screw diad axis parallel to the z axis is one of the elements of symmetry not expressed by the symbol *Cmc*. The symbol 2_1 could have been added after the c.

The point R gives rise to the points S, V, W by the operation of the symmetry elements m and c. Certain additional elements of symmetry not contained in the symbol *Cmc* are thereby generated. Consider the derivation of the point S from the point P. If we take the course P, Q, S, we can write

$$Q = m_1 P,$$

$$S = T_{(\mathbf{a}/2+\mathbf{b}/2)} Q = T_{(\mathbf{a}/2+\mathbf{b}/2)} m_1 P.$$

But

$$T_{(\mathbf{a}/2+\mathbf{b}/2)} = T_{\mathbf{a}/2} T_{\mathbf{b}/2}.$$

Hence

$$S = T_{\mathbf{a}/2} m_1 T_{\mathbf{b}/2} P.$$

The operations $T_{\mathbf{a}/2} m_1$ correspond to reflection in a plane normal to the x axis, cutting it at a distance $\mathbf{a}/4$ from the origin. Combination of this with $T_{\mathbf{b}/2}$ forms a glide plane parallel to (100) gliding in the direction of the y axis. In Fig. 15.4(b), this is shown as a dashed line. By reflection in the m plane, this glide plane also gives a parallel glide plane cutting the x axis at $-\mathbf{a}/4$ or $3\mathbf{a}/4$ from the origin.

Another element of symmetry not contained in the symbols *Cmc* is shown by relating point V to point P. If we follow the course P–R–V, we can write

$$R = T_{\mathbf{a}/2}T_{\mathbf{b}/2}P,$$

$$V = T_{\mathbf{c}/2}m_2R = T_{\mathbf{c}/2}m_2T_{\mathbf{a}/2}T_{\mathbf{b}/2}P.$$

Now $m_2T_{\mathbf{b}/2}$ implies the existence of a glide plane normal to the y axis at a distance $\mathbf{b}/4$ from the origin. $T_{\mathbf{c}/2}T_{\mathbf{a}/2}$ is the same as $T_{(\mathbf{c}+\mathbf{a})/2}$, i.e. it corresponds to a translation along the diagonal of the (010) face of the unit cell, and the amount of the translation is half the length of the whole diagonal. Thus we have

$$V = T_{(\mathbf{c}+\mathbf{a})/2}m_2T_{\mathbf{b}/2}P,$$

which corresponds to the diagonal glide parallel to (010) cutting the y axis at $\mathbf{b}/4$ from the origin, as shown in Fig. 15.4(b).

It can be verified that, starting from any of the equivalent points shown in Fig. 15.4(a), and applying any of the symmetry elements, we always arrive at another one of this group of equivalent points.

Q.15.2. Draw diagrams of the equivalent general positions and of the distribution of symmetry elements in space group *Ccc*2 (C_{2v}^{13}).

Space group $P4_22_12$ (D_4^6)

The 4_2 axes are placed at the corners of the square unit cell (Fig. 15.5(b)). (This is a different arrangement from that of the *International tables for X-ray crystallography*, but it changes only the position of the origin and not the relative disposition of the symmetry elements.) Points P, Q (Fig. 15.5(a)) are at a level + and points R, S are at a level $\frac{1}{2}+$ as is required by the operation of the axis 4_2. Corresponding equivalent points are inserted at the other corners of the unit cell. It is clear that the points R, T, U, V are related by a 4_2 axis, which passes down through the centre of the cell. This is inserted in Fig. 15.5(b). Next the 2_1 axis is to be inserted parallel to the x axis, because it stands next after the 4_2 symbol. The 2_1 axis cannot be inserted at level O passing through the 4_2 axis. If that were attempted it would give a 2 axis parallel to the y axis, and this is not allowed, because the x and y axes must be equivalent.

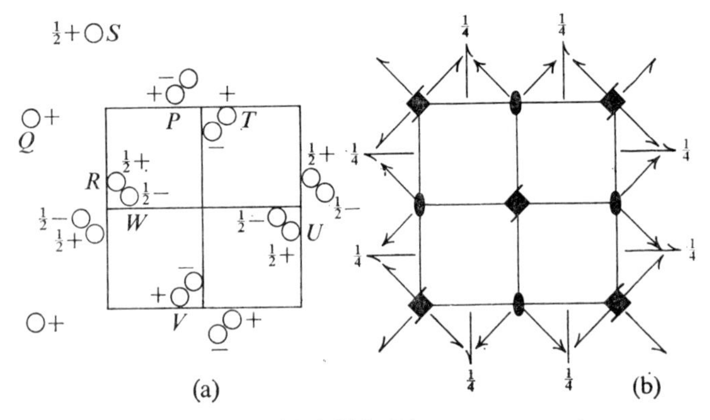

FIG. 15.5. Space group $P4_22_12$ (D_4^6). Diagrams representing
(a) the distribution of equivalent positions,
(b) the distribution of symmetry elements.

The truth of this conclusion can be seen by the drawing in of equivalent points or by the following analysis. First it must be noted that diad axes are necessarily present (compare points P and T) because of the operation of the 4_2 axes. Consider the 2_1 axis to pass at level O through the 4_2 axis at the centre of the cell. The point P would go to a point N (not marked in Fig. 15.5(a)) according to the equation

$$N = T_{\mathbf{a}/2}C_{2x}P.$$

The points P and T are related by the diad axis parallel to z, so we can write

$$P = C_{2z}T,$$

and combining the expressions we have

$$N = T_{\mathbf{a}/2}C_{2x}C_{2z}T.$$

Now $C_{2x}C_{2z} = C_{2y}$, so that

$$N = C_{2y}T_{\mathbf{a}/2}P.$$

Thus the new point N is related to P by a diad axis parallel to the y axis passing through a point $\mathbf{a}/4$ from the origin at level O. This is not allowed, because we have assumed a 2_1 axis parallel to the x axis, and this must apply to the y axis also.

The same argument shows that the 2_1 axis cannot pass through the 4_2 axis at level $\mathbf{c}/4$.

Next we can try putting the 2_1 axis at level O through a point $\mathbf{a}/4$ along the y axis. This results in a point P being changed to a point W related to point R by a diad axis parallel to $(1\bar{1}0)$, i.e. to the 2 in the third place of the space-group symbol. This diagonal diad axis is at height $\mathbf{c}/4$. This is a position for point W consistent with the space-group symmetry. There is another equally consistent possibility, namely that this 2_1 axis should be at level $\mathbf{c}/4$ and distant $\mathbf{a}/4$ from the xz-axial plane. This puts the diagonal diad axis at level O. This is the scheme used in Fig. 15.5. By applying the 4_2 symmetry to the point W, we can find all the remaining equivalent points, as shown in Fig. 15.5(a). The 2_1 and 2 axes of the space-group symbol are shown in Fig. 15.5(b).

Q.15.3. Draw diagrams of the equivalent general positions and of the distribution of symmetry elements in space group $P4_2cm(C_{4v}^3)$.

Space group $P\bar{3}1c$ (D_{3d}^2)

The diagrams of the distribution of the equivalent points and symmetry elements of space group $P\bar{3}1c$ are given in Figs 15.6(a) and (b). The origin is taken at the centre of symmetry associated with

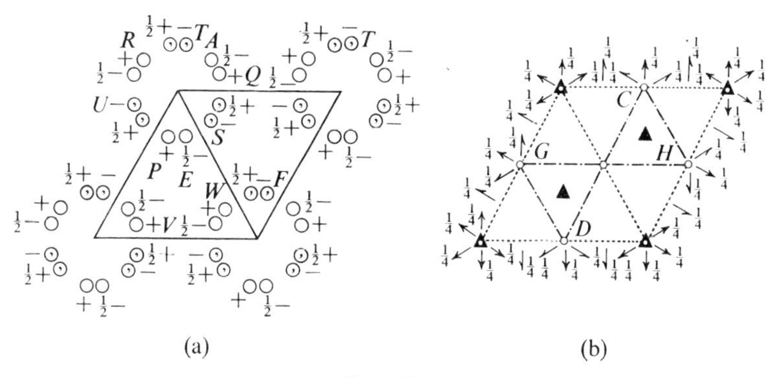

FIG. 15.6. Space group $P\bar{3}1c$ (D_{3d}^2). Diagrams representing
(a) the distribution of equivalent positions,
(b) the distribution of symmetry elements.

the $\bar{3}$ axis. Thus the points P, Q, R are at level $+$ and the points S, T, U are at level $-$ and have a comma. This distribution is repeated at each corner of the unit cell. It can be seen that the points

P, V, W are related by a triad axis. Thus in Fig. 15.6(b) the inversion triad, $\bar{3}$, and the triad, 3, are entered as shown. The place following $\bar{3}$ in the space-group symbol is occupied by 1, which means a one-fold axis is parallel to the x axis. This is ignored in Fig. 15.6(b). The second place following $\bar{3}$ is occupied by c, which means that a glide plane of symmetry, gliding in the z axis direction, is parallel to the horizontal edges of the hexagonal unit cell. In Fig. 15.6(b), this plane is made to pass through the origin, and in this arrangement it produces the distribution of the equivalent points shown in Fig. 15.6(a). If a c-glide parallel to $10\bar{1}0$ were made to pass through the centre of the cell, it would introduce more 3 axes, and this is not allowed. The horizontal diad screw axes that are present among these equivalent points are shown in Fig. 15.6(b). The diagonal glide plane represented by a dot-dash-dot line requires some explanation. Consider the points A and F. They are related because A is at $\frac{1}{2}-$, without a comma, and F is at $-$ with a comma. Thus for these points the line CD (Fig. 15.6(b)) acts as a diagonal glide plane. The same is true of the points E, F, (Fig. 15.6(a)) in relation to the diagonal glide plane GH. The triad axes repeat these glide planes as shown.

Q.15.4. Draw diagrams of the equivalent general positions and of the distribution of symmetry elements in the space group $P\bar{3}m1$ (D^3_{3d}).

Essentially the same principles can be applied to all the 230 space groups. Each combination of symmetry elements must be self-consistent, so that starting from any point and applying all the symmetry elements in turn we reach original point again. It is easy to verify that all space groups satisfy the requirements of a group given in Appendix 4.

In addition to the symmetry operations we have already discussed, it is necessary to include at this stage certain others. With lattices, a translation in one direction has the same result, namely superposition, as a translation of the same amount in the opposite direction. A translation of one cell dimension in one direction is denoted S, and an equal translation in the opposite sense is denoted S^{-1}. A glide plane is associated with a translation which is one-half of the corresponding lattice translation. A glide translation in one sense is, for example, denoted $T_{c/2}$, while the translation in the opposite sense is denoted $T^{-1}_{c/2}$. A glide translation carried out twice leads to a lattice translation. This may be expressed in symbols as

$$T_{c/2}T_{c/2} = S = T^2_{c/2}.$$

Also a glide operation in one direction followed by the same operation in the opposite direction brings the lattice back to its original position. This may be expressed

$$T_{c/2}T_{c/2}^{-1} = E.$$

Further equations expressing the relation between glide translation and lattice translation are

$$T_{c/2}S^{-1} = T_{c/2}^{-1}$$

and

$$T_{c/2}^{-1}S = T_{c/2}.$$

The symmetry elements $T_{c/2}$ and $T_{c/2}^{-1}$ are both present if either of them is present. Similarly, S and S^{-1} are always symmetry elements of a lattice (3, p. 127; 4, p. 199; 7, p. 163; 9, p. 73; 11, p. 89; 14, p. 235).

15.2. The reciprocal lattice

The theory of the electronic nature of conductors and semiconductors has been profoundly influenced during the last half-century by the ideas introduced by Brillouin. In an explanation of these ideas it is a great help to make use of an alternative treatment of lattices. Though the lattices on which crystal structures are normally represented are called Bravais lattices, the alternative description, usually called the reciprocal lattice, was also introduced by Bravais. There is one reciprocal lattice corresponding to each of the fourteen Bravais lattices mentioned in §15.1. The relation between one Bravais lattice and the corresponding reciprocal lattice is shown in Fig. 15.7. The unit cell of the Bravais lattice is the parallelepiped based on the edges OA, OB, OC. The *faces* which contain the points A, B, and C are denoted (100), (010), and (001) respectively. The edges of the reciprocal unit cell are OP, OQ, OR, and the *points* P, Q, R are denoted 100, 010, 001 respectively. (Note the use of brackets for the faces of the Bravais cell.) OP is normal to the face (100) and is of length inversely proportional to the perpendicular distance (spacing) of the two faces of the Bravais cell that are parallel to (100). Similarly OQ and OR are normal to (010) and (001) respectively and of lengths inversely proportional to the spacing of these planes. A lattice is built up on the reciprocal unit cell. The edges OA, OB, OC of the Bravais unit cell are denoted **a**, **b**, and **c**

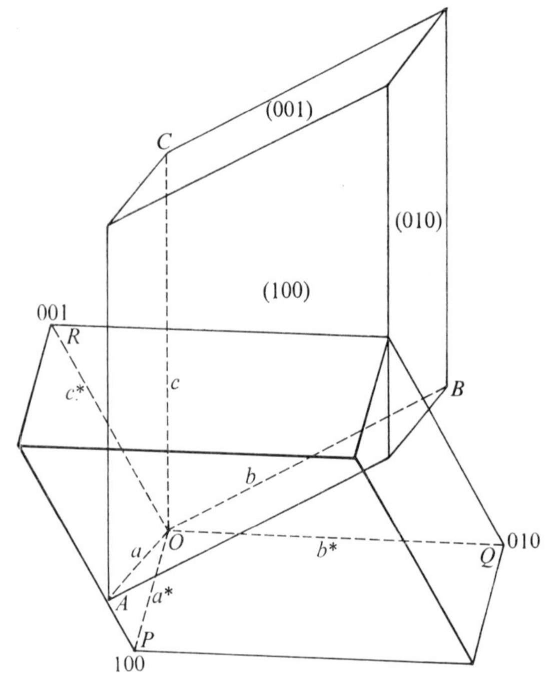

FIG. 15.7. Diagram showing the relation between a unit cell in a Bravais lattice and the corresponding reciprocal-lattice unit cell.

respectively, and the angles BOC, COA, and AOB are denoted α, β, γ respectively. The edges OP, OQ, OR of the reciprocal unit cell are denoted $\mathbf{a}^*$, $\mathbf{b}^*$, $\mathbf{c}^*$ respectively. A plane in the Bravais lattice is defined by Miller indices hkl. Such a plane cuts OA, OB, and OC at distances from O $\mathbf{a}/h$, $\mathbf{b}/k$, and $\mathbf{c}/l$ respectively (Fig. 15.8). The normal to this plane drawn from the origin is the line OS. The point T in the reciprocal lattice corresponds to point S in the Bravais lattice. Relative to the axes OP, OQ, OR, the coordinates of T are $h\mathbf{a}^*$, $k\mathbf{b}^*$, $l\mathbf{c}^*$. In the Bravais lattice there is an infinite set of planes all parallel to the plane hkl and having a spacing equal to the length OS. The point T in the reciprocal lattice corresponds to this infinite set of planes in the Bravais lattice. Such a relationship exists between every set of parallel equispaced planes in the Bravais lattice and a single point in the reciprocal lattice.

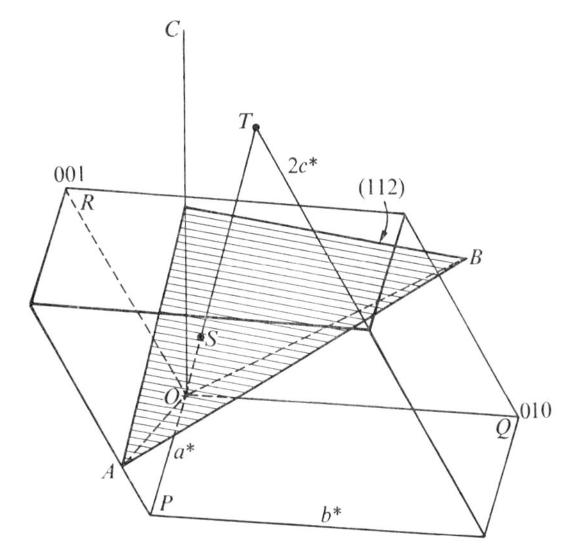

FIG. 15.8. Diagram showing the relation between a plane (*hkl*) cutting off lengths $\mathbf{a}/h$, $\mathbf{b}/k$, $\mathbf{c}/l$ on the axes OA, OB, OC, and the corresponding reciprocal point T having coordinates $h\mathbf{a}^*$, $k\mathbf{b}^*$, $l\mathbf{c}^*$ on the axes of the reciprocal lattice.

Expressed in terms of vectors, this may be written

$$\mathbf{a}^* = \frac{\mathbf{b}\times\mathbf{c}}{[\mathbf{abc}]}, \qquad \mathbf{b}^* = \frac{\mathbf{c}\times\mathbf{a}}{[\mathbf{abc}]}, \qquad \mathbf{c}^* = \frac{\mathbf{a}\times\mathbf{b}}{[\mathbf{abc}]},$$

or

$$\mathbf{a}.\mathbf{a}^* = 1, \qquad \mathbf{a}.\mathbf{b}^* = \mathbf{a}.\mathbf{c}^* = 0, \text{ etc.}$$

The converse statements are equally true, namely

$$\mathbf{a} = \frac{\mathbf{b}^*\times\mathbf{c}^*}{[\mathbf{a}^*\mathbf{b}^*\mathbf{c}^*]}, \quad \mathbf{b} = \frac{\mathbf{c}^*\times\mathbf{a}^*}{[\mathbf{a}^*\mathbf{b}^*\mathbf{c}^*]}, \quad \mathbf{c} = \frac{\mathbf{a}^*\times\mathbf{b}^*}{[\mathbf{a}^*\mathbf{b}^*\mathbf{c}^*]}.$$

The plane *hkl* cuts off a length $\mathbf{a}/h$ on the a axis, and when this is projected on the normal to the same plane the length of the projection is $\mathbf{d}_{hkl}$. Thus we can write

$$\left.\begin{aligned} &\frac{\mathbf{a}}{h}\cdot\mathbf{d}_{hkl} = d_{hkl}^2 \\ \text{and similarly}\quad &\frac{\mathbf{b}}{k}\cdot\mathbf{d}_{hkl} = \frac{\mathbf{c}}{l}\cdot\mathbf{d}_{hkl} = d_{hkl}^2 \end{aligned}\right\} \qquad (15.1)$$

If now we assume that

$$\mathbf{d}^*_{hkl} = h\mathbf{a}^* + k\mathbf{b}^* + l\mathbf{c}^*,$$

then

$$\left.\begin{aligned} \frac{\mathbf{a}}{h}\cdot\mathbf{d}^*_{hkl} &= 1 \qquad && \text{since } \mathbf{a}\cdot\mathbf{a}^* = 1 \\ \frac{\mathbf{b}}{k}\cdot\mathbf{d}^*_{hkl} &= 1 \qquad && \text{since } \mathbf{b}\cdot\mathbf{b}^* = 1 \\ \frac{\mathbf{c}}{l}\cdot\mathbf{d}^*_{hkl} &= 1 \qquad && \text{since } \mathbf{c}\cdot\mathbf{c}^* = 1 \end{aligned}\right\} \qquad (15.2)$$

Comparing equations (16.1) and (16.2), we see that $\mathbf{d}_{hkl}$ and $\mathbf{d}^*_{hkl}$ must be in the same direction, and

$$\mathbf{d}_{hkl} = d^2_{hkl}.\mathbf{d}^*_{hkl},$$

i.e.

$$|d^*_{hkl}| = 1/|d_{hkl}|.$$

Thus the reciprocal *point hkl* having coordinates h, k, l in the reciprocal axes corresponds to the infinite family of planes *hkl* in the Bravais lattice.

Up to this point we have constructed the reciprocal lattice so that the constant k in the expression

$$|d^*_{hkl}| = k/|d_{hkl}|$$

is unity. This is an arbitrary choice, and k may be given any value, since it determines only the scale on which the reciprocal lattice is constructed. In what follows we shall put $k = 2\pi$, as this simplifies some of the expressions used.

Infinite families of lattice planes and infinite sets of parallel wave fronts have this in common. Just as each set of lattice planes can be represented by a reciprocal point having coordinates h, k, l, so every set of wave fronts can be represented by a single reciprocal point, which is on the wave normal and distant from the origin by an amount inversely proportional to the wavelength. In the theory of electronic conduction in crystals, a wave motion is associated with each electron, the wavelength being inversely proportional to its momentum. Such an electron can therefore be represented by a point in reciprocal space, lying on the wave normal at a distance from the origin inversely proportional to the wavelength or directly proportional to the momentum.

There is one great difference between the representations of lattice planes and of electron waves. As the indices h, k, l increase from zero, the spacing d_{hkl} decreases and the corresponding d^*_{hkl} increases. The reciprocal points closest to the origin are therefore usually 100, 010, and 001. Electron waves are not limited in this way. The largest wavelength is determined by the actual size of the crystal and not by the size of the Bravais unit cell. The corresponding reciprocal points for such waves can therefore be very close to the origin or a long way from it. For electron waves the whole of reciprocal space can be regarded as containing reciprocal points distributed on a lattice of very small unit-cell size, this unit cell corresponding to the dimensions of the whole crystal. Since the ratio of the linear dimensions of the crystal and of the Bravais lattice unit cell are of the order of 10^6, we may regard the reciprocal points representing electron waves as forming an almost continuous distribution. It is not quite continuous, because only those waves that form standing waves within the external faces of the crystal have a permanent existence.

The symmetry of given positions within a reciprocal unit cell

The standing waves within the crystal must have a wavelength and wave-normal direction which permit every wave crest to pass through a Bravais lattice point. When this condition is fulfilled, a series of reciprocal lattice points can be used to represent the waves. Thus in

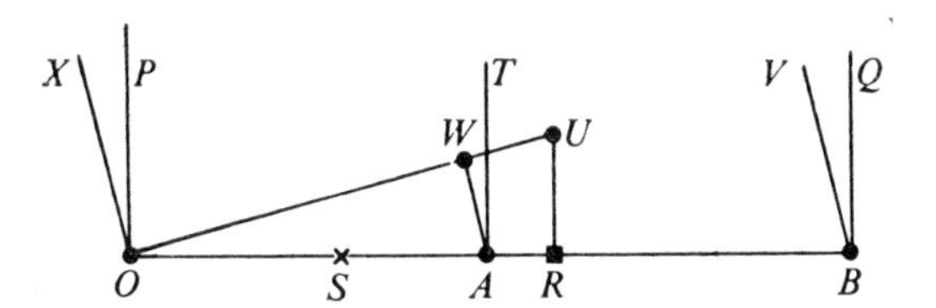

FIG. 15.9. Diagram showing wave fronts OP, AT, BQ, which pass through points along a point row in the Bravais lattice, and the corresponding reciprocal point R.

Fig. 15.9 the waves OP, AT, BQ are represented by the reciprocal point R. If the wave length is OA,

$$OR = 2\pi/OA.$$

Another set of waves such as OX, AW, BV is represented by the reciprocal point U. The line RU is normal to OA, since in the similar triangles OAW, OUR

$$\frac{OA}{OW} = \frac{OU}{OR}.$$

If the direction OP or OB is a two-fold axes of symmetry, the waves and the reciprocal points can satisfy this symmetry, since points such as U can be distributed above and below R in accordance with this symmetry. Thus to particular points in reciprocal space there correspond wave trains of defined wavelength and wave-normal direction, and these obey any symmetry element directed along or normal to the wave normal.

15.3. Wave–vector groups

The wave-vector group at a point $\mathbf{c}^*/2$ *from the origin in space group* $Cmcm$ (D_{2h}^{17})

These relations will now be illustrated by applying them to the wave-vector group associated with a point $\mathbf{c}^*/2$ from the origin in space group $Cmcm$ (D_{2h}^{17}). In Fig. 15.10, the diagram (a) gives the

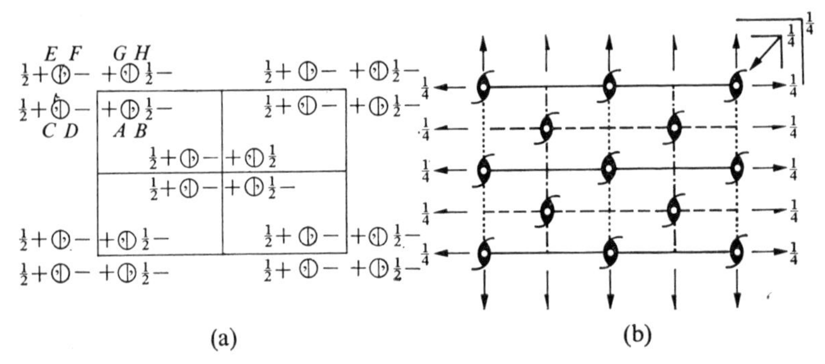

FIG. 15.10. Diagrams showing the distribution of equivalent points and symmetry elements in space group $Cmcm$ (D_{2h}^{17}).

distribution of the equivalent points in the factor group of this space group, and the diagram (b) gives the distribution of the corresponding axes, planes, and centres of symmetry. Table 15.1

TABLE 15.1

Symbols used for the [001] *wave-vector group in Cmcm*

Elements of symmetry in *Cmcm*	Corresponding point in Fig. 15.10 (a)	Symbols used	Corresponding symbols with lattice translation
Identity	A	E	S
Mirror plane normal to [100]	G	m_1	Sm_1
Glide plane normal to [010] gliding parallel to [001]	C	$T_{\mathbf{c}/2}m_2$	
Mirror plane normal to [001] at $\mathbf{c}/4$ and $3\mathbf{c}/4$	B	m_3	Sm_3
Diad axis parallel to [100] at level zero	D $A \rightarrow C \rightarrow D$	$m_3(T_{\mathbf{c}/2}m_2)$	
Diad axis parallel to [010] at level $\mathbf{c}/4$	H	m_3m_1	$Sm_3\,m_1$
Screw diad axis parallel to [001]	E	$T_{\mathbf{c}/2}C_{2z}$	
Centre of symmetry at origin	F $A \rightarrow E \rightarrow F$	$m_3(T_{\mathbf{c}/2}C_{2z})$	

sets out the elements of symmetry in the left-hand column. The letters in the second column show to what letter A is changed by the given symmetry element. The third column gives the symbols used to define the symmetry elements within the factor group, and the fourth column shows the new symbols that arise by addition of a lattice translation (also parallel to [001]). It is not necessary to add a new symbol in the third row of the fourth column. If we write

$$S(T_{\mathbf{c}/2}m_2) = T_{\mathbf{c}/2}^{-1}m_2,$$

or

$$S^{-1}(T_{\mathbf{c}/2}m_2) = T_{\mathbf{c}/2}^{-1}m_2,$$

we see that no new symmetry has been developed, since the symbol $T_{c/2}m_2$ implies the presence of the symbol $T_{c/2}^{-1}m_2$. The same reasoning applies to the fifth, seventh, and eighth rows of the fourth column. It must be noted that the wave-vector group is a finite group. We cannot, for example, include S^2 or S^3 as elements of symmetry within the group. Where S^2 appears in the course of the analysis it must be treated as equal to the identity operation E.

In general, the order in which the products of the symmetry operations are written makes no difference to the result. For example, the operation m_3m_1 corresponds to taking point A to $G(m_1)$ and then to H (m_3 at height c/4). Similarly, the operation m_1m_3 corresponds to taking point A to point B (m_3) and then to H (m_1). Thus we may write

$$m_3m_1 = m_1m_3.$$

By contrast, $m_3(T_{c/2}m_2)$ gives the succession of positions A–C–D, while $(T_{c/2}m_2)m_3$ gives the succession A–B–D^{u}, where D^{u} represents the point one unit cell above point D. The same lack of commutation applies to the product of m_3 and $(T_{c/2}C_{2z})$. It may be shown that all products of symmetry elements involving $T_{c/2}$ and m_3 do not commute with one another.

The classes of symmetry in this wave-vector group

To find which elements of symmetry belong to the same class, we proceed as in §12.2 and conjugate each element with each of the others. To illustrate this process, the following examples are given.

$$m_1^{-1}(T_{c/2}C_{2z})m_1 = A\xrightarrow[m_1]{}G\xrightarrow[(Tc/2C_{2z})]{}C\xrightarrow[m_1^{-1}]{}E = T_{c/2}C_{2z}.$$

$$m_3^{-1}(T_{c/2}C_{2z})m_3 = A\xrightarrow[m_3]{}B\xrightarrow[(Tc/2C_{2z})]{}F^{u}\xrightarrow[m_3^{-1}]{}E^{d} = T_{c/2}^{-1}C_{2z}.$$

(The superscripts 'u' and 'd' to points stand for 'up' and 'down' and refer to corresponding points above and below those points in a cell such as the one shown in Fig. 15.10.) Thus $T_{c/2}C_{2z}$ and $T_{c/2}^{-1}C_{2z}$ belong to the same class. Conjugation of m_1 with any of the elements of symmetry in the group results in m_1 only. Two examples are given,

$$m_3^{-1}m_1m_3 = B\xrightarrow[m_3]{}A\xrightarrow[m_1]{}G\xrightarrow[m_3^{-1}]{}H = m_1.$$

$$(T_{c/2}C_{2z})^{-1}m_1(T_{c/2}C_{2z}) = A\xrightarrow[(Tc/2C_{2z})]{}E\xrightarrow[m_1]{}C\xrightarrow[(Tc/2C_{2z})^{-1}]{}G = m_1.$$

Proceeding in this way, we can show that there are ten classes of symmetry, namely

$\mathscr{C}_1 = E,$ $\mathscr{C}'_1 = m_3(T_{c/2}C_{2z})+m_3(T^{-1}_{c/2}C_{2z}),$

$\mathscr{C}_2 = T_{c/2}C_{2z}+T^{-1}_{c/2}C_{2z},$ $\mathscr{C}'_2 = m_3+Sm_3,$

$\mathscr{C}_5 = m_3m_1+Sm_3m_1,$ $\mathscr{C}'_5 = T_{c/2}m_2+T^{-1}_{c/2}m_2,$

$\mathscr{C}_6 = m_3(T_{c/2}m_2)+m_3(T^{-1}_{c/2}m_2),$ $\mathscr{C}'_6 = m_1,$

$\mathscr{C}_7 = S,$ $\mathscr{C}_9 = Sm_1.$

(The subscripts of the $\mathscr{C}$s are chosen to follow those used by H. Jones (see bibliography).) The class multiplication coefficients, c_{ijs}, are found as described in Chapter 13. From these the determinantal equations can be written down and solved in turn to find the ten roots corresponding to each class of symmetry.

The example of class $\mathscr{C}_2$

As an example the solution for class $\mathscr{C}_2$ is given below. The class multiplication coefficients of the form c_{i2s} are set out in Table 15.2. This matrix is reduced to triangular form by the usual process, which depends on the fact that addition of any multiple of a given row of the determinant to any other row, term by corresponding term, does not alter the value of the determinant. Thus the first row is multiplied by $2/x_2$ and then added, term by term, to the second row. This gives the second row of Table 15.3. The third row of Table 15.2 is transferred to Table 15.3 unchanged. The third row of Table

TABLE 15.2

Matrix containing the class multiplication coefficients of the type c_{i2s}

i \ s	1	2	5	6	7	1′	2′	5′	6′	9
1	$-x_2$	1	·	·	·	·	·	·	·	·
2	2	$-x_2$	·	·	2	·	·	·	·	·
5	·	·	$-x_2$	2	·	·	·	·	·	·
6	·	·	2	$-x_2$	·	·	·	·	·	·
7	·	1	·	·	$-x_2$	·	·	·	·	·
1′	·	·	·	·	·	$-x_2$	2	·	·	·
2′	·	·	·	·	·	2	$-x_2$	·	·	·
5′	·	·	·	·	·	·	·	$-x_2$	2	2
6′	·	·	·	·	·	·	·	1	$-x_2$	·
9	·	·	·	·	·	·	·	1	·	$-x_2$

TABLE 15.3

Matrix of Table 15.2 in triangular form

i \ s	1	2	5	6	7	1′	2′	5′	6′	9
1	$-x_2$	1	·	·	·	·	·	·	·	·
2	·	$\frac{2}{x_2}-x_2$	·	·	2	·	·	·	·	·
5	·	·	$-x_2$	2	·	·	·	·	·	·
6	·	·	·	$\frac{4}{x_2}-x_2$	·	·	·	·	·	·
7	·	·	·	·	$\frac{2x_2}{x_2^2-2}-x_2$	·	·	·	·	·
1′	·	·	·	·	·	$-x_2$	2	·	·	·
2′	·	·	·	·	·	·	$\frac{4}{x_2}-x_2$	·	·	·
5′	·	·	·	·	·	·	·	$-x_2$	2	2
6′	·	·	·	·	·	·	·	·	$\frac{2}{x_2}-x_2$	$\frac{2}{x_2}$
9	·	·	·	·	·	·	·	·	·	$-\frac{x_2(4-x_2^2)}{(2-x_2^2)}$

15.2 is multiplied by $2/x_2$ and added to the fourth row to give the fourth row of Table 15.3. To eliminate the 1 in the second column of row five (Table 15.2) we multiply the term above it in Table 15.3, namely $(2/x_2)-x_2$ by $-x_2/(2-x_2^2)$ and add the second row of Table 15.3 (each term being multiplied by the factor $-x_2/(2-x_2^2)$) to the fifth row of Table 15.2. The procedure follows the same principle in the rest of Table 15.3. The value of the determinant is simply the product of all the diagonal terms, which, on reduction, becomes

$$x_2^2(4-x_2^2)^4.$$

When this expression is equated to zero the ten roots obtained are $2, 2, 2, 2, -2, -2, -2, -2, 0, 0$.

The character table for this wave-vector group

The roots x_j corresponding to all of the classes of symmetry present for the wave group can be found in the manner indicated in the previous paragraph. These roots can then be arranged in columns and rows which satisfy the orthogonality relation that the sum of the roots along any horizontal row except the topmost row must be zero. The characters for the class $\mathscr{C}_1$ must satisfy the relation that the sum of their squares is equal to the total number of symmetry elements. In the example chosen here, there are sixteen elements and ten classes of symmetry. The values of χ_E must therefore be

$$1, 1, 1, 1, 1, 1, 1, 1, 2, 2,$$

since the sum of the squares of these ten numbers is equal to sixteen.

The characters χ_i can now be obtained from the roots x_i, the number of elements in the class, h_i, and the identity character, χ_E, from the equation,

$$x_i\chi_E = h_i\chi_i.$$

The characters must then be arranged to satisfy the orthogonality relations; in particular the sum of the products of the corresponding characters in the χ_E column and any other column must be zero. The resulting character table is given below (Table 15.4).

The eight columns and rows enclosed within the double lines correspond to the characters of the point group *mmm*. Thus only the two extra representations at the bottom of Table 15.4 correspond to the particular wave normal defined by the point at $\mathbf{c}^*/2$ from the origin (10, p. 141).

TABLE 15.4

Character table for the wave-vector group at **c***/2 *from the origin in Cmcm*

	$\mathscr{C}_1$	$\mathscr{C}_2$	$\mathscr{C}_5$	$\mathscr{C}_6$	$\mathscr{C}_{1'}$	$\mathscr{C}_{2'}$	$\mathscr{C}_{5'}$	$\mathscr{C}_{6'}$	$\mathscr{C}_7$	$\mathscr{C}_9$
h_i	1	2	2	2	2	2	2	1	1	1
	1	1	1	1	1	1	1	1	1	1
	1	1	−1	−1	1	1	−1	−1	1	−1
	1	−1	1	−1	1	−1	1	−1	1	−1
	1	−1	−1	1	1	−1	−1	1	1	1
	1	1	1	1	−1	−1	−1	−1	1	−1
	1	1	−1	−1	−1	−1	1	1	1	1
	1	−1	1	−1	−1	1	−1	1	1	1
	1	−1	−1	1	−1	1	1	−1	1	−1
	2	0	0	0	0	0	0	2	−2	−2
	2	0	0	0	0	0	0	−2	−2	2

Q.15.5. Determine the class multiplication coefficients of the type c_{i7s} and find the corresponding roots of x_7 for the group discussed in the preceding paragraphs.

Answers

Q.15.1. See Fig. 15.11 (a), (b).

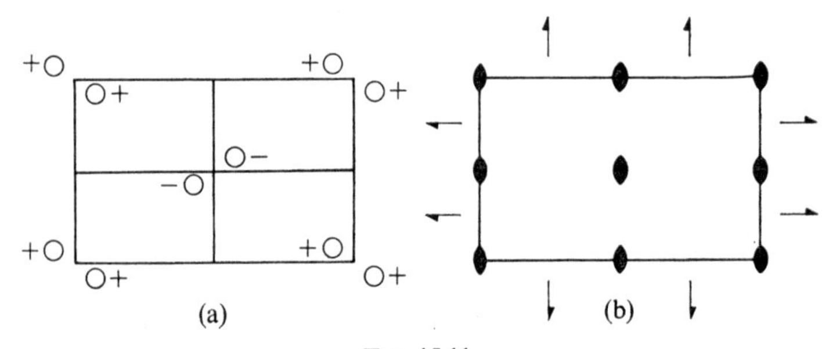

FIG. 15.11.

Q.15.2. See Fig. 15.12 (a), (b).

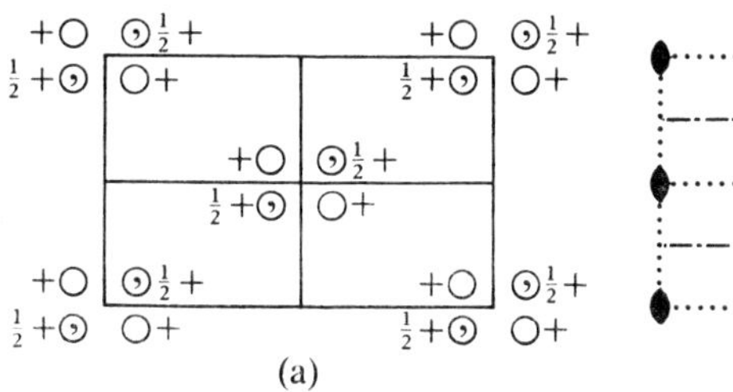

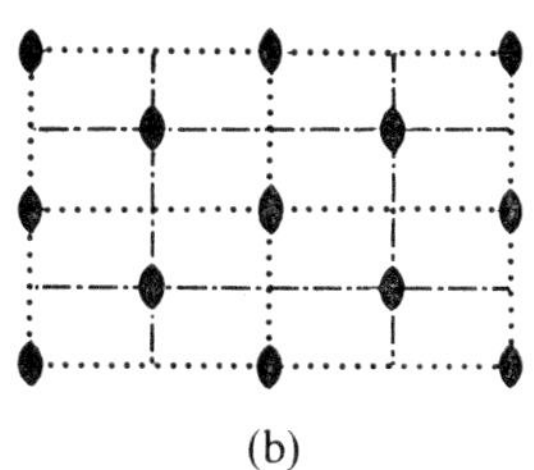

(a) (b)

FIG. 15.12.

Q.15.3. See Fig. 15.13 (a), (b).

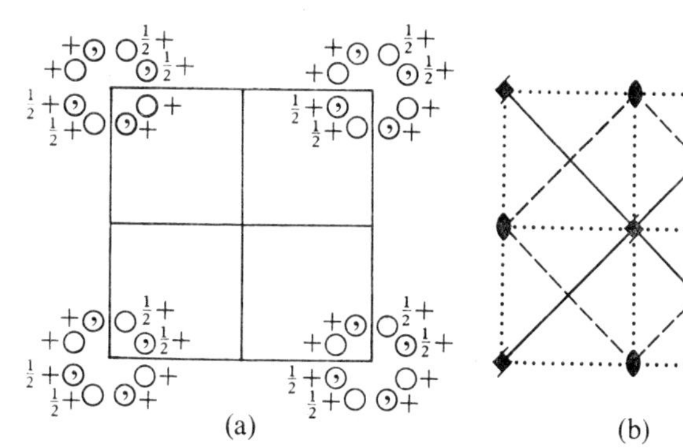

(a) (b)

FIG. 15.13.

Q.15.4. See Fig. 15.14 (a), (b).

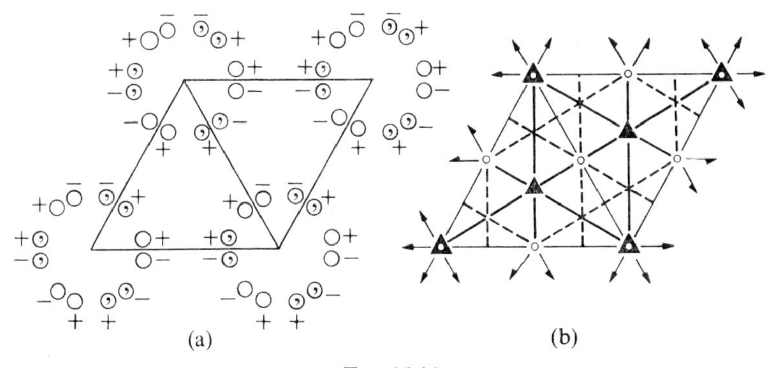

(a) (b)

FIG. 15.14.

Q.15.5.

$\mathscr{C}_1\mathscr{C}_7 = \mathscr{C}_7,$ $\qquad c_{177} = 1.$

$\mathscr{C}_2\mathscr{C}_7 = (T_{c/2}C_{2z}+T_{c/2}^{-1}C_{2z})S$

$= T_{c/2}^{-1}C_{2z}+T_{c/2}C_{2z} = \mathscr{C}_2,$ $\qquad c_{272} = 1.$

$\mathscr{C}_5\mathscr{C}_7 = (m_3m_1+Sm_3m_1)S = Sm_3m_1+m_3m_1 = \mathscr{C}_5,$ $\qquad c_{575} = 1.$

$\mathscr{C}_6\mathscr{C}_7 = \{m_3(T_{c/2}m_2)+m_3(T_{c/2}^{-1}m_2)\}S$

$= m_3(T_{c/2}^{-1}m_2)+m_3(T_{c/2}m_2) = \mathscr{C}_6,$ $\qquad c_{676} = 1.$

$\mathscr{C}_7\mathscr{C}_7 = SS = E = \mathscr{C}_1,$ $\qquad c_{771} = 1.$

$\mathscr{C}'_1\mathscr{C}_7 = \{m_3(T_{c/2}C_{2z})+m_3(T_{c/2}^{-1}C_{2z})\}S$

$= m_3(T_{c/2}^{-1}C_{2z})+m_3(T_{c/2}C_{2z}) = \mathscr{C}'_1,$ $\qquad c_{1'71'} = 1.$

$\mathscr{C}'_2\mathscr{C}_7 = (m_3+m_3S)S = m_3S+m_3 = \mathscr{C}'_2,$ $\qquad c_{2'72'} = 1.$

$\mathscr{C}'_5\mathscr{C}_7 = (T_{c/2}m_2+T_{c/2}^{-1}m_2)S$

$= (T_{c/2}^{-1}m_2)+(T_{c/2}m_2)$

$= \mathscr{C}'_5,$ $\qquad c_{5'75'} = 1.$

$\mathscr{C}'_6\mathscr{C}_7 = m_1S = \mathscr{C}_9,$ $\qquad c_{6'79} = 1.$

$\mathscr{C}_9\mathscr{C}_7 = Sm_1S = m_1 = \mathscr{C}'_6,$ $\qquad c_{976'} = 1.$

TABLE 15.4

Matrix for determining the roots of x_7

i \ s	1	2	5	6	7	1′	2′	5′	6′	9
1	$-x_7$	·	·	·	1	·	·	·	·	·
2	·	$1-x_7$	·	·	·	·	·	·	·	·
5	·	·	$1-x_7$	·	·	·	·	·	·	·
6	·	·	·	$1-x_7$	·	·	·	·	·	·
7	1	·	·	·	$-x_7$	·	·	·	·	·
1′	·	·	·	·	·	$1-x_7$	·	·	·	·
2′	·	·	·	·	·	·	$1-x_7$	·	·	·
5′	·	·	·	·	·	·	·	$1-x_7$	·	·
6′	·	·	·	·	·	·	·	·	$-x_7$	1
9	·	·	·	·	·	·	·	·	1	$-x_7$

When this matrix is made triangular, we obtain the product along the diagonal equal to

$$-x_7(1-x_7)^3\left(\frac{1}{x_7}-x_7\right)(1-x_7)^3(-x_7)\left(\frac{1}{x_7}-x_7\right).$$

On reduction this leads to

$$(1-x_7)^6(1-x_7^2)^2.$$

Equating this expression with zero, we get the roots corresponding to x_7 as

$$1, 1, 1, 1, 1, 1, 1, -1, 1, -1.$$

16

Brillouin zones

16.1. Introduction

IN §15.2 it was shown that trains of parallel plane waves in crystals may be represented by points in a reciprocal lattice. The reciprocal point lies on the wave normal at a distance from the origin of the reciprocal lattice inversely proportional to the wavelength of the waves. Brillouin made use of this concept in explaining the theory of electronic conduction in crystals. Each electron is associated with a train of plane waves and can therefore be represented by a point in the reciprocal lattice. The almost continuous distribution of these reciprocal points can be divided into what are called Brillouin zones. The origin of the reciprocal lattice is joined by straight lines to neighbouring reciprocal points. Usually, 'neighbouring' means 'next-nearest neighbour', but in some cases it refers to lattice points that are not so close to the origin. In Fig. 16.1, a square net is shown which may be regarded as the projection on to the paper of a prim-

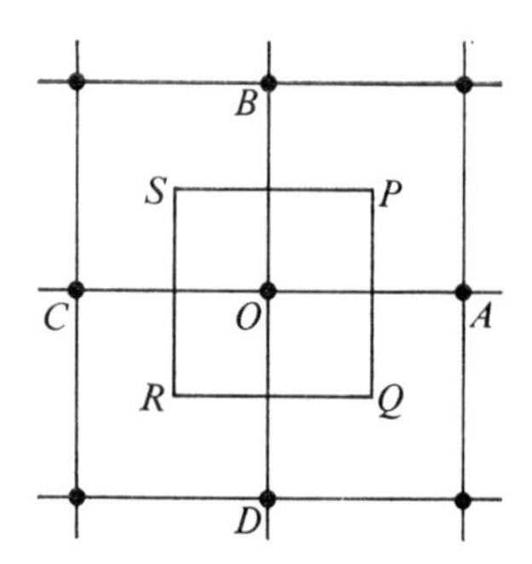

FIG. 16.1. Diagram showing the projection of a primitive cubic lattice of which typical points are *O*, *A*, *B*, *C*, *D* and the corresponding first Brillouin zone *PQRS*.

itive cubic lattice. The lines joining the origin to the nearest neighbours in reciprocal space are OA, OB, OC, OD. To construct the first Brillouin zone, these lines are perpendicularly bisected by planes which project as the lines PQ, PS, RS, and RQ respectively. A second Brillouin zone can be constructed by taking the second-nearest neighbours among the reciprocal points surrounding O, and so on for the higher orders of these zones.

The relation of these zones to electron waves is that waves represented by points on the bounding faces of these zones are reflected by the crystal lattice in the same way that X-rays of corresponding wavelength would be reflected. This can be shown by reference to Fig. 16.2. The reciprocal point Q is joined to the origin O,

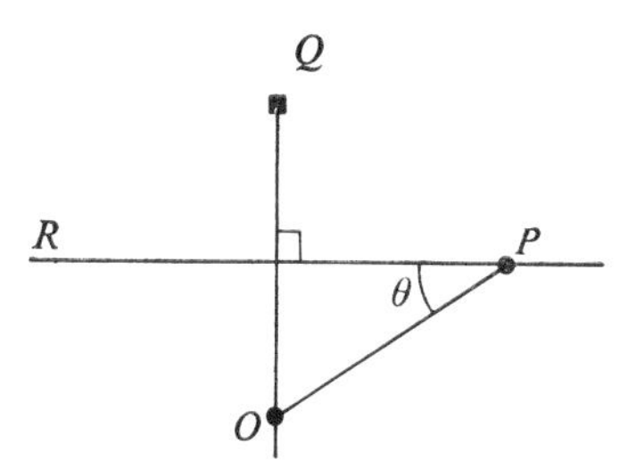

FIG. 16.2. Diagram showing a point P, representing a wave vector OP, lying on a plane PR, which forms a face of a Brillouin zone.

and the plane PR bisects OQ perpendicularly. The angle $\widehat{OPR}$ is denoted θ. The point P corresponds to a wave for which the wave length λ is given by

$$\lambda = 2\pi/OP$$

(with 2π as the constant of proportionality). The spacing, d, of the Bravais lattice planes corresponding to point Q is similarly given by

$$d = 2\pi/OQ.$$

Thus

$$\frac{\lambda}{d} = \frac{OQ}{OP}.$$

From Fig. 16.2 it is clear that this ratio is equal to 2 sin θ. Thus

$$\frac{\lambda}{d} = 2 \sin \theta.$$

This is the usual Bragg relation for reflection at lattice planes. The point P is any point along the plane PR, and therefore the whole plane acts as a reflector for the waves represented by the points lying anywhere in the boundary of the Brillouin zone. The form of Brillouin zones is fundamental to the theory of electron conduction, and in the following paragraphs attention will be given to this geometrical problem. In addition to the external shape of the Brillouin zone, it is necessary for many purposes to know the way in which the energy or momentum of the electrons varies within the zone. This usually involves a knowledge of the requirements of symmetry, and in this connection group theory provides an almost indispensable guide.

16.2. The geometry of Brillouin zones

The cubic system

There are three cubic Bravais lattices, namely primitive, face-centred, and body-centred. The corresponding reciprocal lattices are all based on the cubic reciprocal unit cell, having a side of length a^*. X-ray reflections from a face-centred lattice have indices hkl, which are either all odd or all even so that the corresponding reciprocal lattice has no points with mixed (i.e. odd and even) coordinates h, k, l. Such a lattice may be regarded as body-centred cubic based on a side of length $2a^*$ (Fig. 16.3). The reciprocal lattice for a body-centred Bravais lattice is likewise based on a cell of side $2a^*$, but it is face-centred, since for all the X-ray reflections $h+k+l$ must be even (Fig. 16.4).

Construction of Brillouin zones

We can obtain the forms of Brillouin zones by finding the coordinates (in the reciprocal lattice) of the points in which three planes defining the Brillouin zone meet. An analytical method of finding such points is given below, and is followed by applications to the three types of Brillouin zone of the cubic system.

The vector joining the origin to the point at which three planes intersect is denoted $\mathbf{p}$. The vectors from the origin that are normal to each of the three intersecting planes and end on the planes are denoted $\mathbf{q}$, $\mathbf{r}$, and $\mathbf{s}$ respectively. Then we have

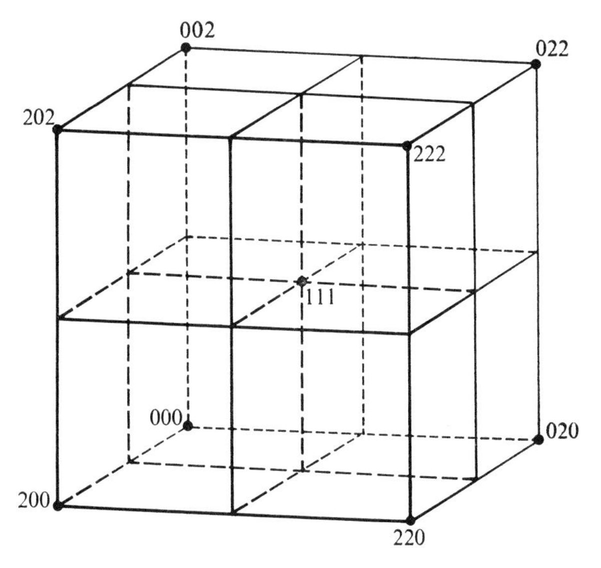

FIG. 16.3. Diagram showing the points of a reciprocal lattice corresponding to a face-centred cubic Bravais lattice.

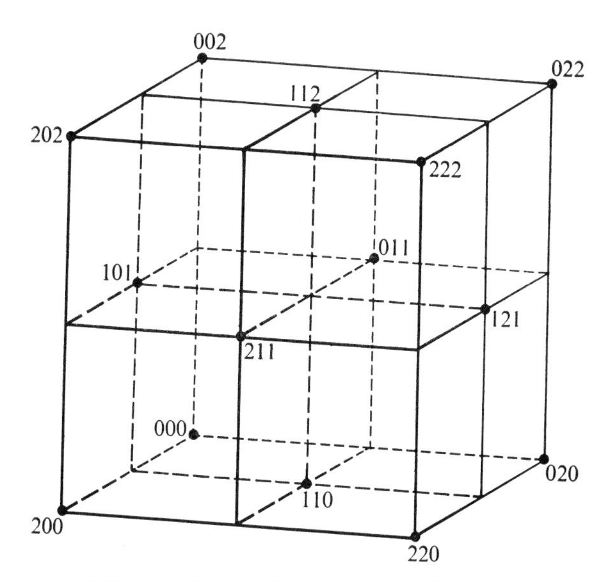

FIG. 16.4. Diagram showing the points of a reciprocal lattice corresponding to a body-centred cubic Bravais lattice.

$$\mathbf{p}\cdot\mathbf{q} = |q|^2,$$
$$\mathbf{p}\cdot\mathbf{r} = |r|^2,$$
$$\mathbf{p}\cdot\mathbf{s} = |s|^2.$$

The components of these vectors parallel to the X, Y, Z axes are denoted by subscripts 1, 2, 3 respectively. Thus

$$p_1q_1+p_2q_2+p_3q_3 = |q|^2,$$
$$p_1r_1+p_2r_2+p_3r_3 = |r|^2,$$
$$p_1s_1+p_2s_2+p_3s_3 = |s|^2,$$

and hence

$$p_1 = \frac{\begin{vmatrix} |q^2| & q_2 & q_3 \\ |r^2| & r_2 & r_3 \\ |s^2| & s_2 & s_3 \end{vmatrix}}{\begin{vmatrix} q_1 & q_2 & q_3 \\ r_1 & r_2 & r_3 \\ s_1 & s_2 & s_3 \end{vmatrix}}$$

and similar expressions give p_2 and p_3.

Brillouin zone of a primitive cubic lattice

The six reciprocal points nearest to the origin have indices 100, 010, 001, $\bar{1}00$, $0\bar{1}0$, $00\bar{1}$. The corresponding coordinates are therefore as follows.

$$q_1 \quad q_2 \quad q_3 = \tfrac{1}{2} \quad 0 \quad 0,$$
$$r_1 \quad r_2 \quad r_3 = 0 \quad \tfrac{1}{2} \quad 0,$$

and
$$s_1 \quad s_2 \quad s_3 = 0 \quad 0 \quad \tfrac{1}{2};$$

hence

$$p_1 = \begin{vmatrix} \frac{1}{4} & 0 & 0 \\ \frac{1}{4} & \frac{1}{2} & 0 \\ \frac{1}{4} & 0 & \frac{1}{2} \end{vmatrix} \Bigg/ \tfrac{1}{8} = 2\begin{vmatrix} 1 & 0 & 0 \\ 1 & \frac{1}{2} & 0 \\ 1 & 0 & \frac{1}{2} \end{vmatrix} = \tfrac{1}{2}.$$

Also $p_1 = p_2 = p_3 = \frac{1}{2}$.

There are eight points having each of their coordinates $\pm\frac{1}{2}$, and the lines joining these points form the cube which is the first Brillouin zone.

Brillouin zone of a face-centred cubic Bravais lattice

In Fig. 16.3, the first reciprocal point along the x axis is 200, along the [110] direction it is 220, but along the body diagonal it is [111]. We shall first determine the coordinates of point A (Fig. 16.5), the meeting place of planes defined by the reciprocal points 200, 111 and $11\bar{1}$.

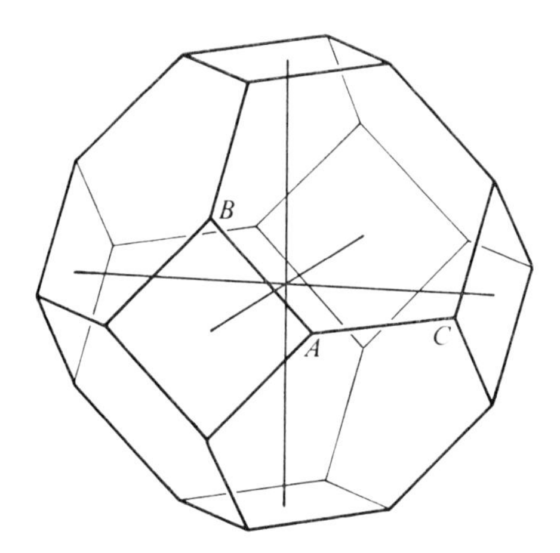

FIG. 16.5. Diagram showing the cubo-octahedral form of the first Brillouin zone corresponding to a face-centred cubic Bravais lattice.

We have, for the components of the normals to these planes,

$$q_1\,q_2\,q_3 = 1 \quad 0 \quad 0, \quad |q|^2 = 1;$$

$$r_1\,r_2\,r_3 = \tfrac{1}{2} \quad \tfrac{1}{2} \quad \tfrac{1}{2}, \quad |r|^2 = \tfrac{1}{4}+\tfrac{1}{4}+\tfrac{1}{4} = \tfrac{3}{4};$$

$$s_1\,s_2\,s_3 = \tfrac{1}{2} \quad \tfrac{1}{2} \quad \bar{\tfrac{1}{2}}, \quad |s|^2 = \tfrac{1}{4}+\tfrac{1}{4}+\tfrac{1}{4} = \tfrac{3}{4}.$$

Then
$$p_1 = \frac{\begin{vmatrix} 1 & 0 & 0 \\ \frac{3}{4} & \frac{1}{2} & \frac{1}{2} \\ \frac{3}{4} & \frac{1}{2} & \bar{\frac{1}{2}} \end{vmatrix}}{\begin{vmatrix} 1 & 0 & 0 \\ \frac{1}{2} & \frac{1}{2} & \frac{1}{2} \\ \frac{1}{2} & \frac{1}{2} & \bar{\frac{1}{2}} \end{vmatrix}} = 1, \quad p_2 = \frac{\begin{vmatrix} 1 & 1 & 0 \\ \frac{1}{2} & \frac{3}{4} & \frac{1}{2} \\ \frac{1}{2} & \frac{3}{4} & \bar{\frac{1}{2}} \end{vmatrix}}{-\frac{1}{2}} = \tfrac{1}{2},$$

and
$$p_3 = \frac{\begin{vmatrix} 1 & 0 & 1 \\ \frac{1}{2} & \frac{1}{2} & \frac{3}{4} \\ \frac{1}{2} & \frac{1}{2} & \frac{3}{4} \end{vmatrix}}{-\frac{1}{2}} = 0.$$

The coordinates of point A are therefore $1, \frac{1}{2}, 0$. Similarly the co-ordinates of points B and C are respectively $1, 0, \frac{1}{2}$ and $\frac{1}{2}, 1, 0$. Other corresponding points may be inserted from the cubic symmetry. When these points are joined by lines the final Brillouin zone is a cubo-octahedron (Fig. 16.5).

Brillouin zone of a body-centred cubic Bravais lattice

The rule for the X-ray reflections from a body-centred cubic, Bravais lattice is that $(h+k+l)$ must be even. Thus along the [100] [110], and [111] directions the first reflecting planes are 200, 110, and 222 respectively.

The reciprocal points of the type 110 are closer to the origin than any others and hence they alone form the first Brillouin zone. For the point A, Fig. 16.6, the indices of the planes meeting at that point are 110, 011, 101. Thus we have for the components of the normals to the bounding planes,

$$q_1 \quad q_2 \quad q_3 = \tfrac{1}{2} \quad \tfrac{1}{2} \quad 0, \qquad |q|^2 = \tfrac{1}{2};$$
$$r_1 \quad r_2 \quad r_3 = 0 \quad \tfrac{1}{2} \quad \tfrac{1}{2}, \qquad |r|^2 = \tfrac{1}{2};$$
$$s_1 \quad s_2 \quad s_3 = \tfrac{1}{2} \quad 0 \quad \tfrac{1}{2}, \qquad |s|^2 = \tfrac{1}{2};$$

thus
$$p_1 = \begin{vmatrix} \frac{1}{2} & \frac{1}{2} & 0 \\ \frac{1}{2} & \frac{1}{2} & \frac{1}{2} \\ \frac{1}{2} & 0 & \frac{1}{2} \end{vmatrix} \Bigg/ \begin{vmatrix} \frac{1}{2} & \frac{1}{2} & 0 \\ 0 & \frac{1}{2} & \frac{1}{2} \\ \frac{1}{2} & 0 & \frac{1}{2} \end{vmatrix} = \tfrac{1}{8}/\tfrac{1}{4} = \tfrac{1}{2}.$$

Similarly $p_2 = p_3 = \frac{1}{2}$.

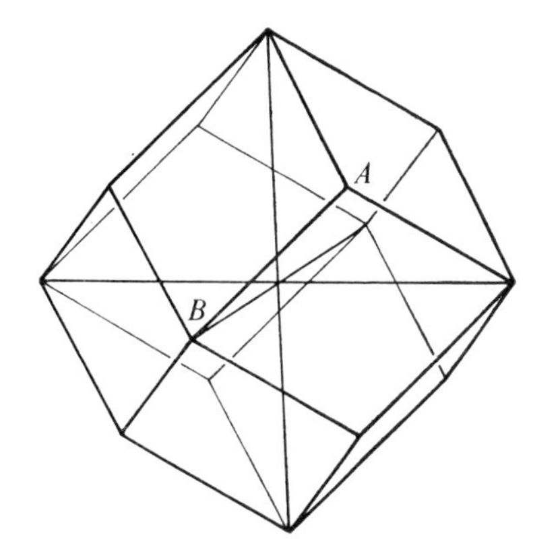

FIG. 16.6. Diagram showing the rhombic-dodecahedral form of the first Brillouin zone corresponding to a body-centred cubic Bravais lattice.

Thus the coordinates of *A* are $\frac{1}{2}, \frac{1}{2}, \frac{1}{2}$. The planes 110, 101, $1\bar{1}0$, $10\bar{1}$ all pass through the point having coordinates 1, 0, 0. This is the point marked *B*. When corresponding points to *A* and *B* are inserted and joined together by lines, we obtain a rhombic dodecahedron as the first Brillouin zone, Fig. 16.6 (6, p. 110; 10, p. 32, 116, 121).

Brillouin zones of non-cubic systems

The Brillouin zones of metals belonging to the non-cubic systems have their own characteristic shapes. These can be derived from the reciprocal points nearest to the origin in the manner described above. Some of the forms are simple, some are more complicated figures. For Brillouin zones for graphite and bismuth see Figs 16.7 and 16.8.

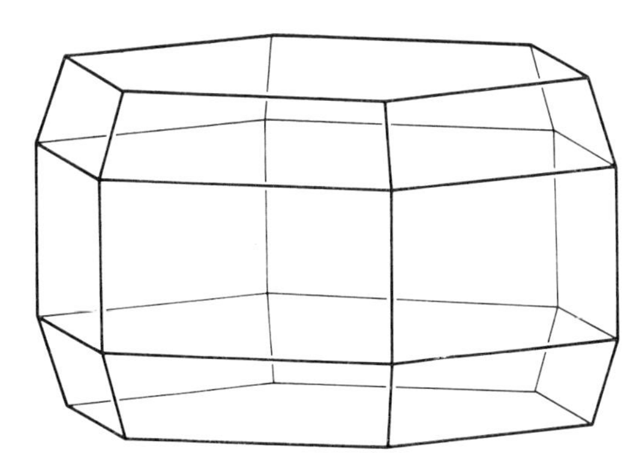

FIG. 16.7. Diagram of the first Brillouin zone of graphite.

Large Brillouin zones

For the discussion of many electrical, magnetic, and elastic properties of metallic and semi-metallic crystals, it is necessary to

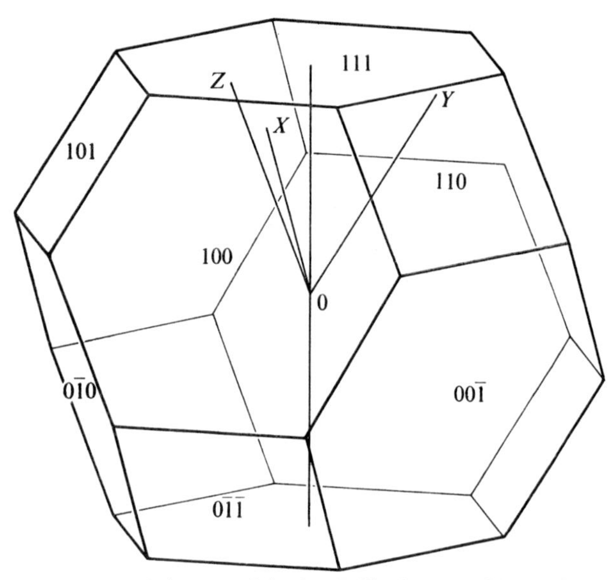

FIG. 16.8. Diagram of the first Brillouin zone of bismuth.

consider not only the first or second Brillouin zones but also larger zones. These are constructed in the same way as those described above, except that the chosen reciprocal points are further from the origin and correspond to X-ray reflections of relatively large intensity. As explained in § 16.1, electron waves are also strongly reflected from the same lattice planes as are the X-rays. The large zones are usually important in connection with the electron:atom ratio in the crystal. The electron:atom ratio is the ratio of the number of valency electrons to the number of atoms per unit cell. This ratio often has a close connection with the type of crystal structure shown by the crystal. For example, the γ-brass Cu_5Zn_8 structure is associated with a 21:13 electron:atom ratio. On the electron theory of metals, the number of electrons depends on the volume of the Brillouin zone. The large zone for γ-brass is $22\frac{1}{2}$ times the volume of its first Brillouin zone, with which one electron is associated. The (Bravais) unit cell contains 13 atoms. The fact that $22\frac{1}{2}$ is greater than 21 implies that the large zone is not completely filled. However, the electron:atom ratio of 21:13 shows that the large zone, which corresponds to the strong X-ray reflections 330 and 411, has a close relation to the electronic characteristics of the structure (10, p. 173).

Q.16.1. The large Brillouin zone of γ-brass, Cu_5Zn_8, (cubic) corresponds to the strong X-ray reflections 411 and 330. Determine the coordinates of the points of intersection of the following planes, (a) 411, $4\bar{1}1$, $41\bar{1}$, $4\bar{1}\bar{1}$; (b) 411, $41\bar{1}$, 330; (c) 330, 303, 033. Determine the coordinates of the points of intersection for the sets of faces corresponding to the reciprocal points given under (a), (b), and (c). Draw on clinographic axes the Brillouin zone, using the points (a), (b), and (c) and those derived from them by operating with the full cubic symmetry $m3m$. The clinographic axes may be copied from Fig. 16.3.

16.3. The symmetry associated with particular points within the first Brillouin zone

The geometrical form of the first Brillouin zone of any crystal is determined by the shape of the unit cell of the crystal lattice. This is the same for many different metals, and their differences in physical properties are associated with the different distributions of the momenta and energies of the electrons within the Brillouin zone. To explain the differences in conductivity, in thermo-electric properties, and in many other physical properties, it is necessary to know the forms of the surfaces drawn through points corresponding to constant electron energies. These are difficult to calculate, and it is necessary to reduce the labour of calculation as much as possible by finding out first what is the effect of symmetry on the energy distributions. The problem can be simplified by a consideration of reciprocal points which lie along certain lines or on certain planes. At the origin, the symmetry of the wave-vector group must apply to waves travelling in all directions. This, in the absence of screw axes or glide planes of symmetry, is the full point-group symmetry and the corresponding character table is one of the tables in Appendix 1. At every other reciprocal point, other than reciprocal-lattice points, it is necessary to consider which elements of symmetry of the point group belong to the wave-vector group. If screw axes or glide planes are present in the space group, the wave-vector group may include elements of symmetry that do not occur in the point group (see § 15.1). The three types of cubic lattice, primitive, face-centred, and body-centred, will be considered in turn, and the symmetry associated with particular points of the first Brillouin zone determined.

The wave-vector group symmetry in cubic lattices

Primitive lattice. It is here assumed that no screw axes or glide planes are present in the space group. Figure 16.9 shows the first

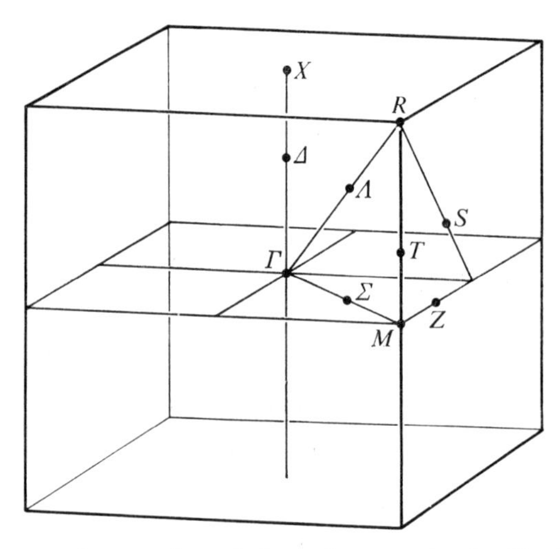

FIG. 16.9. Diagram of particular points associated with the first Brillouin zone of a primitive cubic lattice.

Brillouin zone of a primitive cubic lattice. It is a cube of side a^*, each face being $a^*/2$ from the origin. The letters denote certain important points on or within the zone and the symmetry associated with each is discussed below.

i. Point Γ: this is the centre of the Brillouin zone and represents waves of very large wavelength, travelling in all directions. The symmetry operating among such waves is the full point-group symmetry.

ii. Point R: we are here considering a reciprocal lattice as a geometrical array in which the environment of any one reciprocal lattice point is just the same as that of any other. Any one point may therefore be chosen as origin. The point R therefore can be joined to eight cube centres, not merely to the one shown in Fig. 16.9. The wave normals are directed along normals to octahedral faces, and the symmetry associated with point R is again the full point-group symmetry.

iii. Point X: this can be joined only to the centre of the cube above it, and the symmetry associated with it is $4/mmm$. The first m refers to the plane passing through X normal to ΓX, the second m is parallel to (100), and the third to (110).

iv. Point M: this can be joined to four surrounding cube centres, so the line MR has tetrad symmetry. The horizontal plane ΓMZ and the vertical planes $MR\Gamma$, RMZ give to point M the same symmetry as point X, namely, $4/mmm$.

v. Points Δ and T: these have no horizontal plane of symmetry but they have tetrad axes and two vertical planes of symmetry passing through them. The symmetry associated with these points is therefore $4mm$.

vi. Point Λ: the symmetry operative at this point is a triad axis along ΓR and three planes of symmetry intersecting in it. Thus the symmetry is that of point group $3m$.

vii. Points Σ and S: the line $\Gamma\Sigma M$ is a diad axis and the planes ΓMR and ΓMZ are planes of symmetry intersecting in the diad axis. Thus the point group for Σ is $2mm$. Similarly, the line RS is a diad axis, and two mutually perpendicular planes of symmetry intersect in it, giving point-group symmetry $2mm$ to point S.

viii. Point Z: the line MZ is a diad axis, as we can see by adding a Brillouin zone on the right of the one shown in Fig. 16.9. There is a vertical, and also a horizontal, plane of symmetry intersecting in the line MZ. Thus the point-group symmetry applicable to Z is $2mm$.

Face-centred cubic Bravais lattice. Figure 16.10 shows the form of the first Brillouin zone of a face-centred cubic Bravais lattice,

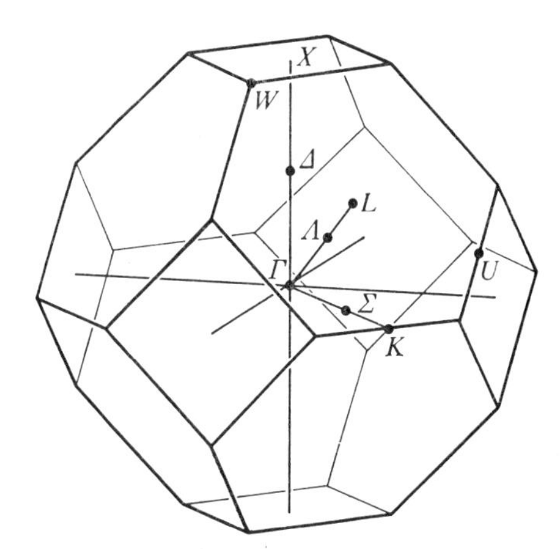

FIG. 16.10. Diagram of particular points associated with the first Brillouin zone of a face-centred cubic Bravais lattice.

and certain important points are marked by letters. The symmetry associated with the wave groups at Γ, X, Δ is the same as that of the corresponding points in the primitive Bravais lattice.

i. Point L: the line ΓL is a triad axis, and plane $X\Gamma L$ is a plane of symmetry. There are therefore three planes of symmetry intersecting in the line ΓL. The point L is midway between the centres of two zones which are in contact on the hexagonal face WKU passing through L. The wave directed along ΓL must therefore be associated with one directed from L towards Γ. Thus the point L has a centre of symmetry, and the point-group symmetry of the wave-vector group is $\bar{3}m$.

ii. Point W: in Fig. 16.11 the environment of point W has been indicated by drawing in parts of the surrounding zones. The sym-

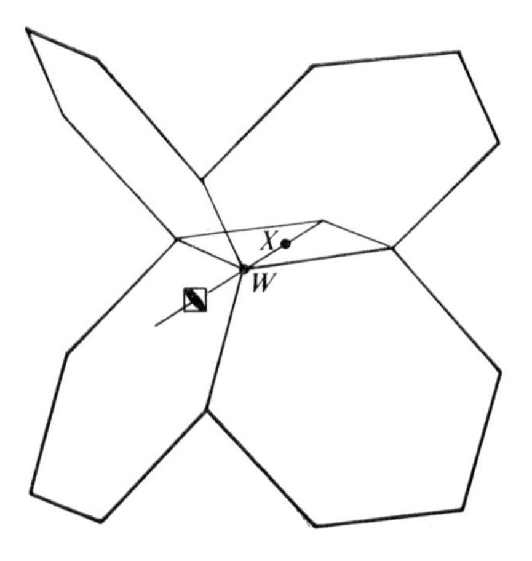

FIG. 16.11. Diagram showing the symmetry at point W in Fig. 16.10.

metry along XW may be seen to be $\bar{4}$, and intersecting in this line are two mutually perpendicular planes of symmetry. The point-group symmetry corresponding to that of point W is thus $\bar{4}2m$.

Q.16.2. What point-group symmetries correspond to those of points K and Σ?

Body-centred cubic Bravais lattice. Figure 16.12 shows the rhombic dodecahedron which is the form of the first Brillouin zone of the body-centred cubic Bravais lattice. The symmetry associated with the groups of the waves that correspond to points marked Γ, Δ, Λ,

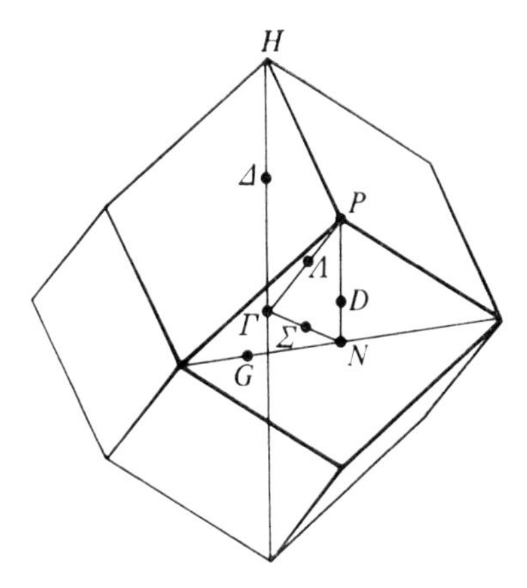

Fig. 16.12. Diagram of particular points associated with the first Brillouin zone of a body-centred cubic Bravais lattice.

and Σ may be seen to be the same as for the corresponding points in the primitive lattice.

i. Point H: six rhombic dodecahedra meet at point H if we take account of neighbouring zones. Thus the symmetry at H is the full cubic point-group symmetry.

ii. Point N: the face PN is shared between two neighbouring zones and is thus a plane of symmetry. The planes $PN\Gamma$ and $GN\Gamma$ are also planes of symmetry. Thus the symmetry associated with point N is mmm.

iii. Point P: Fig. 16.13 shows that, when surrounding zones are taken into account, a $\bar{4}$ axis passes through P in the direction NP. The planes $HP\Gamma$ and PNG are mutually perpendicular planes of symmetry. Four rhombic dodecahedra meet at P, and from the centre of each a triad axis passes to P. Thus the point-group symmetry of point P is $\bar{4}3m$ (10, p. 101).

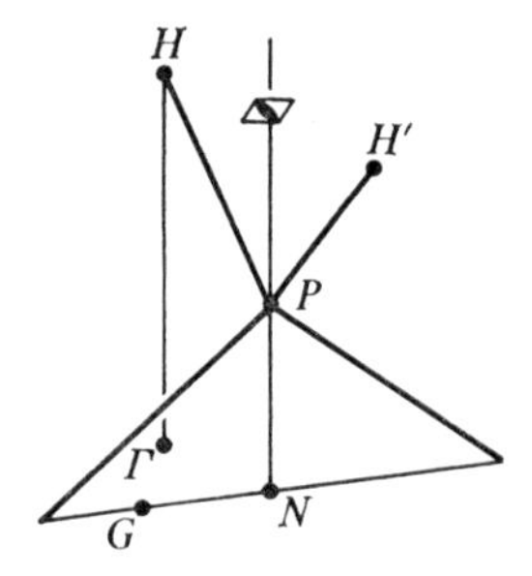

FIG. 16.13. Diagram showing the $\bar{4}$ axis *NP* of Fig. 16.12.

Q.16.3. What symmetry is associated with points *D* and *G* in Fig. 16.12?

16.4. Compatibility relations

The symmetry of the wave-vector groups changes as we pass outwards from the centre of a Brillouin zone in any direction. At the centre it is the full point-group symmetry when screw axes or glide planes are absent, but elsewhere inside the zone it may have fewer elements of symmetry. When screw axes or glide planes are present, the number of elements of symmetry at the centre of the zone is greater than the number in the point group (see § 15.1). Each point group has its own set of irreducible representations, and, in connection with the groups of symmetry elements belonging to waves, the character associated with a given symmetry element cannot suddenly change in moving outwards from the centre of the zone. In other words, the effect of a given element of symmetry on the wave must be continuous between the centre and the boundary of the first Brillouin zone.

To illustrate this, we may consider the line ΓX in Fig. 16.9, corresponding to wave normals parallel to a tetrad axis in a primitive cubic Bravais lattice. At Γ the symmetry is $m3m$, at Δ it is $4mm$, and at X it is $4/mmm$. The character tables for $4mm$ and $m3m$ are given in Tables 16.1 and 16.2 respectively. Corresponding elements of symmetry have been arranged in the same vertical columns to facilitate comparison between them. It will be seen that the four representations in $4mm$ and $m3m$ listed below have the same characters for corresponding elements of symmetry. These representations are said

TABLE 16.1

Character table of point group 4mm

	E	C_2	$2C_4$	$2\sigma_v$	$2\sigma_d$
A_1	1	1	1	1	1
A_2	1	1	1	-1	-1
B_1	1	1	-1	1	-1
B_2	1	1	-1	-1	1
E	2	-2	0	0	0

TABLE 16.2

Character table of point group m3m

	E	$8C_3$	$3C_2$	$6C_4$	$6C_2$	I	$8S_6$	$3\sigma_h$	$6S_4$	$6\sigma_d$
A_{1g}	1	1	1	1	1	1	1	1	1	1
A_{2g}	1	1	1	-1	-1	1	1	1	-1	-1
E_g	2	-1	2	0	0	2	-1	2	0	0
T_{1g}	3	0	-1	1	-1	3	0	-1	1	-1
T_{2g}	3	0	-1	-1	1	3	0	-1	-1	1
A_{1u}	1	1	1	1	1	-1	-1	-1	-1	-1
A_{2u}	1	1	1	-1	-1	-1	-1	-1	1	1
E_u	2	-1	2	0	0	-2	1	-2	0	0
T_{1u}	3	0	-1	1	-1	-3	0	1	-1	1
T_{2u}	3	0	-1	-1	1	-3	0	1	1	-1

to be compatible, and are the only ones that can be used in the further study of the wave functions of the electron waves.

4*mm*	*m*3*m*
A_1	A_{1g}
A_2	A_{1u}
B_1	A_{2g}
B_2	A_{2u}

The E representation of $4mm$ is not compatible with any representation of $m3m$ and so could not correspond to the symmetry of any wave-vector group along the line ΓX (10, p. 108).

Q.16.4. Determine which representations at point X, Fig. 16.9, are compatible with representations valid at point Δ.

Answers

Q.16.1.	(a) $2\frac{1}{4}$	0	0,
	(b) 2	1	0,
	(c) $1\frac{1}{2}$	$1\frac{1}{2}$	$1\frac{1}{2}$.

See Fig. 16.14.

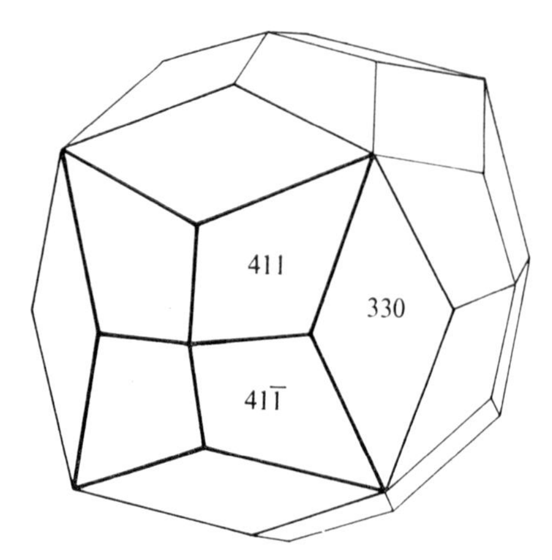

FIG. 16.14. Diagram showing the large Brillouin zone for γ-brass bounded by the planes of indices {411} and {330}.

Q.16.2.	K, $2mm$;	Σ, $2mm$.
Q.16.3.	D, $2mm$;	G, $2mm$.
Q.16.4.	ΔA_1,	$X A_{1g}$ and A_{2u};
	A_2,	A_{2g} and A_{1u};
	B_1,	B_{1g} and B_{2u};
	B_2,	B_{2g} and B_{1u};
	E,	E_g and E_u.

17

The number of constants required to determine a given physical property of a crystal

17.1. Introduction

So far in the study of group theory, we have considered mainly representations that are connected with the geometry of crystals. Thus many representations have been expressed in terms of the x, y, z coordinates of equivalent points in a point group. Also, in the study of wave-vector groups we met representations that are related to lattice translations. Our main study has been in connection with the irreducible representations. An infinite number of representations can be devised, depending simply on how many quantities are represented. In this chapter we shall consider the constants representing the physical properties of crystals. These properties, as we have seen in Chapter 1, are subject to Neumann's Principle, which requires compliance on the part of the physical properties with the symmetry of the crystal. Each physical constant must satisfy every element of symmetry in the point group of the crystal to which it belongs. Thus, if a crystal has three principle thermal conductivities, each one of them must satisfy the symmetry of the crystal. The first row of any character table for one of the thirty-two point groups has the character unity for each class of symmetry. Each conductivity, in the example here chosen, must therefore correspond to this first-row representation. If a representation is constructed that contains all the relations between each of the principal coefficients of conductivity and all the classes of symmetry in the point group, the total of the characters representing them all will be three times the number for one of them. This can be expressed in formal terms as follows.

If h_j is the number of elements of symmetry in class j and χ_j is the corresponding character, then, for the first row of the character table,

$$\sum_j h_j \chi_j = g,$$

where g is the total number of symmetry elements present, and the summation is carried out over all the classes of symmetry. When a reducible representation is formed in any way whatever, let the character be denoted $\chi(R)$. Then we have

$$\sum_j h_j \chi(R)_j = ng,$$

where n is the number of times the irreducible representation (having characters χ_j) is contained in the reducible representation (having characters $\chi(R)_j$). The above expression will be much used in this chapter and can be more conveniently written

$$n = \frac{1}{g} \sum_j h_j \chi(R)_j. \tag{17.1}$$

17.2. The character of a matrix associated with a rotation about an axis

We suppose the coordinate axes are rotated through an angle ϕ about the z axis, as shown in Fig. 17.1. Then the relation between

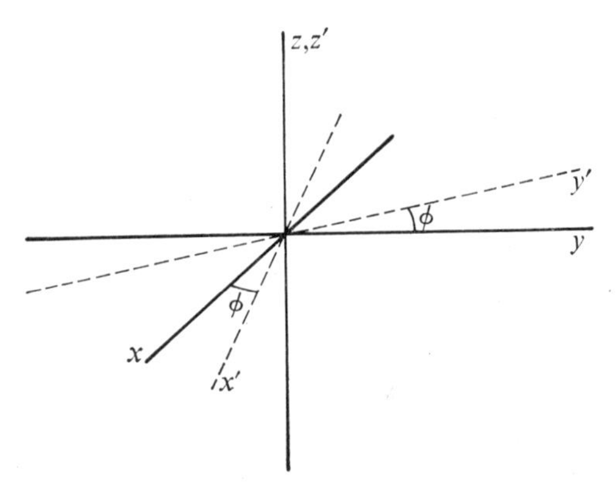

FIG. 17.1. Coordinate axes rotated through ϕ about the z axis.

x', y', z' and x, y, z is given by the following matrix.

	x	y	z
x'	$\cos\phi$	$\sin\phi$	0
y'	$-\sin\phi$	$\cos\phi$	0
z'	0	0	1.

The character of this matrix is $2\cos\phi+1$.

A table for all the crystallographic rotation axes of symmetry can be drawn up as follows (Table 17.1).

TABLE 17.1

Characters $\chi(R)$ of the matrices for rotation about crystallographic axes

	E	C_2	C_3	C_4	C_6
ϕ	0	$2\pi/2$	$2\pi/3$	$2\pi/4$	$2\pi/6$
$(2\cos\phi+1)$	3	-1	0	1	2

17.3. The character of a matrix associated with a reflection across a plane

We can suppose the plane to be perpendicular to the z axis, when the transformation matrix becomes

	x	y	z
x'	1	0	0
y'	0	1	0
z'	0	0	-1.

If a rotation is followed by a reflection in the 001 plane, then the matrix is the product of the above two matrices, namely

	x	y	z
x'	$\cos\phi$	$\sin\phi$	0
y'	$-\sin\phi$	$\cos\phi$	0
z'	0	0	-1.

Thus the effect of a plane of symmetry at right-angles to the rotation axis is to change $(2\cos\phi+1)$ into $(2\cos\phi-1)$. Thus, Table 17.2 shows the effect for the following symmetry elements, S_n.

TABLE 17.2

Characters for rotation followed by reflection in a plane perpendicular to the rotation axis

	S_1	$S_2 = I$	S_3	S_4	S_6
ϕ	2π	$2\pi/2$	$2\pi/3$	$2\pi/4$	$2\pi/6$
$(2\cos\phi-1)$	1	-3	-2	-1	0

17.4. Consideration of a vector property in a crystal having point-group symmetry 32

A vector can be defined by the coordinates x, y, z. The effect of application of the symmetry of the crystal is to rotate the vector about the axis C_3 and about each of the diad axes C_{2x}, C_{2y}, C_{2u}. We have seen in § 17.2 that the character for such rotations is given by $(2\cos\phi+1)$. We denote the character of the vector $\chi(R)$. It must be noted that the character $\chi(R)$ for rotation about any axis is the same as the corresponding character for rotation about the z axis. If the x or y axis, for example, is called the z axis, then the same matrix could be written down as we have used for rotation about the z axis. The classes of symmetry are E, $2C_3$, and $3C_2$ so the value of $\chi(R)$ for each class is given in Table 17.3.

TABLE 17.3

Characters $\chi(R)$ for classes of symmetry in point group 32 for a vector property

	E	$2C_3$	$3C_2$
h	1	2	3
ϕ	0	$2\pi/3$	$2\pi/2$
$\chi(R)$	3	0	-1

Applying the relation (17.1) we obtain

$$n = \tfrac{1}{6}(1 \times 3 + 2 \times 0 + 3 \times -1) = 0.$$

In other words, no vector property, such as pyroelectricity, can exist in a crystal belonging to point group 32.

17.5. A vector property in point group $3m$

In the point group $3m$ the corresponding characters are given in Table 17.4.

TABLE 17.4

Characters $\chi(R)$ for classes of symmetry in point group $3m$ for a vector property (treating σ_v as equivalent to S_1)

	E	$2C_3$	$3\sigma_v$
h	1	2	3
ϕ	0	$2\pi/3$	0
$\chi(R)$	3	0	1

Applying the relation (17.1) we obtain

$$n = \tfrac{1}{6}(1 \times 3 + 2 \times 0 + 3 \times 1) = 1.$$

Thus one coefficient is sufficient to define a vectorial property in this point group.

Q.17.1. How many coefficients are required to define a vectorial property (e.g. pyroelectricity) in the point group $\bar{4}3m$?

17.6. The character for a second-order tensor

A second-order tensor, k, such as that for thermal conductivity, can be transformed from one set of Cartesian axes to another set, according to the relation obtained in Chapter 2, namely

$$k'_{pq} = c_{pr}c_{qs}k_{rs}. \tag{17.2}$$

A representation of the thermal conductivity can be made by drawing up a matrix as shown in Table 17.5, in which columns refer to undashed ks and rows refer to dashed ks. This 6×6 matrix contains all the information concerning the property of thermal conductivity in a given crystal. Since $k_{pq} = k_{qp}$ they are put together in the same column. Because the character of the representation is required, only the diagonal terms need be evaluated.

TABLE 17.5

The diagonal terms in the matrix for a second-order tensor

	k_{11}	k_{12}, k_{21}	k_{13}, k_{31}	k_{22}	k_{23}, k_{32}	k_{33}
k'_{11}	c_{11}^2					
k'_{12}		$c_{11}c_{22}+c_{12}c_{21}$				
k'_{13}			$c_{11}c_{33}+c_{13}c_{31}$			
k'_{22}				c_{22}^2		
k'_{23}					$c_{22}c_{33}+c_{23}c_{32}$	
k'_{33}						c_{33}^2

We now consider a rotation of the coordinate axes through an angle ϕ about the z axis. In this case, the expressions for the c_{ik}s are given by the matrices

$$\begin{matrix} c_{11} & c_{12} & c_{13} \\ c_{21} & c_{22} & c_{23} \\ c_{31} & c_{32} & c_{33} \end{matrix} \quad \text{and} \quad \begin{matrix} \cos\phi & \sin\phi & 0 \\ -\sin\phi & \cos\phi & 0 \\ 0 & 0 & 1. \end{matrix}$$

Applying this to the diagonal terms in Table 17.5, we obtain the character of the 6×6 matrix, which is

$$\begin{aligned} \chi(R) &= \cos^2\phi+\cos^2\phi-\sin^2\phi+\cos\phi+\cos^2\phi+\cos\phi+1 \\ &= 4\cos^2\phi+2\cos\phi. \end{aligned}$$

For reflection across a plane perpendicular to the z axis combined with a rotation through an angle ϕ about that axis, the only change is that c_{33} becomes -1. The value of $\chi(R)$ in this case is

$$4\cos^2\phi-2\cos\phi.$$

This result applies to any plane of symmetry, since the z axis may, for the purpose of finding $\chi(R)$, be chosen normal to the plane. We can now apply this character $\chi(R)$ to any point group and obtain a reducible representation. For example, in the point group *mmm*, we have Table 17.6.

TABLE 17.6

Characters $\chi(R)$ for classes of symmetry in point group mmm for a second-order tensor

	E	C_{2z}	C_{2y}	C_{2x}	I	σ_z	σ_y	σ_x
h	1	1	1	1	1	1	1	1
ϕ	0	$2\pi/2$	$2\pi/2$	$2\pi/2$	$2\pi/2$	0	0	0
(R)	6	2	2	2	6	2	2	2

Applying equation (17.1) we obtain for the number of coefficients of thermal conductivity, since $g = 8$,

$$n = \tfrac{1}{8}(6+2+2+2+6+2+2+2) = 3.$$

These coefficients correspond to k_{11}, k_{22}, k_{33} discussed in Chapter 2 (1, p. 82).

Q.17.2. Determine the number of coefficients defining a second-order polar tensor in point groups $\bar{1}$ and 23.

17.7. The character for a third-order tensor

The property of piezoelectricity relates a stress or strain applied to a crystal to the electric moment developed within it. The relation is thus between a second-order tensor and a vector or first-order tensor. The transformation of the coefficients of piezoelectricity is given by the equation

$$d'_{ikl} = c_{im}c_{kn}c_{lo}d_{mno}. \tag{17.3}$$

A representation may be formed by writing down the matrix having columns under d_{mno} and rows from d'_{ikl}. The character of this matrix, $\chi(R)$, required for the determination of the number of the non-zero piezoelectric coefficients in any given point group, can be obtained as shown in Table 17.7.

TABLE 17.7

The evaluation of the diagonal terms in the piezoelectric matrix

ikl'	mno	$\cos^3\phi$	$\cos^2\phi$	$\cos\phi$	1
111	111	1			
112	112 $11\times 11\times 22 = c^3$; 121 $11\times 12\times 21 = -cs^2 = c^3-c$	2		-1	
113	113 $11\times 11\times 33 = \pm c^2$; 131 $11\times 13\times 31 = 0$		± 1		
122	122 $11\times 22\times 22 = c^3$	1			
123	123 $11\times 22\times 33 = \pm c^2$; 132 $11\times 23\times 32 = 0$		± 1		
133	133 $11\times 33\times 33$			1	
211	211 $22\times 11\times 11 = c^3$	1			
212	212 $22\times 11\times 22 = c^3$; 221 $22\times 12\times 21 = -cs^2 = c^3-c$	2		-1	
213	213 $22\times 11\times 33 = \pm c^2$; 231 $22\times 13\times 31 = 0$		± 1		
222	222 $22\times 22\times 22 = c^3$	1			
223	223 $22\times 22\times 33 = \pm c^2$; 232 $22\times 23\times 32 = 0$		± 1		
233	233 $22\times 33\times 33 = c$			1	
311	311 $33\times 11\times 11 = \pm c^2$		± 1		
312	312 $33\times 11\times 22 = \pm c^2$; 321 $33\times 12\times 21 = \mp s^2 = \pm c^2 \mp 1$		± 2		
313	313 $33\times 11\times 33 = c$; 331 $33\times 13\times 31 = 0$			1	
322	322 $33\times 22\times 22 = \pm c^2$		± 1		
323	323 $33\times 22\times 33 = c$; 332 $33\times 23\times 32 = 0$			1	
333	333 $33\times 33\times 33 = \pm 1$				
	$\chi(R)$	8	± 8	2	

Thus $\chi(R) = 8\cos^3\phi \pm 8\cos^2\phi + 2\cos\phi = (4\cos^2\phi \pm 2\cos\phi)(2\cos\phi \pm 1)$.

This value of $\chi(R)$ may be applied to quartz (point group 32) to find the number of piezoelectric constants. Table 17.8 may be drawn up.

TABLE 17.8

Characters $\chi(R)$ for classes of symmetry in point group 32 for the property of piezoelectricity

	E	$2C_3$	$3C_2$
h	1	2	3
ϕ	0	$2\pi/3$	$2\pi/2$
$\chi(R)$	18	0	-2

Applying equation 17.1 we obtain

$$n = \tfrac{1}{6}(1\times 18+2\times 0-3\times 2) = 2.$$

These are the two constants denoted d_{111} and d_{123} in Chapter 5.

17.8. The characters of fourth-order and higher-order tensors

The character $\chi(R)$ for the third-order tensor property is found to be simply the product of the corresponding characters for a second-order and a first-order tensor. Similar results hold in the derivation of the characters of the representations of fourth-order and higher-order tensors. However, it is necessary to be careful in cases where the physical property imposes certain relations between the suffixes of the coefficients. Thus photoelasticity relates refractive index (a second-order tensor) to stress or strain (both second-order tensors). Photoelasticity is therefore a fourth-order tensor, and its character is simply the square of the character for a second-order tensor, namely $(4\cos^2\phi\pm 2\cos\phi)^2 = 16\cos^4\phi\pm 16\cos^3\phi+4\cos^2\phi$.

Elasticity relates stress and strain and is also a fourth-order tensor. However, its character is $(16\cos^4\phi\pm 8\cos^3\phi-4\cos^2\phi+1)$, which differs from that for photoelasticity. The reason for this is that, in photoelasticity, the suffixes *iklm* of any coefficient can be interchanged to *kilm*, *ikml*, *kiml* only, whereas the corresponding coefficients for elasticity permit the exchanges of the two suffixes on the right with the two suffixes on the left for each of the four arrangements of *iklm*. As an example, the number of elastic constants that may be present in a crystal having the point-group symmetry *mmm* will be evaluated (Table 17.9).

TABLE 17.9

Characters for the classes of symmetry in point group mmm for the property of elasticity

	E	C_{2x}	C_{2y}	C_{2z}	I	σ_x	σ_y	σ_z
h	1	1	1	1	1	1	1	1
ϕ	0	$2\pi/2$	$2\pi/2$	$2\pi/2$	$2\pi/2$	0	0	0
$\chi(R)$	21	5	5	5	21	5	5	5

Applying equation 17.1 we obtain

$$n = \frac{1}{8}(21\times1+5\times1+5\times1+5\times1+21\times1+5\times1+5\times1+5\times1)$$

$$= 9.$$

This corresponds with nine elastic constants derived by a tensor method in § 7.5.

Q.17.3. Given that the character $\chi(R)$ for photoelasticity is

$$(16\cos^4\phi\pm16\cos^3\phi+4\cos^2\phi),$$

find the number of non-zero coefficients in point group 23.

Q.17.4. Given that the character $\chi(R)$ for elasticity is

$$(16\cos^4\phi\pm8\cos^3\phi-4\cos^2\phi+1),$$

find the number of coefficients of elasticity in point group $4/m$.

The characters $\chi(R)$ of sixth-order and higher-order tensors can be found in the same way as that used for fourth-order tensors. Thus $\chi(R)$ for a sixth-order tensor can be written down as the product of the tensors for the fourth-order and second-order tensors which are combined in the given physical property (1, p. 85).

17.9. The characters of axial tensors

In § 9.2 it was shown that, whereas for polar tensors we can write

$$A'_{ijk\cdots}(x') = c_{ip}c_{jq}c_{kr}\cdots A_{pqr\cdots}(x),$$

no matter whether the transformation corresponds to rotation or to reflection or inversion, with axial tensors the equation is valid only for rotation. For reflection or inversion, a negative sign must be placed before the expression on the right-hand side of the equation. Since the characters $\chi(R)$ derived in §§ 17.2–17.8 depend on the transformations of the axes, the same rule must apply to them as to the axial tensors. Thus, for a first-order axial tensor the value of $\chi(R)$ is $\pm(2\cos\phi\pm 1)$ for a rotation ϕ about any axis, the negative signs being taken when the symmetry class contains a reflection or an inversion. Similarly, for a second-order axial tensor the value is $\pm(4\cos^2\phi\pm 2\cos\phi)$.

Optical activity

The gyration tensor relating to optical activity is a second-order axial tensor, and we may derive the number of coefficients that can be non-zero in various point groups. In class $2mm$, for example, Table 17.10 shows in the first row the symmetry elements present, in the second row the formula applicable to each symmetry element (c standing for $\cos\phi$), in the third row the angle of rotation corresponding to the element of symmetry, and in the fourth row the value of $\chi(R)$ given by the formula in that column.

TABLE 17.10

The characters $\chi(R)$ for the classes of symmetry in point group $2mm$

	E	C_2	σ_x	σ_y
	$4c^2+2c$	$4c^2+2c$	$-(4c^2-2c)$	$-(4c^2-2c)$
ϕ	0	$2\pi/2$	0	0
$\chi(R)$	6	2	-2	-2

Hence from equation 17.1 we obtain

$$n = \frac{1}{4}\{(1\times 6)+(1\times 2)+(1\times -2)+(1\times -2)\}$$

$$= 1.$$

In fact the tensor analysis shows that the only non-zero constant is g_{12}.

Q.17.5. Find the number of non-zero constants for optical activity in quartz (point group 32).

The Hall effect

In § 9.4 we saw that Hall coefficients relate the antisymmetric part, a_{ik}, of the conductivity tensor to the axial vector of the magnetic field, H_p, by the expression

$$a_{ik} = \rho_{ikp}H_p.$$

The transformation of a_{ik} is the same as that of an axial vector, as may be seen in the following analysis.

$$a'_{ik} = c_{ip}c_{kq}a_{pq}$$

and $$a_{11} = a_{22} = a_{33} = 0,$$

$$a_{12} = -a_{21};\ a_{13} = -a_{31};\ a_{23} = -a_{32}.$$

Therefore

$$a'_{12} = c_{11}c_{22}a_{12}+c_{11}c_{23}a_{13}+c_{12}c_{23}a_{23}+$$
$$+c_{12}c_{21}a_{21}+c_{13}c_{21}a_{31}+c_{13}c_{22}a_{32}.$$

We shall first consider a rotation through an angle ϕ about the X_3 axis, so that

$$c_{13} = c_{23} = c_{31} = c_{32} = 0.$$

We shall write $\cos\phi = c = c_{11} = c_{22}$, and $\sin\phi = s = c_{12} = -c_{21}$. Then from the previous equation we have

$$a'_{12} = c^2a_{12}-s^2a_{21} = a_{12}.$$

Similarly, we obtain

$$a'_{13} = c_{11}c_{33}a_{13}+c_{12}c_{33}a_{23}$$
$$= ca_{13}+sa_{23} \qquad \text{(when } c_{33} = +1\text{)},$$

and $$a'_{23} = c_{21}c_{33}a_{13}+c_{22}c_{33}a_{23}$$
$$= -sa_{13}+ca_{23} \qquad \text{(when } c_{33} = +1).$$

To construct a representation for a_{ik} we draw up the following matrix

	a_{12}	a_{13}	a_{23}
a'_{12}	1	.	.
a'_{13}	.	c	s
a'_{23}	.	$-s$	c

The character of this matrix is $1+2c$.

When the symmetry operation is not a pure rotation, but involves a plane or centre or inversion axis, $c_{33} = -1$. In this case the expressions for a'_{13} and a'_{23} are changed. The character of the above matrix then becomes $1-2c$. We may write the character as $\pm(2c\pm1)$, it being understood that the negative signs are used when the symmetry operation involves a plane or centre or inversion axis. Thus the character $\chi(R)$ for a second-order antisymmetric tensor is the same as that for an axial vector.

The Hall coefficients thus relate two axial vectors, and the appropriate $\chi(R)$ is the product of their characters, namely

$$\chi(R) = (4c^2\pm4c+1).$$

The same value of $\chi(R)$ can be derived as follows by drawing up a table of the diagonal elements in the 9×9 matrix for the Hall coefficients. Taking account of the fact that

$$\rho_{pqr} = -\rho_{qpr}$$

and $$\rho_{ppr} = 0,$$

we can write, for the first term on the diagonal of the table,

$$\rho'_{121} = c_{11}c_{22}c_{11}\rho_{121}+c_{12}c_{21}c_{11}\rho_{211}$$
$$=(c_{11}c_{22}c_{11}-c_{12}c_{21}c_{11})\rho_{121}.$$

For rotation through an angle ϕ about an axis denoted X_3 the direction-cosine matrix is

$$\begin{matrix} c & s & 0 \\ -s & c & 0 \\ 0 & 0 & 1, \end{matrix}$$

where $c = \cos\phi$ and $s = \sin\phi$. The tensor is axial, and the corresponding matrix for a reflection in a plane perpendicular to Z is (see Chapter 9)

$$\begin{matrix} -c & -s & 0 \\ s & -c & 0 \\ 0 & 0 & 1. \end{matrix}$$

(Note that $c_{33} = +1$.) The corresponding matrix for a centre of symmetry is the same as for a rotation about Z. Again $c_{33} = +1$.

Thus

$$\begin{aligned} \rho'_{121} &= \{c^2(\pm c)+s^2(\pm c)\}\rho_{121} \\ &= \pm c\rho_{121}, \end{aligned}$$

the negative sign applying when there is a plane of symmetry normal to the axis Z.

Similarly, for the diagonal term of the table only,

$$\begin{aligned} \rho'_{131} &= c_{11}c_{33}c_{11}\rho_{131} \\ &= c^2\rho_{131} \qquad \text{since } c_{33} \text{ is } +1 \text{ in all cases.} \end{aligned}$$

Applying this method to all terms on the diagonal of the ρ_{pqr} matrix, we obtain, as before,

$$\chi(R) = 4c^2 \pm 4c + 1.$$

This result may be applied to bismuth, point group $\bar{3}m$, as shown in Table 17.11.

TABLE 17.11

Characters $\chi(R)$ for the Hall effect in classes of symmetry for point group $\bar{3}m$, $A = 4c^2+4c+1$, $B = 4c^2-4c+1$.

	E	$2C_3$	$3C_2$	I	$2S_6$	$3\sigma_d$
	A	A	A	B	B	B
ϕ	0	$2\pi/3$	$2\pi/2$	$2\pi/2$	$2\pi/6$	0
$\chi(R)$	9	0	1	9	0	1

Applying equation 17.1 we obtain

$$n = \left\{\frac{1}{12}(1\times 9)+(2\times 0)+(3\times 1)+(1\times 9)+(2\times 0)+(3\times 1)\right\}$$
$$= 2.$$

In § 9.4 it was shown that for bismuth the non-zero Hall coefficients are ρ_{123} and ρ_{231} (1, p.89; 2, p. 159).

Q.17.6. Compare the number of non-zero tensor components in each of the point groups $2mm$, $\bar{4}$, and $\bar{4}2m$, for (a) a second order symmetric polar tensor and (b) a second-order symmetric axial tensor.

Answers

Q.17.1. The table giving $\chi(R)$ for $\bar{4}3m$ and a first-order polar tensor is as follows.

	E	$8C_3$	$3C_2$	$6S_4$	$6\sigma_d$
h	1	8	3	6	6
ϕ	0	$2\pi/3$	$2\pi/2$	$2\pi/4$	0
$\chi(R)$	3	0	-1	-1	1

Hence, since $g = 24$,

$$n = \left\{\frac{1}{24}(1\times 3)+(8\times 0)+(3\times -1)+(6\times -1)+(6\times 1)\right\}$$
$$= 0.$$

No pyroelectricity can be shown by a crystal belonging to this point group.

Q.17.2. The $\chi(R)$ table for point group $\bar{1}$ and a second-order polar tensor is as follows.

	E	I
h	1	1
ϕ	0	$2\pi/2$
$\chi(R)$	6	6

and so $q = 2$,

$$n = \frac{1}{2}(6+6) = 6.$$

These six coefficients correspond to three principal conductivities and three angles giving the orientation of the principal directions with respect to the crystallographic axes.

The $\chi(R)$ table for point group 23 and a second-order polar tensor is as follows

	E	$4C_3$	$4C_3^2$	$3C_2$
h	1	4	4	3
ϕ	0	$2\pi/3$	$2\pi/3$	$2\pi/2$
$\chi(R)$	6	0	0	2

Hence, since $g = 12$,

$$n = \frac{1}{12}(6+3\times 2) = 1.$$

Thus the cubic point groups have only one coefficient for a second-order tensor property.

Q.17.3. The table giving characters $\chi(R)$ for point group 23 and photo-elasticity is as follows

	E	$4C_3$	$4C_3^2$	$3C_2$
h	1	4	4	3
ϕ	0	$2\pi/3$	$2\pi/3$	$2\pi/2$
$\chi(R)$	36	0	0	4

and so $g = 12, n = \dfrac{1}{12}(36+12) = 4.$

Q.17.4. The $\chi(R)$ table for the elastic coefficients in point group $4/m$ is as follows.

	E	C_4	C_2	C_4^3	I	S_4^3	σ_h	S_4
h	1	1	1	1	1	1	1	1
	0	$2\pi/4$	$2\pi/2$	$2\pi/4$	$2\pi/2$	$2\pi/4$	0	$2\pi/4$
$\chi(R)$	21	1	5	1	21	1	5	1

and so, $g = 8, n = \dfrac{1}{8}(21+1+5+1+21+1+5+1) = 7.$

Q.17.5. In point group 32, only rotation axes are present, so that the formula for $\chi(R)$ is $4c^2+2c$ for all classes.

	E	$2C_3$	$3C_2$
ϕ	0	$2\pi/3$	$2\pi/2$
$\chi(R)$	6	0	2

Hence,

$$n = \frac{1}{6}\{1\times 6+2\times 0+3\times 2\} = 2.$$

Q.17.6. (a) For a second-order symmetric polar tensor, we have the relation

$$\chi(R) = 4\cos^2\phi \pm 2\cos\phi,$$

the minus sign applying to the $\chi(R)$ corresponding to a plane or centre of symmetry. Thus in $2mm$ we have the following table, giving the elements of symmetry in the point group, the angle ϕ corresponding to each element, and the value of $\chi(R)$.

	E	C_2	σ_x	σ_y
ϕ	0	$2\pi/2$	0	0
$\chi(R)$	6	2	2	2

From this we obtain n, the number of non-zero physical constants, as

$$n = \frac{1}{4}(1\times 6+1\times 2+1\times 2+1\times 2) = 3.$$

In point group $\bar{4}$ we have the following table.

	E	S_4	S_4^3	C_2
ϕ	0	$2\pi/4$	$2\pi/4$	$2\pi/2$
$\chi(R)$	6	0	0	2

$$n = \frac{1}{4}(1\times 6+1\times 2) = 2.$$

In point group $\bar{4}2m$ we have the following table.

	E	S_4	S_4^3	C_2	$2C_2'$	$2\sigma_d$
ϕ	0	$2\pi/4$	$2\pi/4$	$2\pi/2$	$2\pi/2$	0
$\chi(R)$	6	0	0	2	2	2

$$n = \frac{1}{8}(1\times 6+1\times 2+2\times 2+2\times 2)$$
$$= 2.$$

(b) For a second-order symmetric axial tensor, the expression for $\chi(R)$ is the same as for the polar tensor, except that it is prefixed by a negative sign when the element of symmetry involved is a plane or a centre. Thus for point group $2mm$ we have the following table.

	E	C_2	σ_x	σ_y
ϕ	0	$2\pi/2$	0	0
$\chi(R)$	6	2	-2	-2

$$n = \frac{1}{4}(1\times 6+1\times 2-1\times 2-1\times 2) = 1.$$

In point group $\bar{4}$ we have

	E	S_4	S_4^3	C_2
ϕ	0	$2\pi/4$	$2\pi/4$	$2\pi/2$
$\chi(R)$	6	0	0	2

$$n = \frac{1}{4}(1\times 6+1\times 2) = 2.$$

In point group $\bar{4}2m$ we have

	E	S_4	S_4^3	C_2	$2C_2'$	$2\sigma_d$
ϕ	0	$2\pi/4$	$2\pi/4$	$2\pi/2$	$2\pi/2$	0
$\chi(R)$	6	0	0	2	2	-2

$$n = \frac{1}{8}(1\times 6+1\times 2+2\times 2-2\times 2) = 1.$$

18

Vibrations of molecules and radicals

18.1. Normal modes of vibration

IN this chapter we shall discuss the modes of vibration of isolated molecules and also those of groups such as CO_3, SO_4, $PtCl_6$, which occur as units in crystal structures. Thermal vibrations are often regarded as random motions, but the motions of the atoms forming molecules or radicals are, in fact, not at all random. They are completely defined by the symmetry of the molecule when it is isolated, as in a gas. When the molecule or ionic complex is in a crystal, the thermal vibrations are subject to limitations imposed by the symmetry of the crystal at the point where the group is located. Any freely suspended body has a number of modes of vibration that are characteristic of its shape and the material of which it is made. An illustration is given in Fig. 18.1 of three modes that are possible in a rectangular bar which is long in comparison with its width and rather thin. It is true of each of these modes of vibration that for small harmonic oscillations, if the bar is set vibrating in one of them, it will continue vibrating in that mode alone, and other modes will not be excited. Such modes are called *normal modes of vibration.*

Consider the vibration of a CO_3 group. We could imagine the three oxygen atoms to be executing vibrations independently of one another, but like the flat bar discussed above this does not happen. Instead the molecule vibrates in a number of particular ways, in each of which the atoms all vibrate with the same frequency but the frequency varies with the mode of vibration. One mode of vibration is that in which the three oxygens move out or in along the lines joining them to the central carbon atom. Their movements are in phase with one another, so that at one extremity of the vibration the triangle is still equilateral but larger than normal, and at the other extremity of the vibration it is also equilateral but

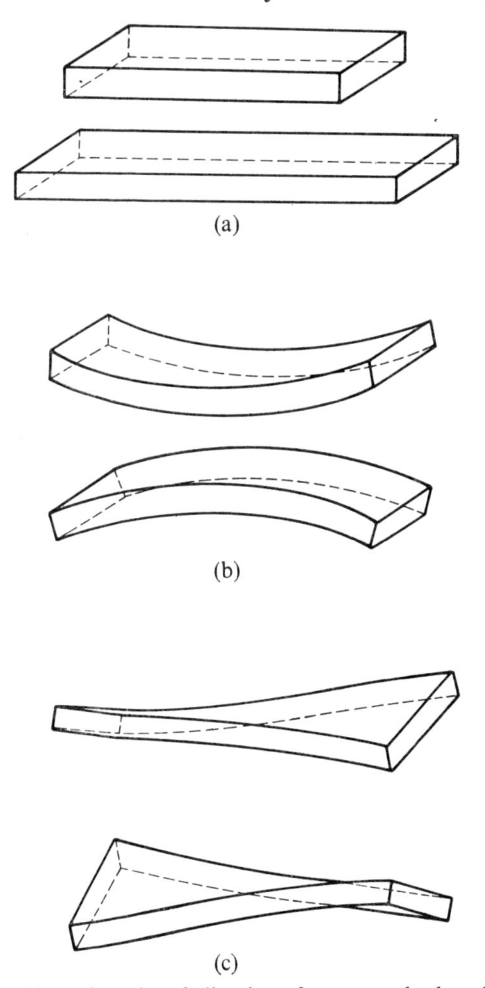

FIG. 18.1. Normal modes of vibration of a rectangular bar. (a) longitudinal, (b) flexural, (c) tortional.

smaller than the mean size. This mode of vibration is called a 'breathing' mode. The symmetry elements of the CO_3 group are the same as those of point group $\bar{6}m2$ (D_{3h}), namely E, $2C_3$, $3C_2$, σ_h, $2S_3$, $3\sigma_v$. During the atomic movements corresponding to the breathing mode, the requirements of every one of these symmetry classes is satisfied. This corresponds to the representation 1, 1, 1, 1, 1, 1 denoted A_1' in the character tables (Appendix 1).

Q.18.1. To which point group, and representation within the point group, do the breathing modes of SO_4 and $PtCl_6$ correspond?

18.2. The number of normal modes of vibration

Each atom has three degrees of freedom, and if there are n atoms in each molecule or ion there are $3n$ degrees of freedom in the whole molecule. Three of these correspond to all the atoms moving together along x or y or z. This is merely bodily translation and does not correspond to a vibration of the kind that interests us here. Similarly, bodily rotations about x or y or z are not of interest in study of vibrations. Thus we are left with $3n-6$ degrees of freedom and there can be one normal made corresponding to each degree of freedom. However, a molecule in which all the atoms lie on a single straight line has $3n-5$ significant degrees of freedom, because rotation about the line passing through all the atoms is not observable.

18.3. Normal modes and irreducible representations

The irreducible representations of any point group contain all the possible relations between the coordinates of any atom (x, y, z) and the classes of symmetry in that point group. The normal modes of vibration also depend on the changes in the coordinates of the atoms constituting any given molecule. Since the irreducible representations cover all possible combinations of the coordinates that are consistent with the symmetry, every normal mode must correspond with one or other of the irreducible representations of the point group to which the molecule belongs. We need to find which modes of vibration are consistent with the symmetry, and to which representations they belong.

18.4. The application of symmetry operations to the vibration of a molecule or ion

The CO_3 ion will be again considered. The breathing mode of vibration has already been discussed—it remains to find $(3n-7)$, i.e. five, other modes. One of these is that in which the carbon atom moves relative to the oxygen atoms along a line perpendicular to the plane containing the oxygen atoms. When the carbon atom is displaced from its equilibrium position, the ion no longer possesses the plane of symmetry σ_h. Alternatively, we can say that the σ_h

transforms the vibration into its opposite. The three diad axes that pass one through each oxygen and the carbon atom in its rest position are lost when the carbon atom is vibrating in the mode under discussion. Similarly the S_3 axes are lost. Again, these elements of symmetry transform the vibration into its opposite. The representation for each of these symmetry elements is therefore -1. This corresponds to the A_2'' representation in the character table for $\bar{6}m2$.

The remaining four normal modes are not so easily determined; they belong in pairs to the E′ representation, which has the character 2 in the E column. This means that a two-dimensional matrix involving x and y is applicable to each element of symmetry. One pair of these normal modes (denoted ν_{3a} and ν_{3b} respectively) is shown in Fig. 18.2. The relative magnitudes of the atomic displacements are indicated by the arrows and the letters a and A, though the actual magnitudes are exaggerated and not relevant here.

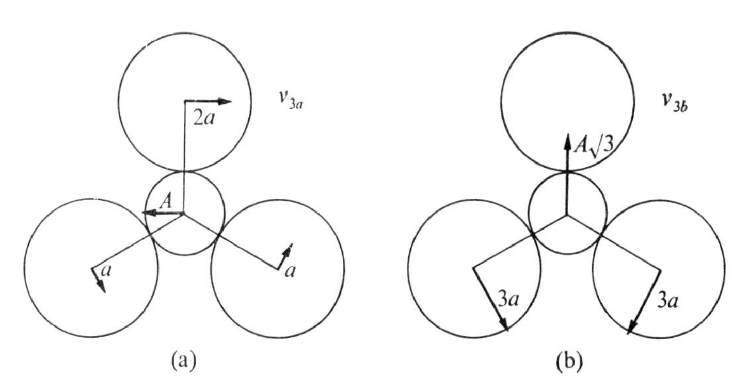

FIG. 18.2. Diagrams showing normal modes of vibration occurring in a CO_3 ion.

If we combine the displacements corresponding to $-\frac{1}{2}$ of those in Fig. 18.2(a) with $+\frac{1}{2}$ of those in Fig. 18.2(b), we obtain the rotation of Fig. 18.2(a) through 120°. This is shown in Fig. 18.3. Thus the operation of C_3^2 on the normal mode ν_{3a} is equivalent to $(-\frac{1}{2}\nu_{3a}+\frac{1}{2}\nu_{3b})$. It may also be shown (see Fig. 18.4) that a similar rotation of ν_{3b} is equivalent to $(-\frac{3}{2}\nu_{3a}-\frac{1}{2}\nu_{3b})$. The transformation may be expressed in the following way.

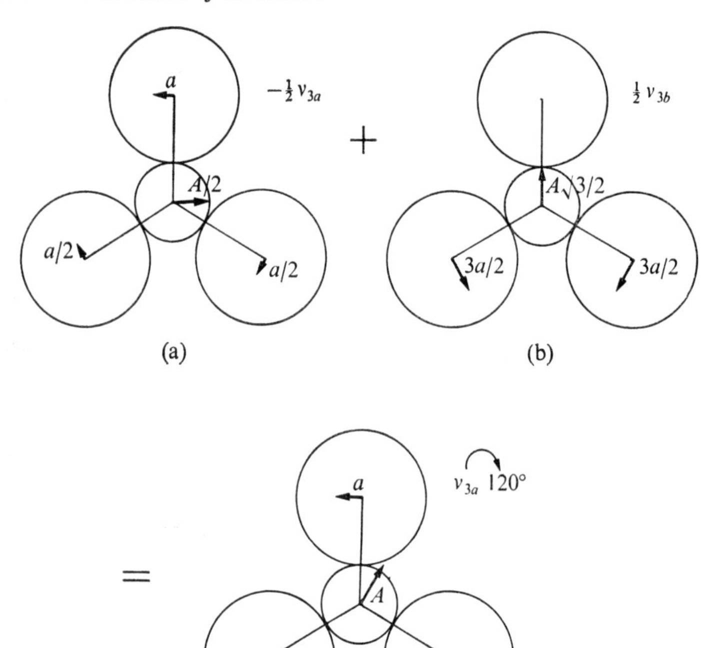

FIG. 18.3. Diagrams showing the addition of two normal modes of vibration of the CO_3 ion $(-\frac{1}{2}, \frac{1}{2})$.

	Fig. 18.2(a)	Fig. 18.2(b)
Fig. 18.3(c)	$-\frac{1}{2}$	$\frac{1}{2}$
Fig. 18.4(c)	$-\frac{3}{2}$	$-\frac{1}{2}$

The character of this matrix is -1, and this is the same value that the E' representation has in the C_3 column of the character table for $\bar{6}m2$.

The transformation corresponding to S_3 gives the same character as that for C_3 since all the atoms lie in the same plane, which is the plane of symmetry.

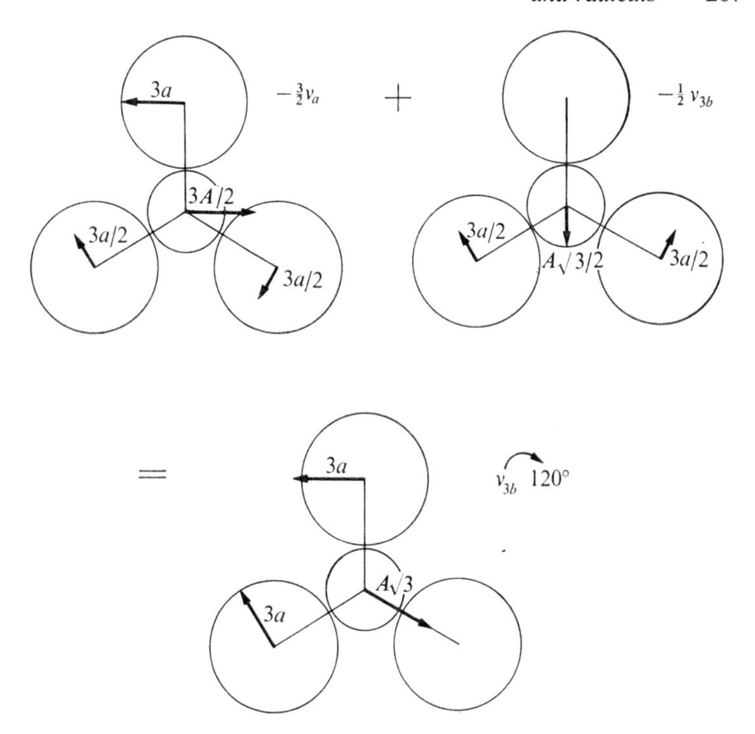

FIG. 18.4. Diagram showing the addition of two normal modes of vibration of the CO_3 ion $(-\frac{3}{2}, -\frac{1}{2})$.

The operation of the diad axis lying in the plane of the CO_3 group and parallel to the long edge of the page produces, from Fig. 18.2(a) and (b), the pattern of vibrations shown in Fig. 18.5(a) and (b). It can be seen that vibrations in Fig. 18.5(a) are the negative of the corresponding vibrations in Fig. 18.2(a), and that Fig. 18.5(b) is the same as Fig. 18.2(b). Thus the matrix for this transformation is as follows.

	Fig. 18.2(a)	Fig. 18.2(b)
Fig. 18.5(a)	−1	0
Fig. 18.5(b)	0	1

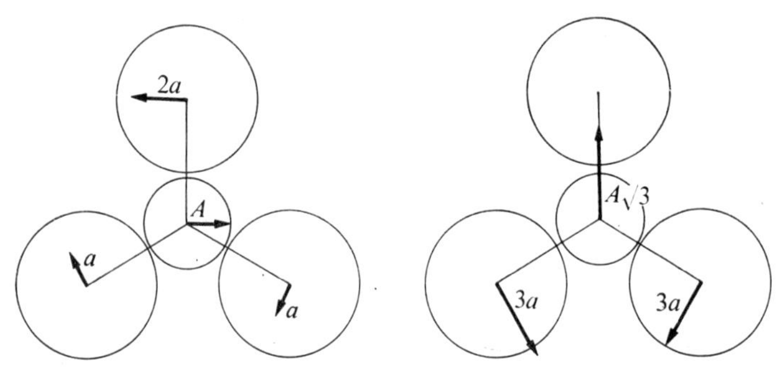

FIG. 18.5. Diagrams showing the operation of C_2 on the vibrations represented in Fig. 18.2.

The character of the transformation is zero, and is thus the same as the character of C_2 in the E' representation of point group $\bar{6}m2$.

The same arguments apply to the operation of the class σ_v having a plane of symmetry perpendicular to the plane of the CO_3 group.

Finally we come to the last class, namely σ_h, and note that, as the movements given in Fig. 18.2(a) and (b) are all in the plane of the CO_3 group, the transformation σ_h leaves the patterns unchanged. The transformation matrix can therefore be written as follows.

	Fig. 18.2(a)	Fig. 18.2(b)
Fig. 18.2(a)′	1	0
Fig. 18.2(b)′	0	1

The character of this matrix is 2. Summarizing our discussion, we have the following character table for the assumed vibrations of the CO_3 ion.

E	C_3	C_2	σ_h	S_3	σ_v
2	-1	0	2	-1	0,

and on comparing this with the character table for $\bar{6}m2$ (Appendix 1) we see that it is the same as for the E' representation.

In most of the applications of group theory it is sufficient to know the characters in the tables given in Appendix 1. This is not the case, however, when two-dimensional or three-dimensional representations have to be analysed in detail. In the above example, we have arrived at complete 2×2 matrices for at least one element in each class of symmetry. The complete set of matrices corresponding to all the symmetry elements is as follows.

$$
\begin{array}{cccccc}
E & C_3 & C_3^2 & C_{2x} & C_{2y} & C_{2u} \\
\begin{matrix} 1 & 0 \\ 0 & 1 \end{matrix} &
\begin{matrix} -\frac{1}{2} & -\frac{1}{2} \\ \frac{3}{2} & -\frac{1}{2} \end{matrix} &
\begin{matrix} -\frac{1}{2} & \frac{1}{2} \\ -\frac{3}{2} & -\frac{1}{2} \end{matrix} &
\begin{matrix} -1 & 0 \\ 0 & 1 \end{matrix} &
\begin{matrix} \frac{1}{2} & \frac{1}{2} \\ \frac{3}{2} & -\frac{1}{2} \end{matrix} &
\begin{matrix} \frac{1}{2} & -\frac{1}{2} \\ -\frac{3}{2} & -\frac{1}{2} \end{matrix}
\end{array}
$$

$$
\begin{array}{cccccc}
\sigma_h & S_3 & S_3^2 & \sigma_x & \sigma_y & \sigma_u \\
\begin{matrix} 1 & 0 \\ 0 & 1 \end{matrix} &
\begin{matrix} -\frac{1}{2} & -\frac{1}{2} \\ \frac{3}{2} & -\frac{1}{2} \end{matrix} &
\begin{matrix} -\frac{1}{2} & \frac{1}{2} \\ -\frac{3}{2} & -\frac{1}{2} \end{matrix} &
\begin{matrix} -1 & 0 \\ 0 & 1 \end{matrix} &
\begin{matrix} \frac{1}{2} & \frac{1}{2} \\ \frac{3}{2} & -\frac{1}{2} \end{matrix} &
\begin{matrix} \frac{1}{2} & -\frac{1}{2} \\ -\frac{3}{2} & -\frac{1}{2} \end{matrix}
\end{array}
$$

It will be seen that they form a group of six different matrices—any pair of matrices multiplied together yields another member of the group, etc.—and that six is the number of classes in the point group. These six matrices form an irreducible representation. It is also noteworthy that classes which, in the character table for $\bar{6}m2$, occur in different columns, here, in some cases, belong to the same matrix, e.g. E and σ_h, C_{2x} and σ_x. We shall see later, in the study of a regular tetrahedral molecule, that very similar relations apply to its E-matrix representation.

18.5. Alternative representations of molecular vibrations

The vibrations of the CO_3 ion have been considered in the previous paragraphs, but the actual modes of vibration were assumed, not derived. This is useful in illustrating the connection between the modes of vibration and the symmetry, but a direct method of obtaining the modes of vibration from the symmetry is required. Another method of analysis will now be considered. A set of axes, x, y, z, is taken to have its origin in each of the four atoms of the CO_3 group. The axes carry subscripts that correspond to the numbers assigned to the corresponding atom (Fig. 18.6).

The symmetry operations of the point group to which the ion CO_3 belongs are E, C_3, C_2, σ_h, S_3, and σ_v, and we shall deal with each of them in turn. Under the operation E, every atom coincides with

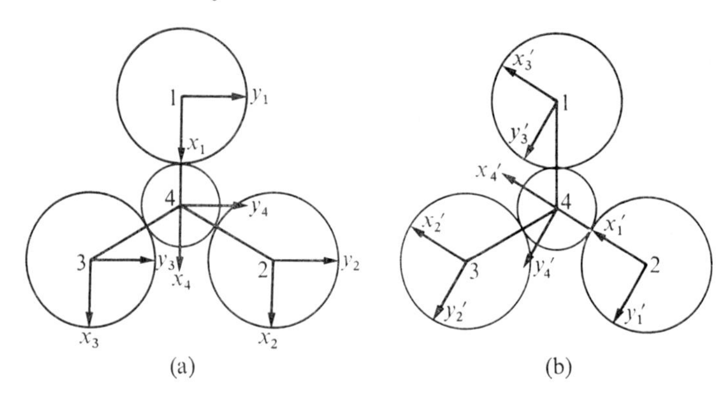

FIG. 18.6. Diagrams showing the system of Cartesian coordinate axes (a) before operation with C_3, (b) after operation with C_3.

itself and every axis coincides with itself. A representation containing $x_1, y_1, z_1, x_2, y_2, \ldots, z_4$ at the head of the twelve columns and $x_1', y_1', z_1', \ldots, z_4'$ on the left-hand side of the twelve rows can be constructed in a similar manner to that used in Table 17.5. For the identity operation, all diagonal terms are unity and off-diagonal terms are zero. The character of the matrix for this representation is thus 12.

The operation C_3 rotates the ion relative to the coordinate axes through $2\pi/3$ in an anticlockwise sense. This relative rotation may be regarded in alternative ways—either as a rotation of the atoms relative to fixed axes or, keeping the atoms fixed, as a rotation of the axes with a transference of the sets of axes as shown in Fig. 18.6(b). Consider the change of axes associated with atom 4. The transformation is as follows.

	x_4	y_4	z_4
x_4'	$-\frac{1}{2}$	$-\frac{\sqrt{3}}{2}$	0
y_4'	$\frac{\sqrt{3}}{2}$	$-\frac{1}{2}$	0
z_4'	0	0	1

The character of this matrix is zero. The element C_3 rotates the Cartesian axes so that the subscripts of x, y, z at each oxygen atom are different from the original subscripts. Therefore, in the 12×12 matrix the contribution from these atoms to the sum of the diagonal terms is zero. Thus the character of the 12×12 matrix for the C_3 operation is zero.

Next, the C_2 symmetry will be considered. In Fig. 18.7, the directions and origins for the dashed axes are shown on the assumption that the diad axis passes through atoms 1 and 4. The diagonal terms

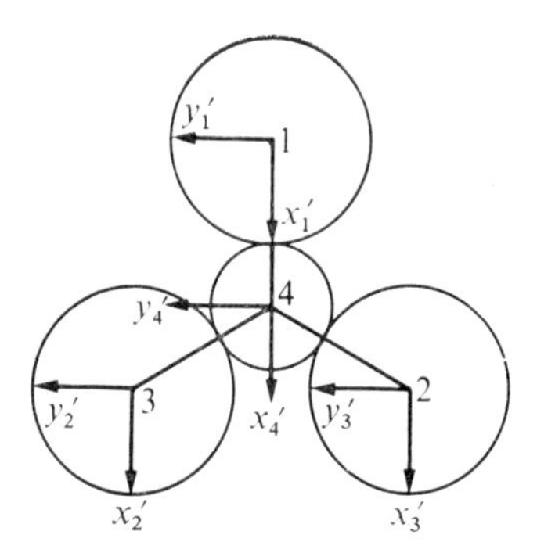

FIG. 18.7. Diagram of change of axes due to operation C_2.

in the 12×12 matrix for atom 1 are $1, -1, -1$, and for atom 4 they are also $1, -1, -1$. Thus the character for this symmetry operation in the 12×12 matrix is -2.

The symmetry σ_h leaves all the atomic positions unchanged, as the atoms all lie in this plane of symmetry. The signs of x and y remain the same, but z becomes $-z$. Thus the diagonal of the 12×12 matrix has the terms $1, 1, -1$, four times. Its character is thus 4.

The symmetry S_3 is similar to that for C_3, except that z changes sign. The only atom that is relevant for determining the character is atom 4, and the character of its transformation matrix is $-\frac{1}{2}-\frac{1}{2}-1 = -2$. Thus the character of the 12×12 matrix for this symmetry operation is -2.

The symmetry σ_v, if we suppose the plane to pass through atoms 1 and 4 leaves x and z unchanged but changes the sign of y. Thus the character for each atom is $+1$ and for the large matrix it is $+2$.

Summarizing these conclusions we obtain the following characters $\chi(R)$ in the 12×12 matrix for the elements of symmetry given below.

	E	C_3	C_2	σ_h	S_3	σ_v
$\chi(R)$	12	0	−2	4	−2	2

This 12×12 representation can be broken down into irreducible representations by making use of the following relation, which is proved in Appendix 7.

$$n_v = \frac{1}{g} \sum_i h_i \chi_i(R) \chi_i^{(v)}, \tag{18.1}$$

where n_v is the number of times the vth irreducible representation is contained in the reducible representation, g is the total number of elements of symmetry in the point group, h_i is the number of elements in class i, $\chi_i(R)$ is the character of the reducible representation in class i, and $\chi_i^{(v)}$ is the character of the irreducible representation in the same class given in the standard table (Appendix 1). The character table for χ_i is reproduced below.

	E	$2C_3$	$3C_2$	σ_h	$2S_3$	$3\sigma_v$
h_i	1	2	3	1	2	3
A'_1	1	1	1	1	1	1
A'_2	1	1	−1	1	1	−1
E'	2	−1	0	2	−1	0
A''_1	1	1	1	−1	−1	−1
A''_2	1	1	−1	−1	−1	1
E''	2	−1	0	−2	1	0
$\chi(R)$	12	0	−2	4	−2	2

Applying equation (18.1) to determine the number of times the representation A'_1 occurs in the 12×12 representation, we obtain

$$n_1 = \tfrac{1}{12}(1\times 12\times 1+2\times 0\times 1+3\times -2\times 1+1\times 4\times 1+2\times -2\times \times 1+3\times 2\times 1)$$

$$= 1.$$

Proceeding in the same way, we may find that the reducible representation $\chi(R)$ is equal to

$$A_1' + A_2' + 3E' + 2A_2'' + E''.$$

From this list we have to eliminate those representations that do not correspond to internal vibration of the ion, but only to bodily translation or rotation. The character table (Appendix 1) shows that A_2'' corresponds to z, i.e. to translation in the z direction, and E' corresponds to the coordinates x, y, i.e. to translation in the xy plane. Rotation about the z axis corresponds in the character table to the A_2' representation, and rotation about the x- and y- axes to the E'' representation. Rejecting these representations, we are left with $A_1' + 2E' + A_2''$, which is the same result as we have already obtained.

18.6. Internal coordinates of vibrations

Up to this point we have considered vibrations in terms of changes in Cartesian coordinates xyz. For some purposes it is better to use internal coordinates such as (a) stretching of the C–O bonds, (b) changes of the O–C–O angles in the plane of the oxygen atoms, and (c) the change of angle between a C–O bond and the plane containing the oxygen atoms.

In Fig. 18.8 is shown the action of the symmetry operation C_3^2 on the bonds denoted a, b, c. All three of them are moved, and a matrix having columns based on abc, and rows on $a'b'c'$ would have

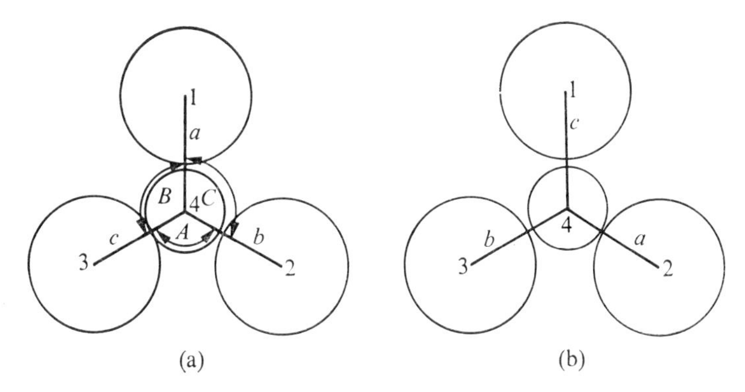

FIG. 18.8. Diagrams showing the internal bonds of the CO_3 group (a) before, (b) after operation with C_3^2.

only off-diagonal non-zero terms. In this respect the internal coordinates behave in the same way as the Cartesian coordinates. The character is thus zero. The same is true for S_3^2.

The symmetry elements E and σ_h leave a, b, and c unchanged, and the character for the whole ion must be 3. The symmetry C_2 leaves only one of the bonds unchanged, and the character is therefore 1. The same is true for the operation σ_v. Thus the characters for the symmetry operations are as follows.

E	C_3	C_2	σ_h	S_3	σ_v
3	0	1	3	0	1.

We now apply equation 18.2 and the data of the character table for $\bar{6}m2$ and obtain irreducible representations

$$A_1' + E'$$

as the components of this reducible representation. Thus the A_1' mode involves stretching and contracting the C–O bonds, giving rise to the breathing normal mode. The E' representation also involves C–O stretching, but, as we shall see in what follows, it may also involve change of the angles O–C–O in the plane of the ion.

The analysis (b) above involves only the three O–C–O angles. By the phase 'coinciding with itself' applied to an angle, we mean that the two lines forming the angle either remain unaffected or are exchanged with one another by the symmetry operation. It must be noted that these are not three independent angles, since their sum must be 2π. Exactly the same arguments can be applied to the angles denoted A, B, C as we have applied to the C–O bonds denoted a, b, c. Thus for C_3^2 and S_3^2 all three angles are moved round, and the characters are zero. For E and σ_h, the angles are not moved, and the characters are 3; and for C_2 and σ_v only one angle coincides with itself, so that the character is 1. From this analysis, the reducible representation can also be broken down into $A_1' + E'$. Now A_1' is a representation that would require all three angles to increase equally at the same time. This is inconsistent with the original definition that the angles lie in the same plane. Thus the A_1' representation must be discarded, and we are left with E' as the representation that corresponds to change of those O–C–O angles. From what was said above, it now appears that E' corresponds both with stretching of the bonds and with changes of the angles between them in the plane of the ion.

Finally the representation A_1'' must correspond with the vibration of the carbon atom in a direction perpendicular to the plane of the oxygen atoms. This follows from the fact that σ_h and S_3 are both -1 in this representation. The results based on internal coordinates can be seen to be the same as those based on Cartesian coordinates by a consideration of Fig. 18.2. In this figure, it can be seen that the E' representation involves both changes in the C–O bond length and also changes in the O–C–O angles.

The study of the molecular vibration using Cartesian coordinates or internal coordinates makes no assumptions about the normal modes and arrives at the irreducible representations that are consistent with the symmetry. The characters of the irreducible representations have been obtained, but the full two-dimensional matrices have not. However, in deciding between various possible alternative normal modes, it is a great help if we know what must be the characters of the corresponding two-dimensional matrices.

The example of the vibration of the CO_3 ion brings out a feature that is of general application. The symmetry elements in the two-dimensional irreducible representations may be given by corresponding matrices that give the transformations of the vibrations taken all together and not separately. Thus in Fig. 18.3 the diagram (a) and the diagram (b) were combined as entities to form diagram (c). The same applies to Fig. 18.5. Later we shall deal with three-dimensional matrices, where also the vibrations are considered *en bloc* (see § 18.8) (2, p. 99; 5, p. 245).

18.7. Vibrations of a tetrahedral molecule or ion

Representation by Cartesian coordinates

We shall now consider a tetrahedral grouping such as occurs in the molecule methane, CH_4, or in the sulphate ion, SO_4. The point group of the molecule is $\bar{4}3m$ (T_d), for which the symmetry elements are E, $8C_3$, $3C_2$, $6S_4$, and $6\sigma_d$ (Appendix 1). A set of x, y, z axes is assigned to each of the five atoms, and a 15×15 matrix representation is made for each element of symmetry in turn.

For the identity operator, E, the character of the matrix is 15.

For C_3, an axis passing through the carbon atom and one hydrogen atom (Fig. 18.9), two atoms remain attached to their original x, y, z-axes but the others do not. For the two atoms which can therefore contribute to the diagonal terms of the 15×15 matrix,

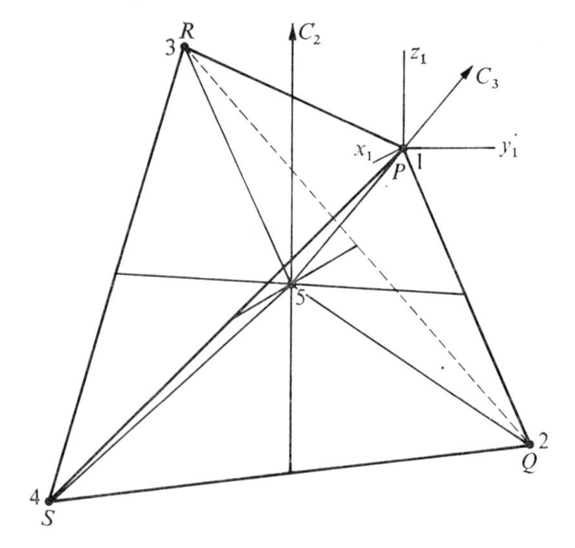

FIG. 18.9. Diagram of Cartesian axes applicable to tetrahedral groups such as CH_4, SO_4, etc.

we have the transformation due to the triad axis, namely $x \to y$, $y \to z$, $z \to x$. The matrix for this transformation is as follows.

	x	y	z
x'	0	1	0
y'	0	0	1
z'	1	0	0

The character of this matrix is zero, and therefore the two atoms that could contribute to the character of the large matrix do not do so, and the character is zero.

For C_2, a diad axis parallel to x passing through the carbon atom and the mid-point of an edge of the tetrahedron, only atom 5 remains attached to its original axes. The character for the diad axis is -1, and this is the character for the 15×15 matrix.

The operation S_4 moves all the Cartesian axes except those passing through the carbon atom, for which the matrix is -1, as indicated below.

	x	y	z
x'	0	-1	0
y'	1	0	0
z'	0	0	-1.

Finally, σ_d leaves three unmoved sets of axes, and the matrix in each case is as follows.

	x	y	z
x'	0	1	0
y'	1	0	0
z'	0	0	1

The character derived from all three sets of axes is thus 3.

Summarizing this analysis, we have for $\chi(R)$ in the 15×15 matrix the characters in the symmetry classes as shown below

	E	$8C_3$	$3C_2$	$6S_4$	$6\sigma_d$
$\chi(R)$	15	0	-1	-1	3

Using the equation (18.1) and the character table for point group $\bar{4}3m$ (Appendix 1), we obtain the reduction of the 15×15 matrix into the irreducible representations

$$A_1 + E + T_1 + 3T_2.$$

It can be seen at once that A_1, which corresponds to a character $+1$ for each symmetry element, is the breathing mode, in which only extensions and contractions of the C–H bonds occur. From the character table it can be seen that T_1 is associated with R_x, R_y, R_z i.e. with bodily rotations of the molecule, and T_2 with (x, y, z), i.e. with bodily translations. Neither of these representations is of interest to us at the moment. Hence the genuine vibrations correspond to representations A_1, E, and $2T_2$. The total number of possible normal vibrations is $3n-6 = 9$, so that, apart from the mode corresponding to A_1, the two-dimensional E representation has two modes, and the three-dimensional T_2 mode has three modes in each of its forms.

Representation by internal coordinates

Using internal coordinates, we can first deal with the four C–H bonds. The identity operator has a character 4, equal to the number of bonds. The triad axis passes along one C–H bond only, so that the character is $+1$. The diad axis moves all four C–H bonds so the character is 0. The same is true of the S_4 axis, for which the character is also zero. The diagonal plane of symmetry, σ_d, contains two C–H bonds, so that its character is 2. Thus we have finally, for the characters in the various classes,

	E	$8C_3$	$3C_2$	$6S_4$	$6\sigma_d$
$\chi(R)$	4	1	0	0	2

This can be seen from the character table to correspond to

$$A_1 + T_2.$$

We have already seen that A_1 corresponds to the mode depending only on simultaneous extension or contraction of all four C–H bonds.

The other internal coordinate relates to the six H–C–H angles. These six angles are not entirely independent, so that some redundancy in the representations is to be expected. The character associated with the identity operator for these six angles is 6. To determine the character associated with C_3, we can use the stereographic projection of the C–H bonds joining the C atoms to atoms P, Q, R, S, shown in Fig. 18.10. Under the operation of C_3, all of these angles,

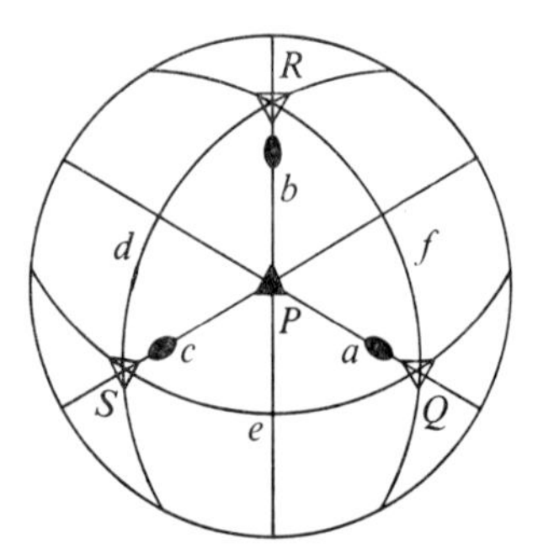

FIG. 18.10. Stereogram showing the directions of the C–H bonds in CH_4 (marked by triangles), and the H–C–H angles (denoted $a, b, \ldots, f$).

a–f, move round, and consequently the character is zero. The stereographic projection of Fig. 18.11 shows that two of the six angles, namely *b* and *e*, superpose on themselves under the action of C_2 a diad axis normal to the paper. The character of the 6×6 matrix for this element of symmetry is therefore 2. Under the operation S_4

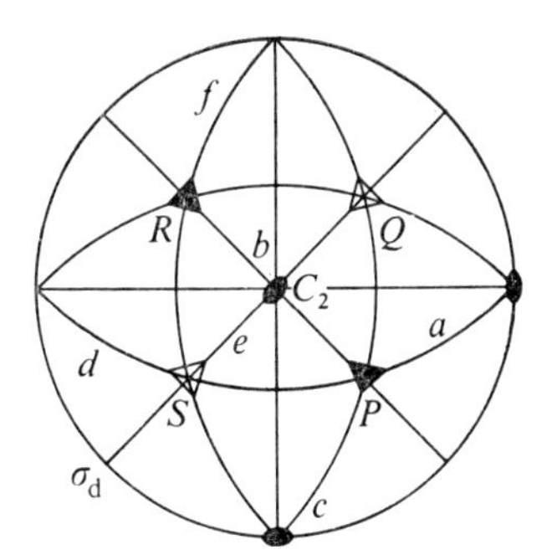

FIG. 18.11. Stereogram of C–H bonds in CH_4 projected on (001).

the angle *b* is superposed on the angle *e*, and all the others are moved to new positions, so that the character is zero. The symmetry plane σ_d parallel to (110) leaves the angles *b* and *e* unchanged, though it changes the others. Thus the character is 2. Summarizing the results for the characters $\chi(R)$ relating to the H–C–H angles, we have

	E	$8C_3$	$3C_2$	$6S_4$	$6\sigma_d$
$\chi(R)$	6	0	2	0	2

This may be broken down by equation (18.1) to

$$\chi(R) = \mathrm{A}_1 + E + T_2.$$

A false assumption has been made in this analysis, namely that all six angles could simultaneously increase. This is what is implied by the presence of A_1 in the expression $A_1 + E + T_2$. Five only of these angles are independent and the sixth is determined by the other five. Thus A_1 must be struck out of the list, so that only E and T_2 are representations associated with changes of angle between the bonds. Thus the A_1 representation is related to bond-stretching, the E representation to change of H–C–H angle, and the T_2 representation is related both to bond-stretching and to change of bond angle.

18.8. Degenerate modes of vibration

The triply degenerate T_2 modes

The above analysis leaves undecided the detailed vibrations associated with the normal modes corresponding to the representations E and T_2. There are two T_2s among the genuine vibrations. The determination of these modes cannot be achieved directly by the simple group-theoretical arguments, but proposed schemes of vibration can be readily tested by these arguments. The representation T_2 has the following characters in the classes specified.

	E	C_3	C_2	S_4	σ_d
$\chi(R)$	3	0	-1	-1	1

In Fig. 18.12 (α), (β), (γ) there is shown a proposed set of diagrams representing T_2 vibrations. We shall test whether these vibration schemes satisfy the symmetry requirements in a manner similar to that used in § 18.4. The clinographic drawing Fig. 18.12 (a) shows the molecule with carbon at point *e* and hydrogens at points *a*, *b*, *c*, and *d*. The other diagrams (α), (β), (γ) show stereograms, having 001 at the centre, at each of the points *a*–*e*. On each stereogram is marked a cross corresponding to the direction of vibration of the atom at that point. Thus in diagram (α) the vibration is along [100], in (β) it is along [010], and in (γ) it is along [001]. It will be noticed that the vibrations of the hydrogens are in each case oppositely directed to that of the carbon. The diagrams marked (1), (2), (3) are clinographic drawings showing the same vibrations as are indicated by the figures (α), (β), (γ). If these vibration schemes are correct, the vibrations of figure (α), *considered as an entity and not as separate vibrations*, must transform into those of figures (β) and (γ) under each symmetry operation of the point group.

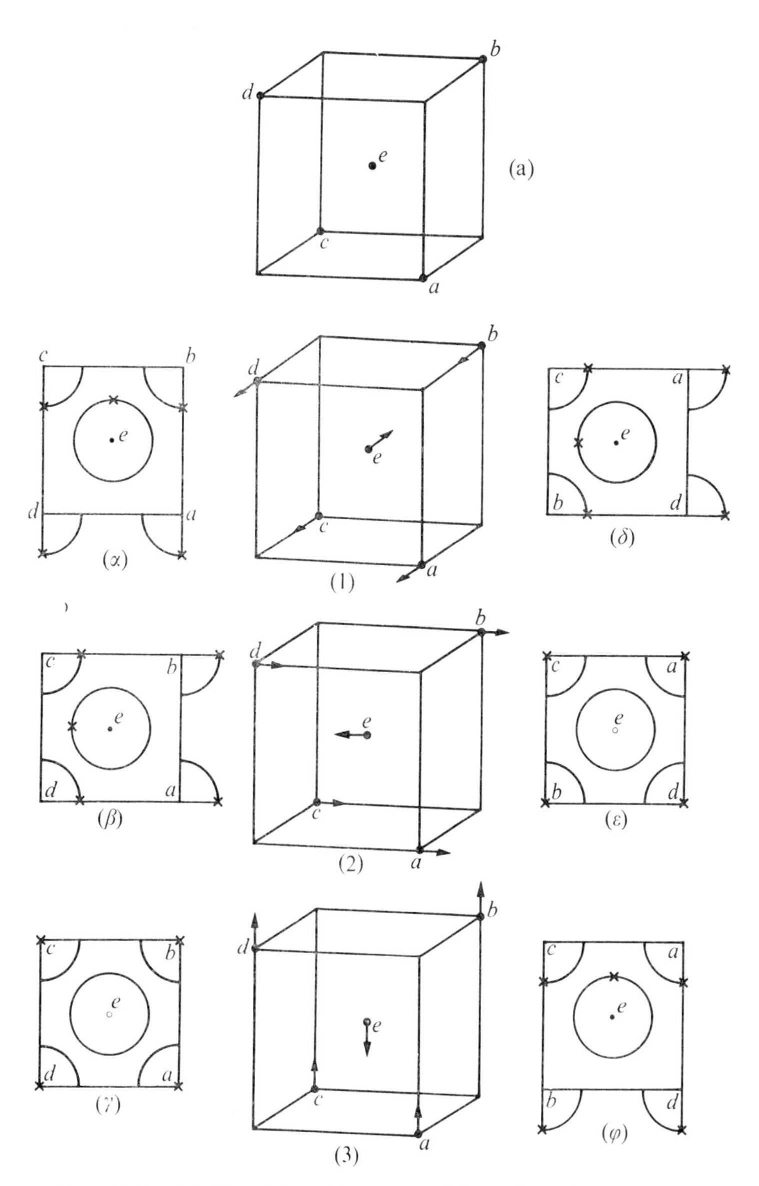

FIG. 18.12. (a) disposition of atoms in CH_4. a, b, c, d = H, e = C; (α), (β), (γ) vibration schemes; (1), (2), (3) clinographic drawings of vibrations; (δ), (ε), (ϕ) transformations of (α), (β), (γ) by operation of C_3.

Consider the operation of the triad axis C_3 on both the atomic centres and the vibration displacements. The matrix for this transformation is

$$\begin{matrix} 0 & 1 & 0 \\ 0 & 0 & 1 \\ 1 & 0 & 0, \end{matrix}$$

which means $x \to y$, $y \to z$, $z \to x$. In Fig. 18.12, it can be seen that atomic displacement in (α) is related to the corresponding displacement in (δ) as x is to y. Similarly, displacements in (β) are related to those in (ε) as y is to z, and, finally, displacements in (γ) and (ϕ) are related as z is to x. Further, except for the exchange of identical hydrogen atoms, (δ) and (β), (ε) and (γ), and (ϕ) and (α) are the same patterns. Thus, these three modes (α), (β), (γ) satisfy the requirement of C_3 symmetry.

A diad axis, C_{2x}, has a transformation matrix

$$\begin{matrix} 1 & 0 & 0 \\ 0 & -1 & 0 \\ 0 & 0 & -1, \end{matrix}$$

which means $x \to x$, $y \to -y$, $z \to -z$. In Fig. 18.12, a diad axis parallel to x leaves (α) unaffected except for the interchange of hydrogen atoms; the scheme (β) is turned into its negative with the arrows pointing in opposite directions to those shown. Since the atoms vibrate equally on either side of their mean centre, this reversal of directions is no change in the vibration scheme. Thus (β) transforms into itself. In the same way (γ) transforms under C_{2x} into itself. The modes (α), (β), (γ) therefore satisfy the C_{2x} requirement. All the symmetry elements of this point group, $\bar{4}3m$, give the same result, namely that they transform one T_2 mode into another T_2 mode. This establishes the proposed vibration schemes of Fig. 18.12 (α), (β), (γ) as the triply degenerate T_2 modes, all having the same frequency of vibration.

Q.18.2. Show that the modes represented in Fig. 18.12 satisfy the requirement of σ_d symmetry.

The second T_2 mode of vibration

The proposed modes of vibration for the second T_2 mode are shown in Fig. 18.13, (α), (β), (γ), (1), (2), (3). The vibration directions

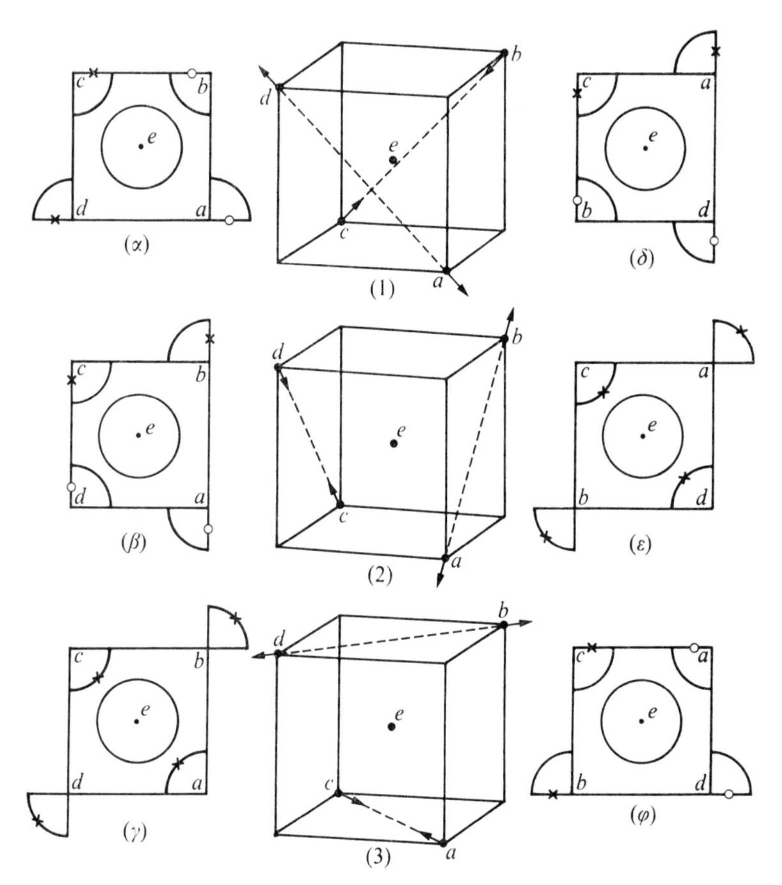

FIG. 18.13. (α), (β), (γ) vibration schemes of second T_2 mode of CH_4; (1), (2), (3) clinographic drawings of (α), (β), (γ); (δ), (ε), (ϕ) transformations of (α), (β), (γ) by the operation of C_3.

are along face-diagonals of the cube, as is shown by the stereograms about the atomic centres in Fig 18.13 (α), (β), and (γ), and by the clinographic drawings in Fig 18.13 (1), (2), (3). The hydrogen atoms are thus vibrating along certain $\langle 110 \rangle$ directions, and the carbon is stationary. To assist in understanding how the ⟨110⟩ directions are changed by the operation of a triad axis [111], Fig. 18.14 shows vibration directions labelled 1, 2, 3 which are related by the rotation about [111]. The trigonal axis considered here passes through the hydrogen atom *c* and the carbon atom *e*, so that *a* goes to *b*, *b* goes to *d*, and *d* goes to *a*, while *c* remains fixed. The operation of the triad axis on the vibration directions is shown, for example, by the

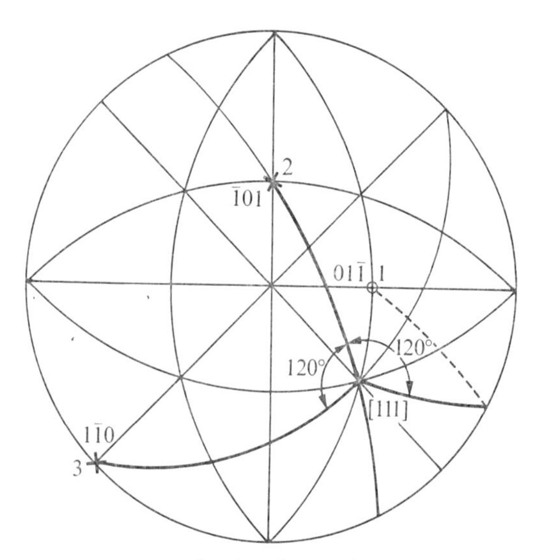

FIG. 18.14. Stereograms showing the rotation of {110} about a triad axis [111].

change of the circle atom *a*, Fig. 18.13(α), to the cross atom *a*, Fig. 18.13(δ). It can be seen that, except for the change of letters attached to the hydrogen atoms $\delta = \beta$, $\varepsilon = \gamma$, and $\phi = \alpha$, so that the modes transform into one another by the operation of the C_3 element of symmetry. We can show by applying in turn all the elements of symmetry of this point group to those modes that they transform into one another. The proposed schemes can therefore be accepted as giving the second T_2 mode.

Q.18.3. Show that the scheme of vibrations proposed for the second T_2 mode satisfies the symmetry requirements of S_4 and σ_d in the point group $\bar{4}3m$.

Since it has already been shown that only two T_2 modes are possible, these, which have been found to satisfy the symmetry requirements, must be the only possible ones.

The doubly degenerate E mode

The remaining mode after the consideration of the A_1 and T_2 modes is the E mode. As may be seen from the character table for this mode, which is reproduced below, the character in the identity-operation column is 2.

	E	C_3	C_2	S_4	σ_d
$\chi(R)$	2	-1	2	0	0

Thus all matrices corresponding to the five symmetry elements are two-dimensional. For our present purpose, it is not enough to know the characters given above. We must know the actual transformation matrices for all the elements of symmetry. Only if we know the complete matrices can we find the transformations of the proposed system of vibrations corresponding to this mode. In § 18.4 a similar problem was discussed in relation to the CO_3 ion. There, two-dimensional matrices were set up for the C_3, S_3, C_2, σ_h, and σ_d elements of symmetry.

The matrices representing in two dimensions the symmetry elements of point group $\bar{4}3m$. For our present purpose the characters of the E representation are not sufficient; we require the complete two-dimensional matrices. The symmetry requirements give rise to a different allocation of the symmetry elements to particular matrices from that in the character table for this point group. There are six two-dimensional matrices, which form a group, and the elements of symmetry are divided between them as shown below (2).

$$\begin{matrix} 1 & 0 \\ 0 & 1 \end{matrix} \qquad\qquad \begin{matrix} 1 & 0 \\ 0 & -1 \end{matrix}$$

$$E,\ C_{2x},\ C_{2y},\ C_{2z} \qquad\qquad \sigma_{(110)},\ \sigma_{(1\bar{1}0)},\ S_{4z},\ S_{4z}^3$$

$$\begin{matrix} -\frac{1}{2} & \frac{\sqrt{3}}{2} \\ -\frac{\sqrt{3}}{2} & -\frac{1}{2} \end{matrix} \qquad \begin{matrix} -\frac{1}{2} & -\frac{\sqrt{3}}{2} \\ \frac{\sqrt{3}}{2} & -\frac{1}{2} \end{matrix}$$

$$C_{3a}, C_{3b}, C_{3c}, C_{3d} \qquad C_{3a}^2, C_{3b}^2, C_{3c}^2, C_{3d}^2$$

$$\begin{matrix} -\frac{1}{2} & -\frac{\sqrt{3}}{2} \\ -\frac{\sqrt{3}}{2} & \frac{1}{2} \end{matrix} \qquad \begin{matrix} -\frac{1}{2} & \frac{\sqrt{3}}{2} \\ \frac{\sqrt{3}}{2} & \frac{1}{2} \end{matrix}$$

$$\sigma_{(011)}, \sigma_{(0\bar{1}1)}, S_{4x}, S_{4x}^3 \qquad \sigma_{(101)}, \sigma_{(101)}, S_{4y}, S_{4y}^3$$

The indices forming subscripts to the σs correspond to the Miller indices of the planes of symmetry.

The normal modes for the E representation. We can now assume the forms of two degenerate modes of vibration and test whether, given the above matrix group, they satisfy the required criterion. This criterion is the same as was applied in § 18.4 to the CO_3 ion, namely that the three-dimensional transformation of the positions of the atoms and their vibration directions due to a given element of symmetry shall give the same pattern of vibrations as the two-dimensional transformation of the vibrations of a given atom for the same element of symmetry.

The two degenerate normal modes of vibration that are assumed are drawn in Fig. 18.15. Figure 18.15 (1), (2) shows clinographic drawings of the atomic positions and vibration directions. Figure 18.15 (α), (β) shows plans on (001) of the drawings (1), (2), with stereograms surrounding each atomic centre to show the vibration directions. The directions $\langle 112 \rangle$, $\langle 110 \rangle$ have the usual significance of zone indices.

Transformation due to C_{3a}. This triad axis passes through atoms c and e; it moves atom a to the position of atom b, atom b to atom d, atom d to atom a and atom c remains unmoved. The two-dimensional matrix for the triad axis is

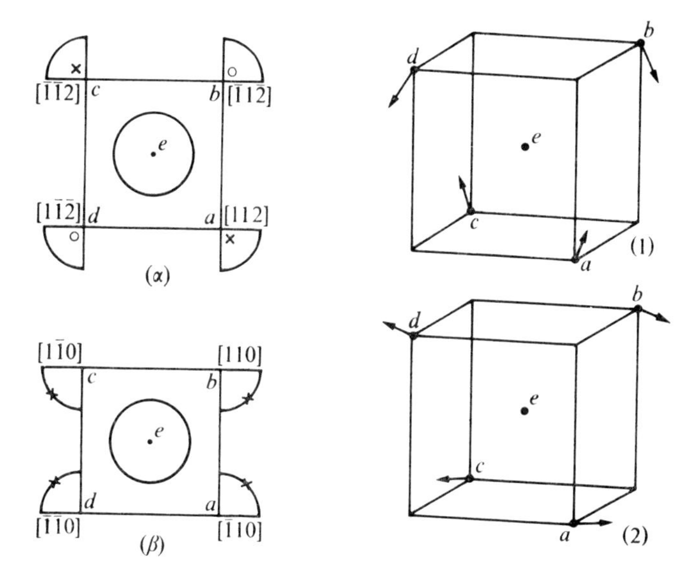

FIG. 18.15. Diagram showing the two modes (α) (β) associated with the E representation. (1), (2) are clinographic drawings of the (α)-, (β)-modes of vibration.

$$\begin{matrix} -\frac{1}{2} & \frac{\sqrt{3}}{2} \\ -\frac{\sqrt{3}}{2} & -\frac{1}{2}. \end{matrix}$$

This transforms the α and β modes to α' and β' modes, according to the scheme

	α	β
α'	$-\frac{1}{2}$	$\frac{\sqrt{3}}{2}$
β'	$-\frac{\sqrt{3}}{2}$	$-\frac{1}{2}$

(The similarity to the table in § 18.4 should be noted.) Consider how this transformation affects the vibrations of the atom b, for which

$$\alpha = \frac{1}{\sqrt{6}}(\bar{1}, 1, \bar{2}), \quad \beta = \frac{1}{\sqrt{2}}[1, 1, 0].$$

The $\sqrt{6}$ and $\sqrt{2}$ are needed to make the amplitude unity in each case. Inserting these values into the matrix, we obtain

$$\alpha' = -\frac{1}{2\sqrt{6}}[\bar{1}, 1, \bar{2}] + \frac{\sqrt{3}}{2\sqrt{2}}[1, 1, 0].$$

Collecting corresponding components together, we get

$$\alpha' = \frac{1}{2\sqrt{6}}[(1+3)+(-1+3)+(2)] = \frac{1}{\sqrt{6}}[2, 1, 1],$$

$$\beta' = \frac{-\sqrt{3}}{2\sqrt{6}}[\bar{1}, 1, \bar{2}] + \frac{-1}{2\sqrt{2}}[1, 1, 0]$$

$$= \frac{1}{2\sqrt{2}}[(1-1)+(-1-1)+(2)] = \frac{1}{\sqrt{2}}[0, \bar{1}, 1].$$

Thus the effect of C_{3a} at the point b is to change a vibration direction $[\bar{1}\ 1\ \bar{2}]$ into $[2\ 1\ 1]$ and a vibration direction $[1\ 1\ 0]$ into $[0\ \bar{1}\ 1]$.

In three dimensions, the vibration of atom a must change according to the scheme

$$x \to y, \quad y \to z, \quad z \to x,$$

which requires $[1\ 1\ 2] \to [2\ 1\ 1]$ and $[\bar{1}\ 1\ 0] \to [0\ \bar{1}\ 1]$. As we have seen, this agrees with the two-dimensional transformation of the C_3 axis and indicates that the original assumption concerning the modes of vibration was correct.

The same test may be applied to each of the symmetry elements in the point group Thus for C_{2z} the two-dimensional matrix gives

	α	β
α'	1	0
β'	0	1

so that for atom *d*

$$\alpha' = \alpha = 1\ \bar{1}\ \bar{2},$$

$$\beta' = \beta = \bar{1}\ \bar{1}\ 0.$$

The atom *b* is turned to atom *d* by the action of C_{2z}, which requires

$$x \to \bar{x}, \quad y \to \bar{y}, \quad z \to z.$$

Thus the vibrations of atom *b* change according to the scheme

$$[\bar{1}\ 1\ \bar{2}] \to [1\ \bar{1}\ \bar{2}]; \qquad [1\ 1\ 0] \to [\bar{1}\ \bar{1}\ 0];$$

which is again consistent with the requirement of the two-dimensional transformation.

It may be shown in a similar manner that with all the two-dimensional matrices applying to the elements of symmetry the two modes α and β transform correctly.

The *E* mode, being two-dimensional, can have only two vibration patterns, and since the proposed α and β schemes satisfy all the elements of symmetry they must be the only possible normal modes.

In conclusion of this discussion of the normal modes, it should be noted that they are determined by the symmetry alone and are not dependent on the inter atomic forces or the sizes of atoms (6, p. 91).

Q.18.4. Find which irreducible representations are associated with the normal modes of the regular octahedral ion $PtCl_6$. This ion belongs to point group $m3m(O_h)$.

Q.18.5. Which representations obtained in Q.18.4 are associated (a) with stretching of the Pt–Cl bonds and (b) with change of angle Cl–Pt–Cl?

Answers

Q.18.1. For SO_4 the point group is $\bar{4}3m$ and the representation is A_1. For $PtCl_6$ the point group is $m3m$ and the representation is A_{1g}.

Q.18.2. The transformation for a plane of symmetry parallel, for example, to $(1\bar{1}0)$ is as follows.

$$\begin{matrix} 0 & 1 & 0 \\ 1 & 0 & 0 \\ 0 & 0 & 1, \end{matrix}$$

which means $x \to y$, $y \to x$, $z \to z$. It can be seen that the displacement (α) is related to the displacement (β) as x is to y. The other requirements are similarly fulfilled.

Q.18.3. The symmetry element S_{4x} exchanges a for b, b for d, d for c, and c for a. Thus the new diagrams derived from Fig. 18.13 (α), (β), (γ), (1), (2), (3), are as shown in Fig. 18.16, namely, (μ), (ν), (ρ), (1), (2), (3). It will be seen that a rotation of these diagrams through 180° about the Z axis would make them the same as (α), (γ), (β) in Fig. 18.13. Such a rotation does not contradict the symmetry requirement, since such a two-fold axis already exists in the molecule. Thus this scheme of vibrations satisfies the symmetry S_{4x}.

If σ_d represents a plane of symmetry parallel to $(1\bar{1}0)$, then $x \to y$, $y \to x$, $z \to z$, $b \to d$, $d \to b$. The patterns (α), (β), (γ) Fig. 18.13 change into (ξ), (ζ), (ω) of Fig. 18.16. It will be seen that (ξ) = (β), (ζ) = (α), (ω) = (γ), so that the proposed vibration scheme satisfies this symmetry requirement.

Q.18.4. In the ion $PtCl_6$, there are seven atoms and $3 \times 7 - 6 = 15$ genuine normal modes. The characters of the 21×21 matrix based on the Cartesian coordinates of the displacements of the seven atoms are as follows.

	E	$8C_3$	$6C_2$	$6C_4$	$3C_2 = C_4^2$	I	$6S_4$	$8S_6$	$3\sigma_h$	$6\sigma_d$
$\chi(R)$	21	0	-1	3	-3	-3	-1	0	5	3

The following notes are intended to help in arriving at the above values of $\chi(R)$.

C_3—only the Pt atom lies on a triad axis, and the character of the axial transformation is zero.

C_2—this is a [110] axis and passes through the Pt atom only.

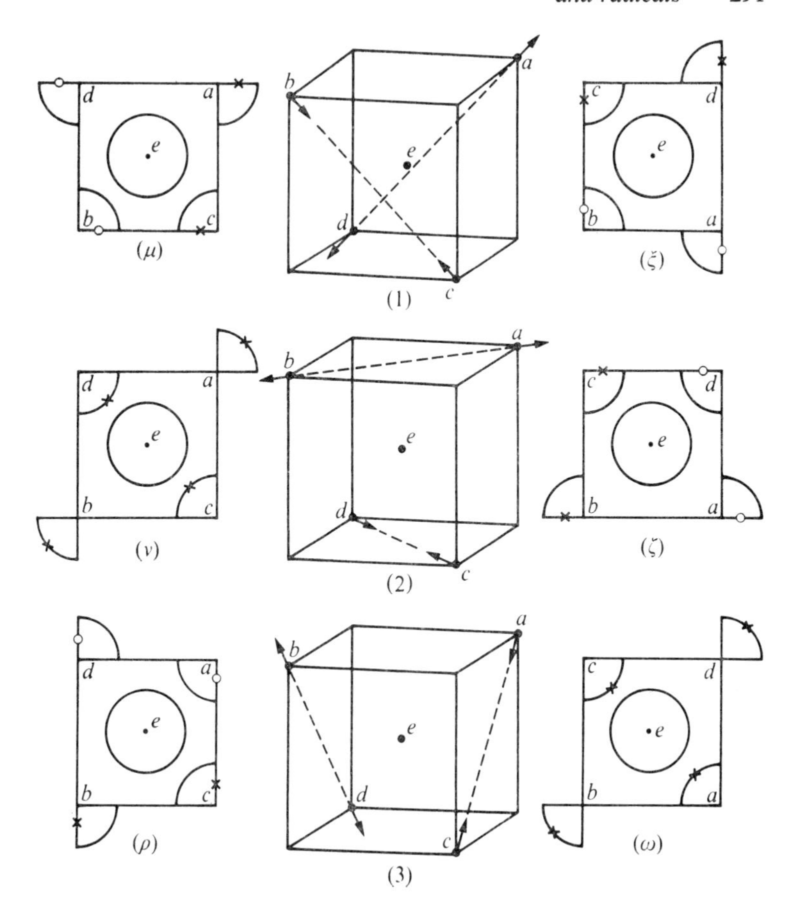

FIG. 18.16. Diagrams (μ), (ν), (ρ) show the transformations of Fig. 18.3 (α), (β), (γ) by the symmetry element S_{4x}. Diagrams (ξ) (ζ) (ω) show the corresponding transformations due to $\sigma_{1\bar{1}0}$.

$C_2 = C_4^2$—three atoms lie on any tetrad axis, its character as a diad axis is -1.

C_4—three atoms lie on any tetrad axis; its character as a tetrad axis is $+1$.

I—inverts all atoms except the Pt atom. Its character is -3.

S_4—passes through three atoms, but inverts the chlorines into one another; character -1.

S_6—the axis passes through the Pt atom only. The character for the S_6 axis is zero.

σ_h—passes through five atoms; character $+1$.

σ_d—the diagonal (110) plane of symmetry passes through three atoms and has a character $+1$.

The number of times the irreducible representations are contained in the $\chi(R)$ representation, n_v, is given by the formula

$$n_v = \frac{1}{g}\sum_i h_i\chi_i(R)\chi_i^{(v)}.$$

Using this expression, we obtain for the representations of point group $m3m\,(O_h)$

$$A_{1g} = 1/48\{(21\times 1)+(6\times -1)+(6\times 3)+(3\times -3)+(1\times -3)+(6\times -1)+ +(3\times 5)+(6\times 3)\}$$

$$= 1;$$

$$A_{2g} = 1/48\{(21\times 1)+(6\times -1\times -1)+(6\times 3\times -1)+(3\times -3)+(1\times -3)+ +(6\times -1\times -1)+(3\times 5)+(6\times 3\times -1)$$

$$= 0;$$

$$E_g = 1/48\{(21\times 2)+(6\times -1\times 0)+(6\times 3\times 0)+(3\times -3\times 2)+ +(1\times -3\times 2)+(6\times -1\times 0)+(3\times 5\times 2)+(6\times 3\times 0)\}$$

$$= 1;$$

$T_{1g} = 1; T_{2g} = 1; A_{1u} = 0; A_{2u} = 0; E_u = 0; T_{1u} = 3; T_{2u} = 1.$

The total representation Γ is thus

$$\Gamma = A_{1g}+E_g+T_{1g}+T_{2g}+3T_{1u}+T_{2u}$$

From the character table for point group $m3m$, it may be seen that, x, y, z, i.e. translations, correspond to T_{1u} and rotations, i.e. R_x, R_y, R_z, to T_{1g}. The genuine normal modes are therefore A_{1g}, E_g, T_{2g}, $2T_{1u}$ and T_{2u}. There are $3\times 7-6 = 15$ normal modes. The A_{1g} is one-dimensional and is the breathing mode in which all the chlorine atoms move out in synchronism and then move in together. E_g is two-dimensional, and each of the Ts is three-dimensional, making up the total of fifteen modes.

Q.18.5. (a) The characters for the stretching of the Pt–Cl bonds are as follows.

	E	$8C_3$	$6C_2$	$6C_4$	$3C_2 = C_4^2$	I	$6S_4$	$8S_6$	$3\sigma_h$	$6\sigma_d$
$\chi(R)$	6	0	0	2	2	0	0	0	4	2

(The $\chi(R)$ numbers correspond to the numbers of bonds unmoved by the corresponding symmetry operation.) Applying the formula for n_v, we obtain the number of irreducible representations contained in the reducible Γ(Pt–Cl) representation as follows.

$$\Gamma(\text{Pt–Cl}) = A_{1g}+E_g+T_{1u}.$$

(b) There are 12 Cl–Pt–Cl angles but they are not all independent. If we carry through the same analysis as for the stretching of the Pt–Cl bonds, there must be some redundancy.

The characters for the change of angle between bonds are as follows.

	E	$8C_3$	$6C_2$	$6C_4$	$3C_2 = C_4^2$	I	$6S_4$	$8S_6$	$3\sigma_h$	$6\sigma_d$
$\chi(R)$	12	0	2	0	0	0	0	0	4	2

(It should be observed that the test of an angle not being moved by the symmetry operation is that the lines defining the angle either stay fixed in position or are exchanged with one another.)

The above $\chi(R)$ values are resolved by the formula for n_v, and we obtain the component irreducible representations as follows.

$$\Gamma(\text{Cl–Pt–Cl}) = A_{1g}+E_g+T_{2g}+T_{1u}+T_{2u}.$$

The redundancy mentioned above relates to A_{1g} and E_g, since these already occur in (a), and the Pt–Cl extensions are quite independent of changes in Cl–Pt–Cl angles. Thus the representations for change of angle between the bonds are T_{2g}, T_{1u} and T_{2u}. These have nine dimensions, which, together with the six for the bond stretching, make up the necessary total of fifteen.

19

Infrared and Raman spectra

19.1. Introduction

In this chapter we shall be concerned with the absorption of infrared radiation in crystals and with the Raman scattering of visible light. Infrared and visible light can be linearly polarized, and the plane of polarization is taken to be the plane containing the direction of the electric vector of the radiation and its direction of travel. Consider a plane-polarized beam of infrared radiation, containing a continuous range of wavelengths from 5 to 15 μm, travelling in a direction parallel to the plane of the CO_3 radical in a calcite crystal. Suppose also that the plane of polarization is perpendicular to the CO_3 group. The spectrum of the radiation transmitted by the calcite crystal has, under these conditions, an absorption band at 11 μm. This arises from the excitation by the radiation of a vibration within the CO_3 group. The carbon can be considered to have a positive charge and the oxygens a negative charge. Under the influence of the electric field in the infrared wave, these charges are displaced in opposite directions. The CO_3 group belongs to the $\bar{6}m2(D_{3h})$ point group. From the character tables it may be seen that the A_2'' representation corresponds to the z coordinate, since the movement of the carbon atom out of the O_3 plane destroys the elements of symmetry C_2, σ_h, and S_3, so that the corresponding characters are all -1.

Suppose now that the plane of polarization is rotated through 90° about the direction of the incident beam, so that the electric vector is parallel to the plane of the CO_3 group. Now two absorption bands are observed, at 7 μm and 15 μm respectively. These correspond to relative displacements of the positive and negative charges in the plane of the CO_3 group. These vibrations are therefore associated with the x and y coordinates. From the character table for $\bar{6}m2$, it may be seen that the corresponding representation for these normal modes is E'. In § 18.4, the actual atomic motions for this representation were evaluated.

The calculation of the wavelengths of the absorption bands depends on a knowledge of the forces tending to restore the vibrating atoms to their equilibrium positions. In some cases, it is easy to see that some modes of vibration must have a higher frequency than others, but we shall not go into this subject here, because it belongs to the study of lattice dynamics rather than to that of group theory.

Raman spectra are produced when visible light is scattered from a crystal and analysed in a spectrometer. An incident beam of monochromatic light produces not only scattered light of the same wavelength as the incident light, but also much fainter beams of light of slightly different wavelengths. Thus in the spectrometer faint satellite lines appear, together with the line due to the unchanged incident light.

The explanation of the production of the Raman lines is as follows. The incident light is absorbed and excites the ion; this excited state decays with the emission of a photon, but its energy might differ from that of the incident light because the ion might be left in a vibrationally excited state (in which case the energy of the scattered photon will be reduced), or the ion might have been in a vibrationally excited state initially and have imparted this energy to the photon during the inelastic collision (in which case the energy of the scattered photon will be increased). If the quantum of vibrational energy involved is $h\nu_0$, then the energies of the incident and scattered photons are related by the equation

$$h\nu = h\nu' \pm h\nu_0 \tag{19.1}$$

(ν is the incident frequency, ν' the scattered frequency). The above expression can be simplified by using wave numbers, ω, (reciprocals of wavelengths) instead of frequencies.

Since

$$\nu = c/\lambda = c\omega,$$

where c is the velocity of light, if we substitute ωs for νs we obtain the expression

$$\omega = \omega' \pm \omega_0.$$

Thus the wave number ω_0 is equal to the difference between the wave numbers of the incident and scattered beams. The data for Raman lines are usually expressed in terms of wave-number differences and in units of cm^{-1}. The reciprocal of these values gives the wavelength corresponding to the quantum of energy involved. This quantum is, of course, just the same as that corresponding to the infrared absorption band. The measurements of these bands are commonly expressed in units of μm. The quanta of infrared radiation have much smaller energies than those of visible light. Thus the wave number for a 10μm radiation is 1000, whereas the corresponding number for blue light of wavelength 4000 Å is 25 000. In the following paragraphs, both wavelengths and wave numbers will be listed for convenience of reference.

The Raman effect depends on the inelastic collision of a photon with the molecule, and therefore on the ability of the electric field of the light to distort the molecule. Therefore one expects the effect to be determined by the polarizability of the molecule or ion, and quantum mechanical calculation confirms that this is so. As we have seen in discussing the magnetic and dielectric properties of crystals (p. 60), the polarizability is in general represented by a triaxial ellipsoid. The equation to such a surface can be written

$$a_{11}x^2+a_{22}y^2+a_{33}z^2+a_{23}yz+a_{31}zx+a_{12}xy,$$

where the a_{ij}s are constants. The only point which concerns us here is that only second-order terms in the coordinates x, y, z are involved. The normal modes of vibration that take part in producing the Raman lines must be compatible with the ellipsoid of polarizability, i.e. they must be determined by second-order terms in the coordinates. Thus, for infrared absorption, representations of the possible modes of vibration are related to first powers of the coordinates, but for Raman lines only the representations involving second-order terms are of interest.

There are many examples among the point groups where a given representation is related to both first-order and second-order terms in the coordinates. In such cases, the representation can apply both to infrared and to Raman lines. However, when a centre of symmetry is present in the point group, a given representation can never be associated with both infrared and Raman lines. This follows from a consideration of the action of a centre of symmetry on first-order and second-order terms in x, y, z. A centre converts $+x$, $+y$, $+z$ into $-x$, $-y$, $-z$; but xy, yz, zx, x^2, y^2, z^2 are unchanged in sign. Thus all representations depending on first powers of the coordinates in a centrosymmetric point group must be the 'u' (*ungerade*) type and all of those depending on the second powers must be of the 'g' (*gerade*) type. These two types must always be distinct from one another. Thus we arrive at the following selection rules.

Selection rule 1

Representations in the character tables corresponding to first powers of x, y, z can be associated with normal modes of vibration giving infrared absorption lines.

Selection rule 2

Representations in the character tables corresponding to squares or products of two coordinates can be associated with normal modes related to Raman lines.

Selection rule 3

Centrosymmetric point groups never contain representations that correspond to both infrared and Raman lines.

19.2. The infrared absorption bands and the Raman lines of flat pyramidal BX_3 groups

Some examples of the infrared absorption bands and Raman lines of flat pyramidal BX_3 groups are given in Table 19.1.

TABLE 19.1

Absorption bands and Raman lines of certain flat pyramidal BX_3 groups

BX_3	E		A_1		A_1		E	
	cm^{-1}	μm	cm^{-1}	μm	cm^{-1}	μm	cm^{-1}	μm
SO_3	470		602		976		—	
ClO_3	480	25·0	625	16·0	933	10·6	963	
BrO_3	356		450	23·0	800	12·2	835	
IO_3	330	—	357	27·0	780	12·8	826	

Raman spectra from *Tables of physical constants*, H. Landolt, Berlin (1950). Infrared spectra from F. I. G. Rawlins and A. M. Taylor, *Infrared analysis of molecular structure*, University Press, Cambridge (1929).

The reciprocal relation between the Raman wave number and the absorption-band wavelength is approximately fulfilled. The precision obtained with the sharp Raman lines is considerably higher than that obtainable with the relatively broad absorption bands.

The symmetry of this flat pyramidal group is that of the point group $3m$ (C_{3v}). The character table for this point group shows that a displacement z corresponds to the A_1 representation. In Table 19.1, the absorption band in the second A_1 column has the values 10·6, 12·2. and 12·8 respectively for the chlorates, bromates, and iodates. Compared with the corresponding changes in the other columns, these values are not very different from one another. This must imply that only oxygen atoms are involved in the 10·6 μm vibration. A vibration of the oxygens along the medians of the triangle which they form would be a breathing mode for the O_3 group. The frequency of this vibration would be, to a large extent, independent of the mass of the halogen atom. This type of vibration involves both change of bond length and change of bond angle. The E modes are associated with movements in the O_3 plane and are similar to those already discussed in connection with the vibrations of the CO_3 group (Fig. 18.5 (a), (b) (2, p. 140; 5, p. 264; 6, p. 113).

Q.19.1. Using Cartesian coordinates drawn through each atom, or otherwise, find which representations of the character table for point group $3m$, correspond to normal modes of vibration of the flat pyramidal BX_3 group. Find how these representations are divided between changes of bond length and change of bond angle.

19.3. The infrared absorption bands and Raman lines of tetrahedral BX_4 groups

In Table 19.2 the infrared and Raman data for some BX_4 groups are given.

The point group to which the tetrahedral BX_4 groups belong is $\bar{4}3m$ (T_d), and in § 18.7 it was shown that the normal modes are given by the A_1, E, and T_2 representations. It can be seen from the character table that only T_2 modes, being associated with x, y, z, can give rise to infrared absorption, and that Raman lines, being associated with second-order terms in x, y, z, may be related to the A_1, E, and T_2 modes. The A_1 mode is the breathing mode, in which the oxygens vibrate along the lines joining them to the central atom.

TABLE 19.2

Absorption bands and Raman lines of some tetrahedral BX_4 *groups*

BX_4	E cm^{-1}	μm	T_2 cm^{-1}	μm	A_1 cm^{-1}	μm	T_2 cm^{-1}	μm
PO_4	420		550		935		1080	
SO_4	450		611	16	983		1105	9
ClO_4	460		626		932		1110	
IO_4	280		340		795		840	

Raman spectra from H. Landolt, *Tables of physical constants*, Berlin (1950). Infrared spectra from F. I. G. Rawlins and A. M. Taylor, *Infrared analysis of molecular structure*. University Press Cambridge (1929).

The E mode is doubly degenerate, and the T_2 modes are triply degenerate (see § 18.8). In the first T_2 mode, Fig. 18.12, applied to the BX_4 group, the central B atom moves, say, along [100], while the oxygens move along [$\bar{1}$00], i.e. in the opposite direction. Thus a dipole is created, which can react with the incident infrared radiation. In the second T_2 modes, Fig. 18.13, two oxygens move inwards together along a face-diagonal of the cube, and the other two move outwards along a perpendicular face-diagonal on the opposite face of the cube. The electric moment induced on the B atom by the opposed movements of the oxygens is thus capable of producing a dipole having its axis normal to the face-diagonal along which the vibrations are occurring.

19.4. The influence of the symmetry of a particular site

In the preceding paragraphs we have to a large extent ignored the fact that the CO_3 group in calcite or the SO_4 group in $SrSO_4$ (celestite) is in a particular environment of surrounding atoms and ions. We have considered the symmetry of the radical as though it were isolated and independent of its neighbours. This assumption is largely justified, as may be seen by the relative constancy of the wavelengths of infrared absorption bands for the same radical in different crystal structures. However, it is not altogether possible to neglect the influence of the environment of the radical or molecule.

The CO_3 group in calcite is in a trigonal environment, but the same group in aragonite is in an orthorhombic environment. Although the CO_3 group in aragonite is no longer trigonal, the only difference between the Raman lines given by the two crystals is that for aragonite one of the vibrations in which the atoms move in the plane of the CO_3 group is split into two peaks differing by 14 cm^{-1} in wave number, i.e. by roughly 1 per cent. A further effect of the different environment in the lower symmetry of aragonite is to make infrared active certain modes of vibration which were inactive in the higher symmetry of calcite. From the character tables, it can be seen that, in the point group *m* which applies to the position of the CO_3 ion in the unit cell of aragonite, the A' representation (top line) relates to x, y as well as x^2, y^2. In calcite the CO_3 ion has three diad axes passing through it and a trigonal axis passing through the carbon atom, so its site symmetry corresponds that of point group 32. The A_1 representation corresponds only to second-order terms in the coordinates. This difference in the environment is responsible for the fact that the symmetric stretching mode, A_1, does not appear in the infrared spectrum of calcite but does appear as a weak line in that of aragonite.

Another example of the influence of the site symmetry on the Raman spectrum is afforded by some sulphates, the data for which are given in Table 19.3.

In the SO_4 group isolated from its environment the E mode is doubly degenerate and the two vibrations have the same frequency.

In the minerals given in Table 19.3, the environment impresses a lower symmetry on the SO_4 group, and the E mode breaks into two frequencies close together. The two T_2 modes are triply degenerate in the isolated state of the SO_4 group, but, in the lower symmetry of the environment, the single frequency is resolved into two, three, or four frequencies all close to the original frequency. In celestite, the point group of the crystal is *mmm*, and the representations to which Raman lines can correspond, because they are related to second powers of the coordinates, are four in number (2, p. 152; 5, p. 267).

TABLE 19.3

The Raman frequencies in some mineral sulphates

Mineral	Site symmetry of SO_4	E cm^{-1}	T_2 cm^{-1}	A_1 cm^{-1}	T_2 cm^{-1}
Anhydride $CaSO_4$	2 mm	416, 495	604, 626	1015	1107, 1125, 1157
Gypsum $CaSO_4 \cdot 2H_2O$	2	415, 492	618, 622, 672	1006	1115, 1136, 1143
Celestite $SrSO_4$	m	453, 458	617, 624, 637, 656	999	1094, 1103, 1159, 1185

From H. Landolt, *Tables of physical constants*, Berlin (1950).

Answers

Q.19.1. The character table for point group $3m$ is reproduced below.

	E	$2C_3$	$3\sigma_v$	
A_1	1	1	1	z
A_2	1	1	-1	R_z
E	2	-1	0	$(x, y)\ (R_x, R_y)$

A 12×12 matrix is constructed from the x, y, z coordinates of the four atoms in the BX_3 group. For the E matrix, the character is 12.

The symmetry element C_3 moves all three sets of axes passing through the oxygens, and the character $(1+2\cos 2\pi/3)$ for the remaining Cl, Br, atom is zero. Thus for C_3 the $\chi(R)$ is zero.

The element σ_v leaves coordinate axes through two atoms unchanged. The character for a plane of symmetry is $(-1+2\cos 2\pi 0)$, i.e. $+1$. Thus the character for $\chi(R)$ is $+2$. Summarizing, for $\chi(R)$ we have the characters 12, 0, 2 in the columns for E, $2C_3$ and $3\sigma_v$ respectively. Applying the formula for n_v (equation (18.1)), we obtain for A_1

$$n_v = \frac{1}{6}(12 \times 1 + 2 \times 3) = 3;$$

for A_2

$$n_v = \frac{1}{6}(12 \times 1 + 2 \times 3 \times -1) = 1;$$

for E

$$n_v = \frac{1}{6}(12 \times 2) = 4.$$

Thus $\chi(R) = 3A_1 + A_2 + 4E$.

Elimination of one A_1 and one E is required to get rid of pure bodily translations, and one A_2 and E to avoid pure rotations. This leaves only two A_1 and two E. The As are one-dimensional and the Es two-dimensional, so the total number of dimensions included in these representations is six. This is the number to be expected from the number of atoms involved.

We may also use internal coordinates, namely change of bond length and change of bond angle. The values of $\chi(R)$ for change of bond length under column E are 3, under C_3, 0, and under σ_v, 1.

Applying the usual formula for n_v we obtain for A_1,

$$n_v = \frac{1}{6}(3\times 1+1\times 3) = 1;$$

for A_2,

$$n_v = \frac{1}{6}(3\times 1-1\times 3) = 0;$$

for E,

$$n_v = \frac{1}{6}(3\times 2) = 1.$$

Thus,

$$\chi(R) = A_1+E.$$

For a change of bond angle also the characters for E, C_3, and σ_v are respectively, 3, 0, and 1. This again gives

$$\chi(R) = A_1+E.$$

Thus the representations are divided equally between changes of bond length and changes of bond angle.

Appendices

A.1. Character tables and stereograms for the thirty–two point groups

Meaning of the symbols used

The point-group symbols are quoted first according to the Hermann–Mauguin system adopted by the *International tables for X-ray crystallography*, Vol. 1, and, in brackets, also in the Schönfliess notation. In the left-hand column of each character table, A stands for a one-dimensional representation which is symmetric, i.e. $+1$, for rotation about a principal axis of symmetry; and B stands for a one-dimensional representation which is antisymmetric, i.e. -1, for rotation about a principal axis of symmetry.

The subscripts 'g' and 'u', which occur in all centrosymmetric classes, indicate representations that are respectively symmetric and antisymmetric with respect to a centre of symmetry.

Numerical subscripts to As and Bs indicate representations related in different ways to axes of symmetry that are normal to the main axis or to a vertical plane of symmetry. A prime and a double prime indicate representations that are respectively symmetric and antisymmetric with respect to a horizontal plane of symmetry (σ_z, σ_h).

Along the top row of each character table, the classes of symmetry elements present in each point group are given. The number of such elements in a given class is indicated by the number preceding the symbol. E is the identity operator; C_n is a vertical n-fold axis; C_{2x}, C_{2y}, C_{2z} are diad axes parallel respectively to the X, Y, Z axes; I is a centre of symmetry, σ_v is a vertical plane of symmetry, σ_z, σ_h are horizontal planes of symmetry, σ_d is a diagonal plane of symmetry. Subscripts give the number of operations carried out with a particular axis of symmetry. S_n is an n-fold axis involving reflection in a horizontal plane of symmetry after a rotation of $2\pi/n$ about the main (vertical) axis. x, y, z are coordinates of equivalent points. R_x, R_y, R_z are infinitesimally small rotations about axes X, Y, Z respectively. E stands for a two-dimensional representation (and has no connection, in spite of the symbol, with the identity operator). T stands for a three-dimensional representation and occurs only in the cubic system.

Triclinic system

Point group 1 (C_1)

	E	
A	1	All functions of coordinates, R_x, R_y, R_z

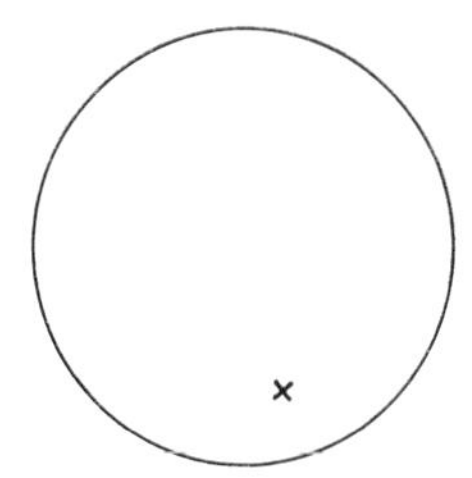

Fig. A.1.1

Point group $\bar{1}$ (C_i)

	E	I		
A_g	1	1	R_x, R_y, R_z	x^2, y^2, z^2, xy, xz, yz
A_u	1	-1	x, y, z	

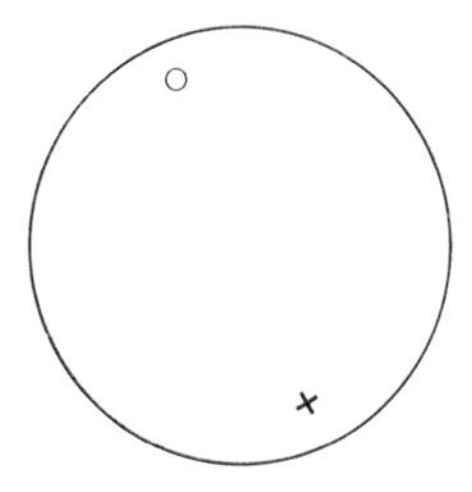

Fig. A.1.2

Monoclinic system

Point group 2 (C_2)

	E	C_{2z}		
A	1	1	z, R_z	x^2, y^2, z^2, xy
B	1	-1	x, y, R_x, R_y	yz, xz

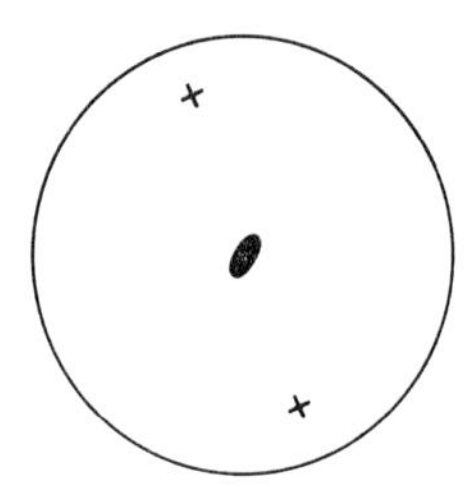

FIG. A.1.3

Point group m (C_s)

	E	σ_z		
A'	1	1	x, y, R_z	x^2, y^2, z^2, xy
A''	1	-1	z, R_x, R_y	yz, xz

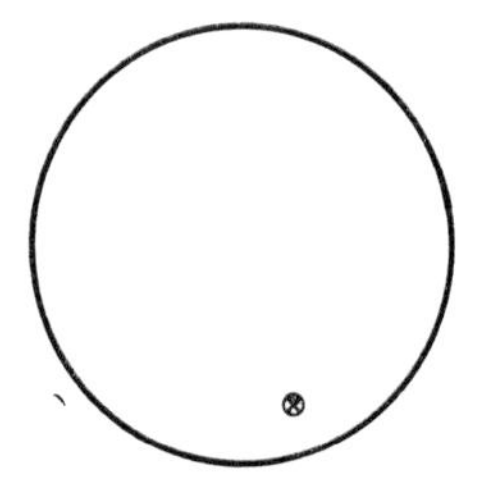

FIG. A.1.4

Point group $2/m$ (C_{2h})

	E	C_{2z}	I	σ_z		
A_g	1	1	1	1	R_z	x^2, y^2, z^2, xy
B_g	1	-1	1	-1	R_x, R_y	xz, yz
A_u	1	1	-1	-1	z	
B_u	1	-1	-1	1	x, y	

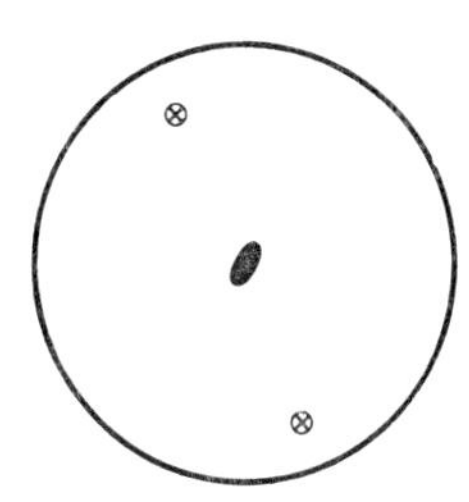

FIG. A.1.5

Orthorhombic system

Point group 222 (D_2)

	E	C_{2z}	C_{2y}	C_{2x}		
A	1	1	1	1		x^2, y^2, z^2
B_2	1	-1	1	-1	y, R_y	xz
B_1	1	1	-1	-1	z, R_z	xy
B_3	1	-1	-1	1	x, R_x	yz

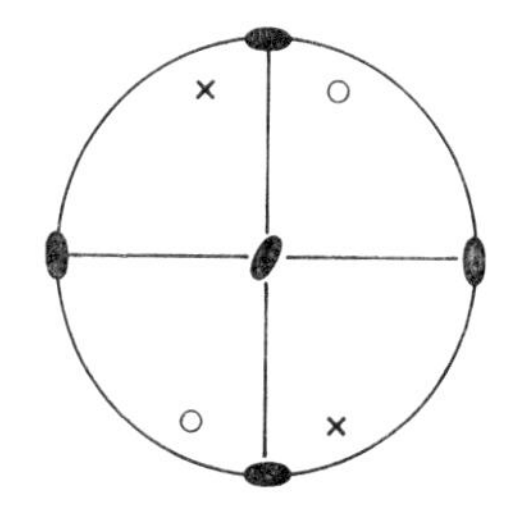

FIG. A.1.6

Point group mm2 (C_{2v})

	E	C_{2z}	σ_y	σ_x		
A_1	1	1	1	1	z	x^2, y^2, z^2
B_1	1	−1	1	−1	x, R_y	xz
A_2	1	1	−1	−1	R_z	xy
B_2	1	−1	−1	1	y, R_x	yz

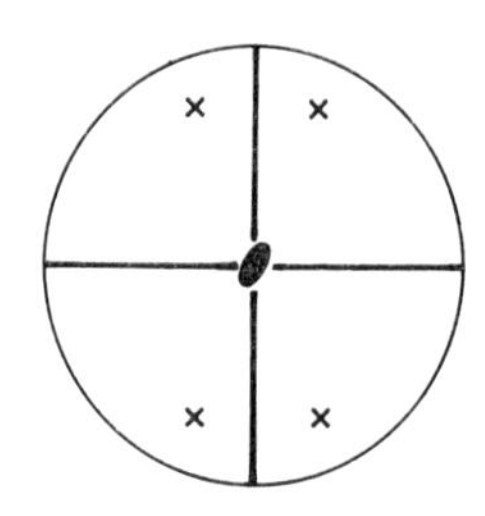

FIG. A.1.7

Point group mmm (D_{2h})

	E	C_{2z}	C_{2y}	C_{2x}	I	σ_z	σ_y	σ_x		
A_g	1	1	1	1	1	1	1	1		x^2, y^2, z^2
B_{2g}	1	−1	1	−1	1	−1	1	−1	R_y	xz
B_{1g}	1	1	−1	−1	1	1	−1	−1	R_z	xy
B_{3g}	1	−1	−1	1	1	−1	−1	1	R_x	yz
A_u	1	1	1	1	−1	−1	−1	−1		xyz
B_{2u}	1	−1	1	−1	−1	1	−1	1	y	
B_{1u}	1	1	−1	−1	−1	−1	1	1	z	
B_{3u}	1	−1	−1	1	−1	1	1	−1	x	

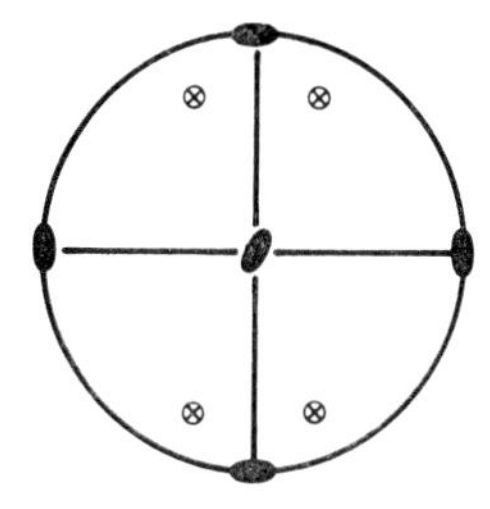

FIG. A.1.8

Tetragonal system

Point group 4 (C_4)

	E	C_4	C_2	C_4^3		
A	1	1	1	1	z, R_z	x^2+y^2, z^2
B	1	-1	1	-1		$(x^2-y^2), xy$
E	$\begin{cases} 1 \\ 1 \end{cases}$	$\begin{matrix} i \\ -i \end{matrix}$	$\begin{matrix} -1 \\ -1 \end{matrix}$	$\left.\begin{matrix} -i \\ i \end{matrix}\right\}$	(x, y) (R_x, R_y)	(yz, xz)

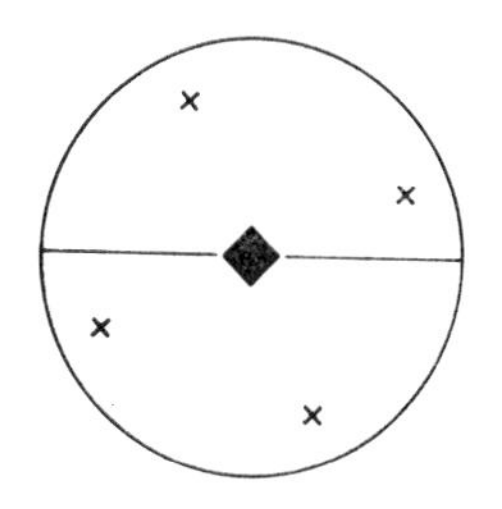

FIG. A.1.9

Point group $\bar{4}$ (S_4)

	E	S_4	C_2	S_4^3		
A	1	1	1	1	R_z	x^2+y^2, z^2
B	1	-1	1	-1	z	x^2-y^2, xy
E	1	i	-1	$-$i	(x, y) (R_x, R_y)	(xz, yz)
	1	$-$i	-1	i		

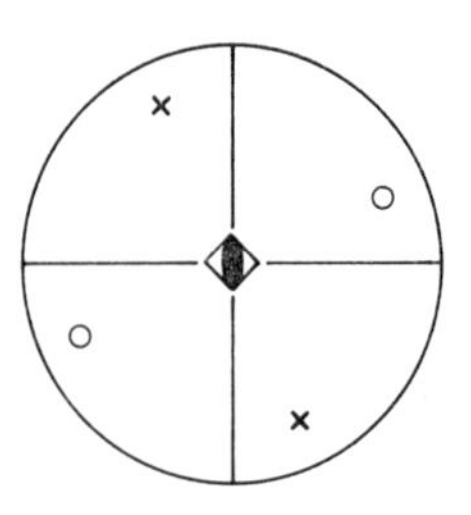

FIG. A.1.10

Point group $4/m$ (C_{4h})

	E	C_4	C_2	C_4^3	I	S_4^3	σ_h	S_4		
A_g	1	1	1	1	1	1	1	1	R_z	x^2+y^2, z^2
B_g	1	-1	1	-1	1	-1	1	-1		x^2-y^2, xy
E_g	1	i	-1	$-$i	1	i	-1	$-$i	R_x, R_y	xz, yz
	1	$-$i	-1	i	1	$-i$	-1	i		
A_u	1	1	1	1	-1	-1	-1	-1	z	
B_u	1	-1	1	-1	-1	1	-1	1		
E_u	1	i	-1	$-$i	-1	$-i$	1	i	(x, y)	
	1	$-$i	-1	i	-1	i	1	$-$i		

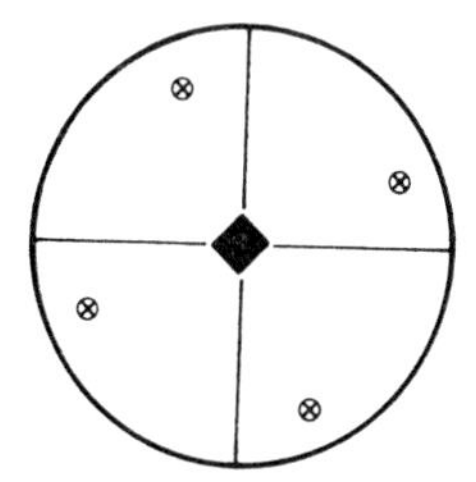

FIG. A.1.11

Point group 422 (D_4)

	E	$2C_4$	$C_2(C_4^2)$	$2C_2'$	$2C_2''$		
A_1	1	1	1	1	1		x^2+y^2, z^2
A_2	1	1	1	-1	-1	z, R_z	
B_1	1	-1	1	1	-1		x^2-y^2
B_2	1	-1	1	-1	1		xy
E	2	0	-2	0	0	(x, y) (R_x, R_y)	(xz, yz)

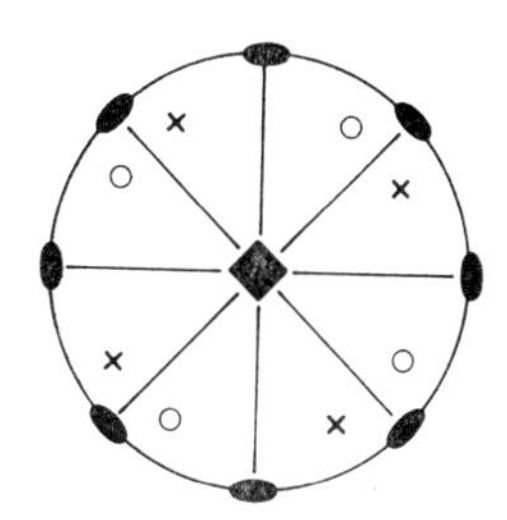

FIG. A.1.12

Point group 4mm (C_{4v})

	E	$2C_4$	C_2	$2\sigma_v$	$2\sigma_d$		
A_1	1	1	1	1	1	z	x^2+y^2, z^2
A_2	1	1	1	-1	-1	R_z	
B_1	1	-1	1	1	-1		x^2-y^2
B_2	1	-1	1	-1	1		xy
E	2	0	-2	0	0	(x, y) (R_x, R_y)	(xz, yz)

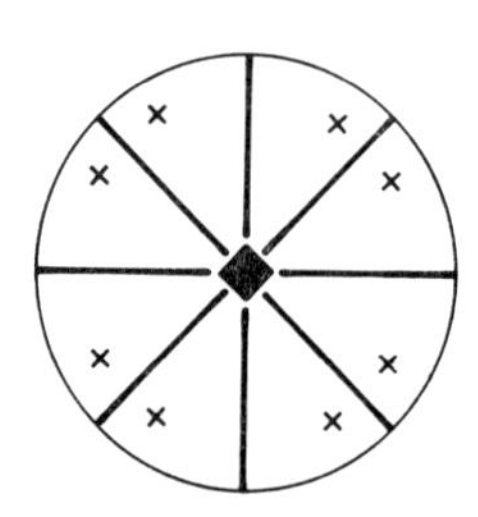

FIG. A.1.13

Point group $\bar{4}2m$ (D_{2d})

	E	$2S_4$	C_2	$2C_2'$	$2\sigma_d$		
A_1	1	1	1	1	1		x^2+y^2, z^2
A_2	1	1	1	-1	-1	R_z	
B_1	1	-1	1	1	-1		x^2-y^2
B_2	1	-1	1	-1	1	z	xy
E	2	0	-2	0	0	(x, y) (R_x, R_y)	(xz, yz)

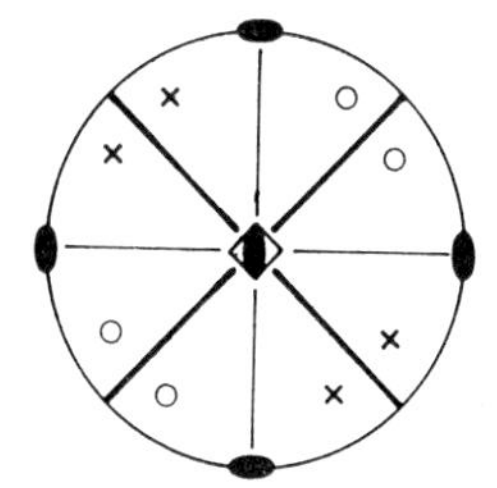

FIG. A.1.14

Point group $4/mmm$ (D_{4h})

	E	$2C_4$	C_2	$2C_2'$	$2C_2''$	I	$2S_4$	σ_z	$2\sigma_v$	$2\sigma_d$		
A_{1g}	1	1	1	1	1	1	1	1	1	1		x^2+y^2, z^2
A_{2g}	1	1	1	−1	−1	1	1	1	−1	−1	R_z	
B_{1g}	1	−1	1	1	−1	1	−1	1	1	−1		x^2-y^2
B_{2g}	1	−1	1	−1	1	1	−1	1	−1	1		xy
E_g	2	0	−2	0	0	2	0	−2	0	0	(R_x, R_y)	(xz, yz)
A_{1u}	1	1	1	1	1	−1	−1	−1	−1	−1		
A_{2u}	1	1	1	−1	−1	−1	−1	−1	1	1	z	
B_{1u}	1	−1	1	1	−1	−1	1	−1	−1	1		
B_{2u}	1	−1	1	−1	1	−1	1	−1	1	−1		
E_u	2	0	−2	0	0	−2	0	2	0	0	(x, y)	

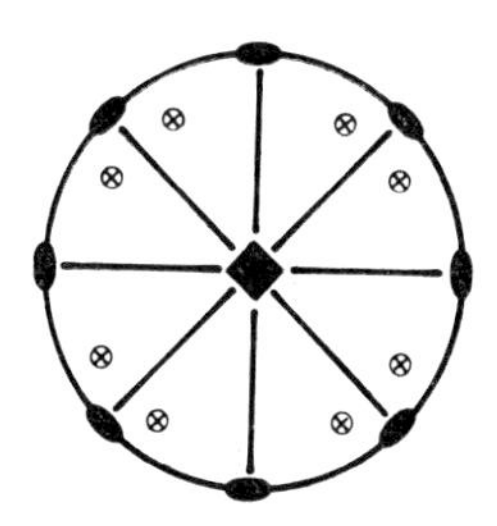

FIG. A.1.15

Trigonal system

Point group 3 (C_3)

	E	C_3	C_3^2		$\varepsilon = \exp(2\pi i/3)$
A	1	1	1	z, R_z	x^2+y^2, z^2
E	1	ε	ε^*	$(x, y)\ (R_x, R_y)$	$(x^2-y^2, xy)\ (yz, xz)$
	1	ε^*	ε		

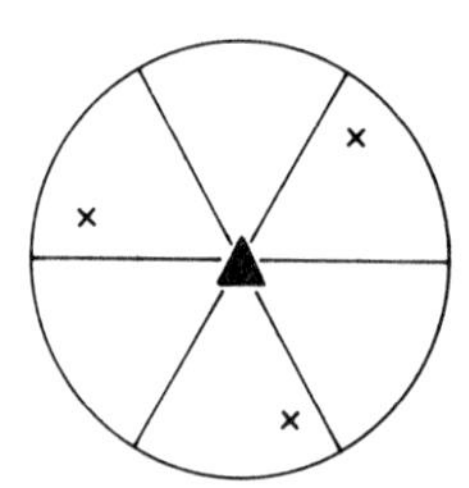

FIG. A.1.16

Point group $\bar{3}$ (S_6)

	E	C_3	C_3^2	I	S_6^5	S_6		$\varepsilon = \exp(2\pi i/3)$
A_g	1	1	1	1	1	1	R_z	x^2+y^2, z^2
E_g	1	ε	ε^*	1	ε	ε^*	(R_x, R_y)	$(x^2-y^2, xy)\ (xz, yz)$
	1	ε^*	ε	1	ε^*	ε		
A_u	1	1	1	-1	-1	-1	z	
E_u	1	ε	ε^*	-1	$-\varepsilon$	$-\varepsilon^*$	(x, y)	
	1	ε^*	ε	-1	$-\varepsilon^*$	$-\varepsilon$		

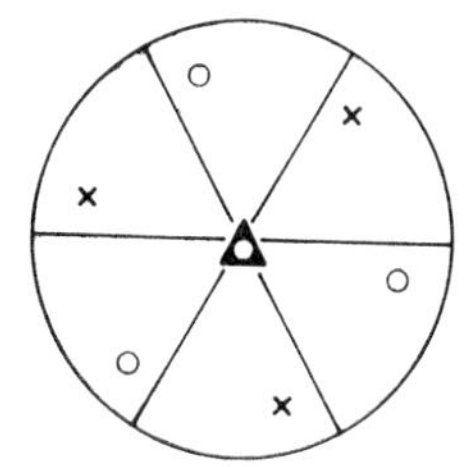

FIG. A.1.17

Point group 32 (D_3)

	E	$2C_3$	$3C_2$		
A_1	1	1	1		x^2+y^2, z^2
A_2	1	1	-1	z, R_z	
E	2	-1	0	(x, y) (R_x, R_y)	(x^2-y^2, xy) (xz, yz)

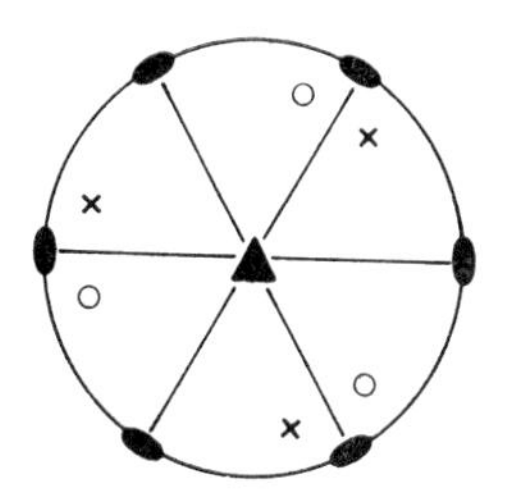

FIG. A.1.18

Point group 3m (C_{3v})

	E	$2C_3$	$3\sigma_v$		
A_1	1	1	1	z	x^2+y^2, z^2
A_2	1	1	-1	R_z	
E	2	-1	0	$(x, y)\ (R_x, R_y)$	$(x^2-y^2, xy)\ (xz, yz)$

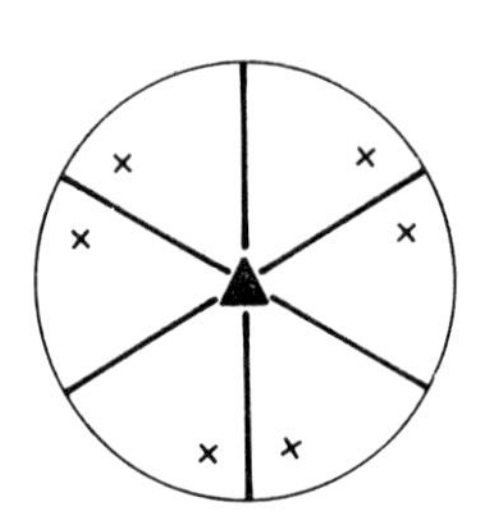

FIG. A.1.19

Point group $\bar{3}m$ (D_{3d})

	E	$2C_3$	$3C_2$	I	$2S_6$	$3\sigma_d$		
A_{1g}	1	1	1	1	1	1		x^2+y^2, z^2
A_{2g}	1	1	-1	1	1	-1	R_z	
E_g	2	-1	0	2	-1	0	R_x, R_y	$(x^2-y^2, xy)\ (xz, yz)$
A_{1u}	1	1	1	-1	-1	-1		
A_{2u}	1	1	-1	-1	-1	1	z	
E_u	2	-1	0	-2	1	0	(x, y)	

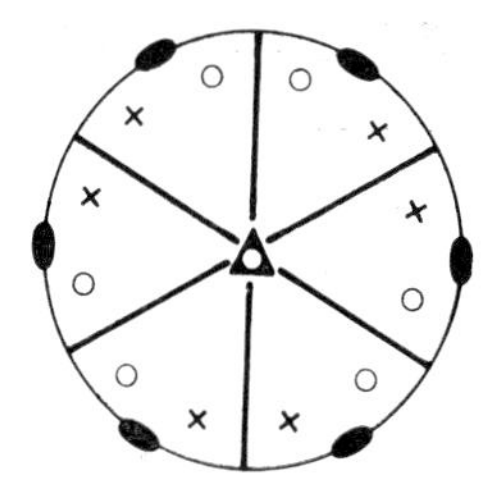

FIG. A.1.20

Hexagonal system

Point group 6 (C_6)

	E	C_6	C_3	C_2	C_3^2	C_6^5		$\varepsilon = \exp(2\pi i/6)$
A	1	1	1	1	1	1	z, R_z	x^2+y^2, z^2
B	1	-1	1	-1	1	-1		
E_1	1	ε	$-\varepsilon^*$	-1	$-\varepsilon$	ε^*	(x, y) (R_x, R_y)	(xz, yz)
	1	ε^*	$-\varepsilon$	-1	$-\varepsilon^*$	ε		
E_2	1	$-\varepsilon^*$	$-\varepsilon$	1	$-\varepsilon^*$	$-\varepsilon$		(x^2-y^2, xy)
	1	$-\varepsilon$	$-\varepsilon^*$	1	$-\varepsilon$	$-\varepsilon^*$		

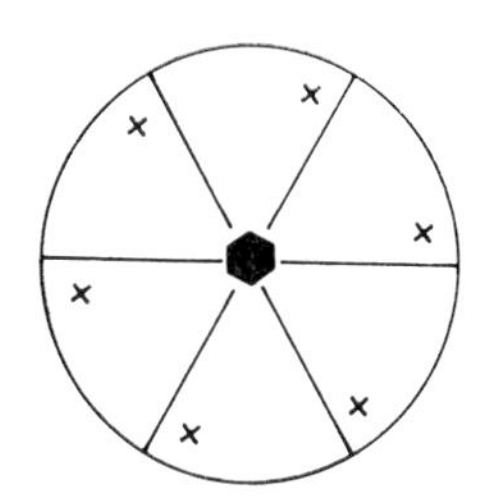

FIG. A.1.21

Point group $\bar{6}$ (C_{3h})

	E	C_3	C_3^2	σ_h	S_3	S_3^2		$\varepsilon = \exp(2\pi i/3)$
A'	1	1	1	1	1	1	R_z	x^2+y^2, z^2
E'	1	ε	ε^*	1	ε	ε^*	(x, y)	(x^2-y^2, xy)
	1	ε^*	ε	1	ε^*	ε		
A''	1	1	1	-1	-1	-1	z	
E''	1	ε	ε^*	-1	$-\varepsilon$	$-\varepsilon^*$	(R_x, R_y)	(xz, yz)
	1	ε^*	ε	-1	$-\varepsilon^*$	$-\varepsilon$		

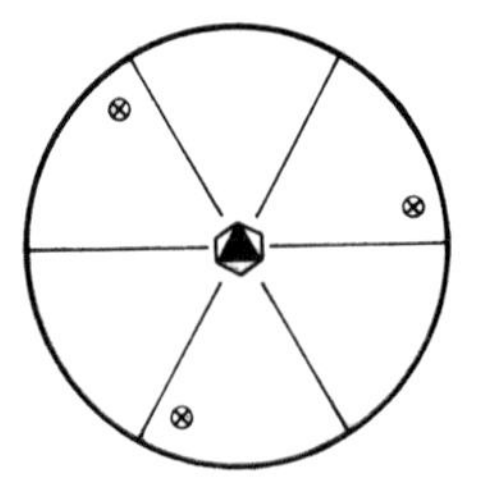

FIG. A.1.22

Point group $6/m$ (C_{6h})

	E	C_6	C_3	C_2	C_3^2	C_6^5	I	S_3^2	S_6^5	σ_h	S_6	S_3		$\varepsilon = \exp(2\pi i/6)$
A_g	1	1	1	1	1	1	1	1	1	1	1	1	R_z	z^2+y^2, z^2
B_g	1	-1	1	-1	1	-1	1	-1	1	-1	1	-1		
E_{1g}	1	ε	$-\varepsilon^*$	-1	$-\varepsilon$	ε^*	1	ε	$-\varepsilon^*$	-1	$-\varepsilon$	ε^*	(R_x, R_y)	(xz, yz)
	1	ε^*	$-\varepsilon$	-1	$-\varepsilon^*$	ε	1	ε^*	$-\varepsilon$	-1	$-\varepsilon^*$	ε		
E_{2g}	1	$-\varepsilon^*$	$-\varepsilon$	1	$-\varepsilon^*$	$-\varepsilon$	1	$-\varepsilon^*$	$-\varepsilon$	1	$-\varepsilon^*$	$-\varepsilon$		(x^2-y^2, xy)
	1	$-\varepsilon$	$-\varepsilon^*$	1	$-\varepsilon$	$-\varepsilon^*$	1	$-\varepsilon$	$-\varepsilon^*$	1	$-\varepsilon$	$-\varepsilon^*$		
A_u	1	1	1	1	1	1	-1	-1	-1	-1	-1	-1	z	
B_u	1	-1	1	-1	1	-1	-1	1	-1	1	-1	1		
E_{1u}	1	ε	$-\varepsilon^*$	-1	$-\varepsilon$	ε^*	-1	$-\varepsilon$	ε^*	1	ε	$-\varepsilon^*$	(x, y)	
	1	ε^*	$-\varepsilon$	-1	$-\varepsilon^*$	ε	-1	$-\varepsilon^*$	ε	1	ε^*	$-\varepsilon$		
E_{2u}	1	$-\varepsilon^*$	$-\varepsilon$	1	$-\varepsilon^*$	$-\varepsilon$	-1	ε^*	ε	-1	ε^*	ε		
	1	$-\varepsilon$	$-\varepsilon^*$	1	$-\varepsilon$	$-\varepsilon^*$	-1	ε	ε^*	-1	ε	ε^*		

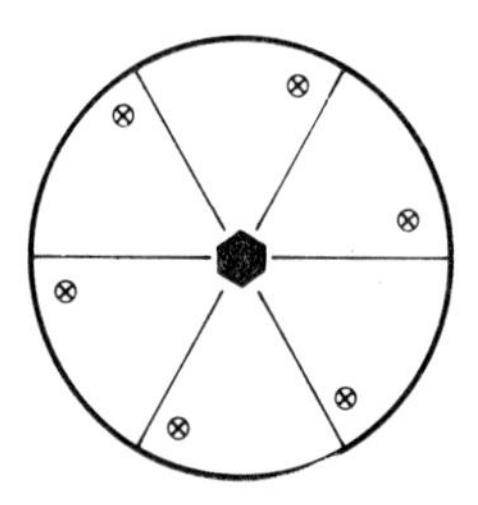

FIG. A.1.23

Point group 622 (D_6)

	E	$2C_6$	$2C_3$	C_2	$3C_2'$	$3C_2''$		
A_1	1	1	1	1	1	1		x^2+y^2, z^2
A_2	1	1	1	1	-1	-1	z, R_z	
B_1	1	-1	1	-1	1	-1		
B_2	1	-1	1	-1	-1	1		
E_1	2	1	-1	-2	0	0	(x, y) (R_x, R_y)	(xz, yz)
E_2	2	-1	-1	2	0	0		(x^2-y^2, xy)

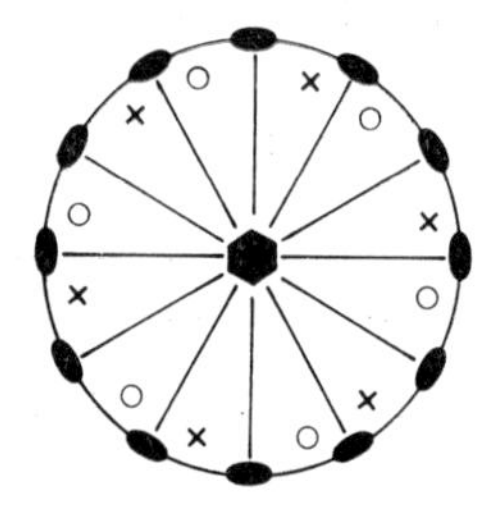

FIG. A.1.24

Point group 6mm (C_{6v})

	E	$2C_6$	$2C_3$	C_2	$3\sigma_v$	$3\sigma_d$		
A_1	1	1	1	1	1	1	z	x^2+y^2, z^2
A_2	1	1	1	1	-1	-1	R_z	
B_1	1	-1	1	-1	1	-1		
B_2	1	-1	1	-1	-1	1		
E_1	2	1	-1	-2	0	0	(x, y) (R_x, R_y)	(xz, yz)
E_2	2	-1	-1	2	0	0		(x^2-y^2, xy)

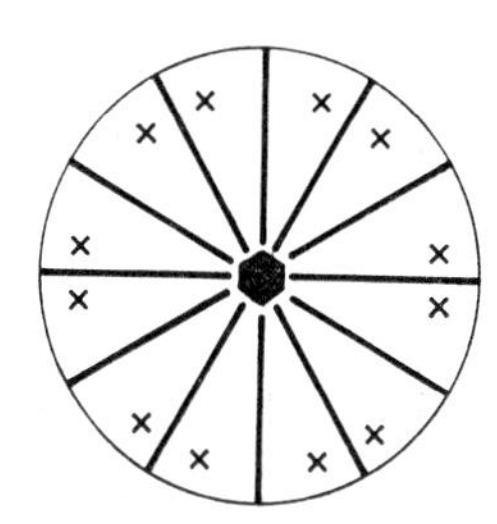

FIG. A.1.25

Point group $\bar{6}m2$ (D_{3h})

	E	$2C_3$	$3C_2$	σ_h	$2S_3$	$3\sigma_v$		
A_1'	1	1	1	1	1	1		x^2+y^2, z^2
A_2'	1	1	-1	1	1	-1	R_z	
E'	2	-1	0	2	-1	0	(x, y)	(x^2-y^2, xy)
A_1''	1	1	1	-1	-1	-1		
A_2''	1	1	-1	-1	-1	1	z	
E''	2	-1	0	-2	1	0	(R_x, R_y)	(xz, yz)

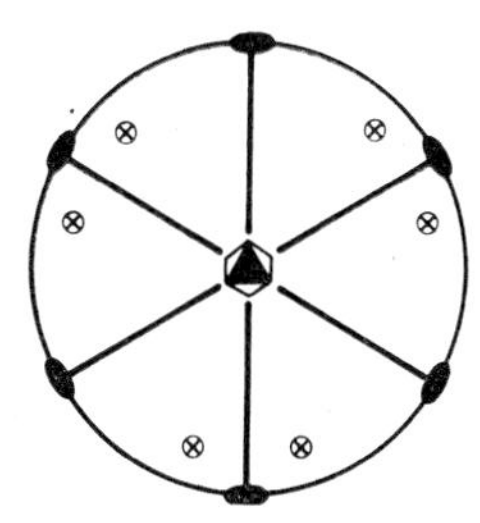

FIG. A.1.26

Point group 6/mmm (D_{6h})

	E	$2C_6$	$2C_3$	C_2	$3C_2'$	$3C_2''$	I	$2S_3$	$2S_6$	σ_h	$3\sigma_d$	$3\sigma_v$		
A_{1g}	1	1	1	1	1	1	1	1	1	1	1	1		x^2+y^2, z^2
A_{2g}	1	1	1	1	−1	−1	1	1	1	1	−1	−1	R_z	
B_{1g}	1	−1	1	−1	1	−1	1	−1	1	−1	1	−1		
B_{2g}	1	−1	1	−1	−1	1	1	−1	1	−1	−1	1		
E_{1g}	2	1	−1	−2	0	0	2	1	−1	−2	0	0	(R_x, R_y)	(xz, yz)
E_{2g}	2	−1	−1	2	0	0	2	−1	−1	2	0	0		(x^2-y^2, xy)
A_{1u}	1	1	1	1	1	1	−1	−1	−1	−1	−1	−1		
A_{2u}	1	1	1	1	−1	−1	−1	−1	−1	−1	1	1	z	
B_{1u}	1	−1	1	−1	1	−1	−1	1	−1	1	−1	1		
B_{2u}	1	−1	1	−1	−1	1	−1	1	−1	1	1	−1		
E_{1u}	2	1	−1	−2	0	0	−2	−1	1	2	0	0	(x, y)	
E_{2u}	2	−1	−1	2	0	0	−2	1	1	−2	0	0		

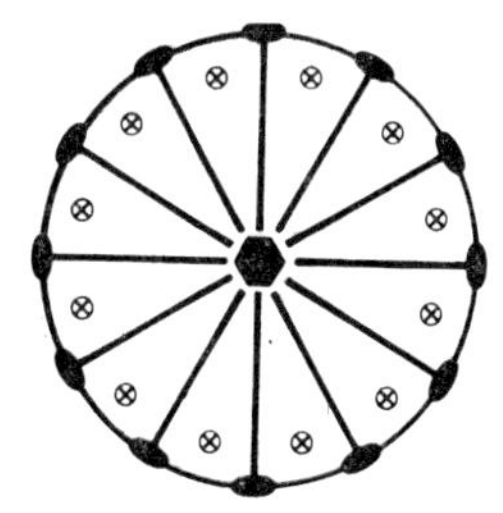

FIG. A.1.27

Cubic system

Point group 23 (T)

	E	$4C_3$	$4C_3^2$	$3C_2$	$\varepsilon = \exp(2\pi i/3)$	
A	1	1	1	1		$x^2+y^2+z^2$
E	1	ε	ε^*	1		$x^2-y^2, (2z^2-x^2-y^2)$
	1	ε^*	ε	1		
T	3	0	0	−1	$(x, y, z), (R_x, R_y, R_z)$	(yz, zx, xy)

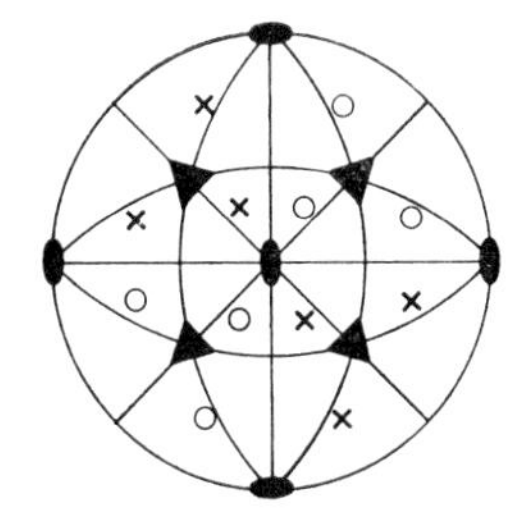

FIG. A.1.28

Point group m3 (T_h)

	E	$4C_3$	$4C_3^2$	$3C_2$	I	$4S_6$	$4S_6^5$	$3\sigma_h$	$\varepsilon = \exp(2\pi i/3)$
A_g	1	1	1	1	1	1	1	1	$x^2+y^2+z^2$
E_g	1	ε	ε^*	1	1	ε	ε^*	1	$(x^2-y^2), (2z^2-x^2-y^2)$
	1	ε^*	ε	1	1	ε^*	ε	1	
T_g	3	0	0	−1	3	0	0	−1	$(xy, xz, yz), (R_x, R_y, R_z)$
A_u	1	1	1	1	−1	−1	−1	−1	
E_u	1	ε	ε^*	1	−1	$-\varepsilon$	$-\varepsilon^*$	−1	
	1	ε^*	ε	1	−1	$-\varepsilon^*$	$-\varepsilon$	−1	
T_u	3	0	0	−1	−3	0	0	1	(x, y, z)

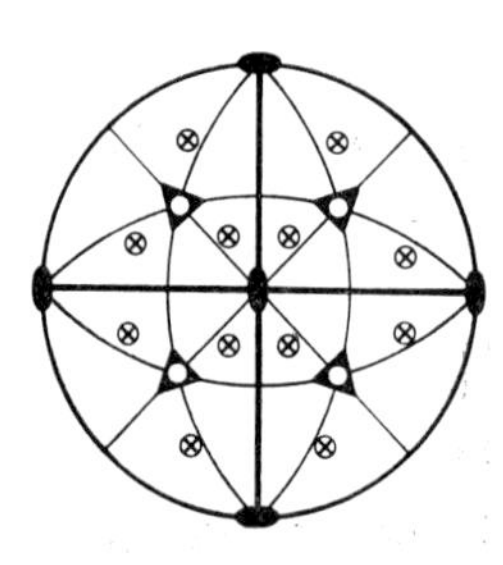

FIG. A.1.29

Point group 432 (O)

	E	$8C_3$	$3C_2$	$6C_4$	$6C_2$		
A_1	1	1	1	1	1		$x^2+y^2+z^2$
A_2	1	1	1	−1	−1		
E	2	−1	2	0	0		$x^2-y^2, 2z^2-x^2-y^2$
T_1	3	0	−1	1	−1	$(x, y, z), (R_x, R_y, R_z)$	
T_2	3	0	−1	−1	1		yz, zx, xy

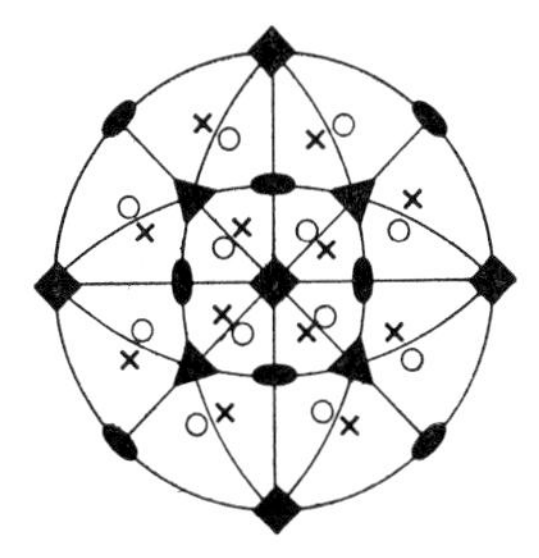

FIG. A.1.30

Point group $\bar{4}3m$ (T_d)

	E	$8C_3$	$3C_2$	$6S_4$	$6\sigma_d$		
A_1	1	1	1	1	1		$x^2+y^2+z^2$
A_2	1	1	1	−1	−1		
E	2	−1	2	0	0		$(2z^2-x^2-y^2,\ x^2-y^2)$
T_1	3	0	−1	1	−1	$(R_x,\ R_y,\ R_z)$	
T_2	3	0	−1	−1	1	$(x,\ y,\ z)$	$(xy,\ xz,\ yz)$

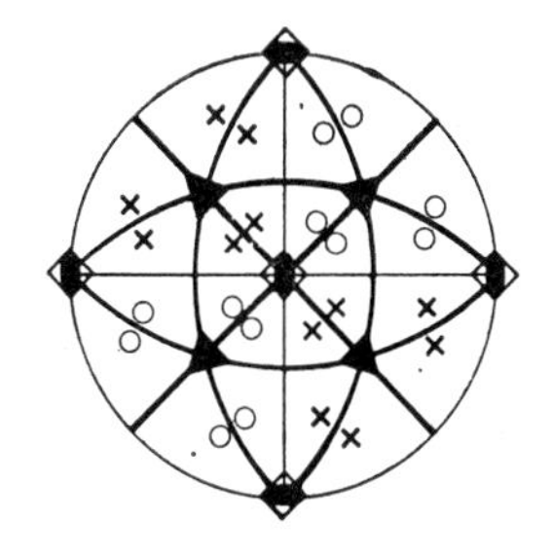

FIG. A.1.31

Point group m3m (O_h)

	E	$8C_3$	$3C_2(=C_4^2)$	$6C_4$	$6C_2$	I	$8S_6$	$3\sigma_h$	$6S_4$	$6\sigma_d$		
A_{1g}	1	1	1	1	1	1	1	1	1	1		$x^2+y^2+z^2$
A_{2g}	1	1	1	−1	−1	1	1	1	−1	−1		
E_g	2	−1	2	0	0	2	−1	2	0	0		$(2z^2-x^2-y^2, x^2-y^2)$
T_{1g}	3	0	−1	1	−1	3	0	−1	1	−1	(R_x, R_y, R_z)	
T_{2g}	3	0	−1	−1	1	3	0	−1	−1	1		(xz, yz, zy)
A_{1u}	1	1	1	1	1	−1	−1	−1	−1	−1		
A_{2u}	1	1	1	−1	−1	−1	−1	−1	1	1		
E_u	2	−1	2	0	0	−2	1	−2	0	0		
T_{1u}	3	0	−1	1	−1	−3	0	1	−1	1	(x, y, z)	
T_{2u}	3	0	−1	−1	1	−3	0	1	1	−1		

(3, p. 57; 5, Appendix II)

FIG. A.1.32

A.2. Point–group symbols in Hermann–Mauguin (international) and in Schoenfliess notation

Crystal system	*Class symbol*	
	International	*Schoenfliess*
Triclinic	1	C_1
	$\bar{1}$	C_i
Monoclinic	m	C_s
	2	C_2
	$2/m$	C_{2h}
Orthorhombic	$2mm$	C_{2v}
	222	D_2
	mmm	D_{2h}
Tetragonal	4	C_4
	$\bar{4}$	S_4
	$4/m$	C_{4h}
	$4mm$	C_{4v}
	$\bar{4}2m$	D_{2d}
	422	D_4
	$4/mmm$	D_{4h}
Rhombohedral	3	C_3
	$\bar{3}$	C_{3i}
	$3m$	C_{3v}
	32	D_3
	$\bar{3}m$	D_{3d}
Hexagonal	$\bar{6}$	C_{3h}
	6	C_6
	$6/m$	C_{6h}
	$\bar{6}m2$	D_{3h}
	$6mm$	C_{6v}
	622	D_6
	$6/mmm$	D_{6h}

Cubic	23	T
	$m3$	T_h
	$\bar{4}3m$	T_d
	432	O
	$m3m$	O_h

(9, p. 44.)

A.3. A theorem on the transformation of tensors

It is required to prove that the transformation of second order tensors takes place according to the relation

$$p'_{ik} = c_{il}c_{km}p_{lm}.$$

We shall prove this for a particular example, because it is rather easier to follow than the general case.

Suppose the relation between two vector quantities A, B is given by

$$A_i = p_{ik}B_k,$$

then $A'_2 = p'_{21}B'_1 + p'_{22}B'_2 + p'_{23}B'_3$

and, transforming to X'_1, X'_2, X'_3 from X_1, X_2, X_3, we have

$$B'_1 = c_{11}B_1 + c_{12}B_2 + c_{13}B_3,$$
$$B'_2 = c_{21}B_1 + c_{22}B_2 + c_{23}B_3,$$
$$B'_3 = c_{31}B_1 + c_{32}B_2 + c_{33}B_3,$$

	X_1	X_2	X_3
X'_1	c_{11}	c_{12}	c_{13}
X'_2	c_{21}	c_{22}	c_{23}
X'_3	c_{31}	c_{32}	c_{33}

where c_{ik} are the direction cosines defining the transformed axes in their relation to the original axes. Hence

$$A'_2 = p'_{21}(c_{11}B_1 + c_{12}B_2 + c_{13}B_3) +$$
$$+ p'_{22}(c_{21}B_1 + c_{22}B_2 + c_{23}B_3) +$$
$$+ p'_{23}(c_{31}B_1 + c_{32}B_2 + c_{33}B_3),$$

or $A'_2 = p'_{2i}c_{ik}B_k.$ (1)

We also have the relation

$$A'_2 = c_{21}A_1 + c_{22}A_2 + c_{23}A_3$$

and
$$A_1 = p_{11}B_1 + p_{12}B_2 + p_{13}B_3,$$
$$A_2 = p_{21}B_1 + p_{22}B_2 + p_{23}B_3,$$
$$A_3 = p_{31}B_1 + p_{32}B_2 + p_{33}B_3.$$

$$\text{Hence } A'_2 = c_{21}(p_{11}B_1+p_{12}B_2+p_{13}B_3)+$$
$$+\ c_{22}(p_{21}B_1+p_{22}B_2+p_{23}B_3)+$$
$$+\ c_{23}(p_{31}B_1+p_{32}B_2+p_{33}B_3),$$

or $$A'_2 = c_{2i}p_{ik}B_k. \tag{2}$$

Both of the expressions for A'_2 in equations (1) and (2) may be arranged as the sum of three terms containing B_1, B_2, B_3. Since B is a vector which has no particular direction, B_1, B_2, and B_3 are independent, and the coefficients of each in expressions (1) and (2) must be equal. Hence

$$p'_{21}c_{11}+p'_{22}c_{21}+p'_{23}c_{31} = c_{21}p_{11}+c_{22}p_{21}+c_{23}p_{31},$$
$$p'_{21}c_{12}+p'_{22}c_{22}+p'_{23}c_{32} = c_{21}p_{12}+c_{22}p_{22}+c_{23}p_{32},$$
$$p'_{21}c_{13}+p'_{22}c_{23}+p'_{23}c_{33} = c_{21}p_{13}+c_{22}p_{23}+c_{23}p_{33}.$$

We may solve these three equations to obtain p'_{21}, p'_{22} and p'_{23} in terms of the c_{ik}s and the p_{ik}s. If we multiply the first equation by c_{11}, the second by c_{12} and the third by c_{13}, we obtain

$$p'_{21}c_{11}c_{11}+p'_{22}c_{11}c_{21}+p'_{23}c_{11}c_{31} = c_{11}c_{21}p_{11}+c_{11}c_{22}p_{21}+c_{11}c_{23}p_{31},$$
$$p'_{21}c_{12}c_{12}+p'_{22}c_{12}c_{22}+p'_{23}c_{12}c_{32} = c_{12}c_{21}p_{12}+c_{12}c_{22}p_{22}+c_{12}c_{23}p_{32},$$
$$p'_{21}c_{13}c_{13}+p'_{22}c_{13}c_{23}+p'_{23}c_{13}c_{33} = c_{13}c_{21}p_{13}+c_{13}c_{22}p_{23}+c_{13}c_{23}p_{33}.$$

If these three equations are added, then because

$$c_{11}^2+c_{12}^2+c_{13}^2 = 1,$$
$$c_{11}c_{21}+c_{12}c_{22}+c_{13}c_{23} = 0,$$
$$c_{11}c_{31}+c_{12}c_{32}+c_{13}c_{33} = 0,$$

$$p'_{21} = c_{11}c_{21}p_{11}+c_{11}c_{22}p_{21}+c_{11}c_{23}p_{31}+$$
$$+\ c_{12}c_{21}p_{12}+c_{12}c_{22}p_{22}+c_{12}c_{23}p_{32}+$$
$$+\ c_{13}c_{21}p_{13}+c_{13}c_{22}p_{23}+c_{13}c_{23}p_{33}$$
$$= c_{2i}c_{1k}p_{ik}\ .$$

This result may be derived quite generally by employing the tensor notation, as follows. Starting from the relations

$$A_i = p_{ik}B_k, \quad A'_i = c_{im}A_m, \text{ and } A_n = c_{on}A'_o,$$

we have

$$A'_i = c_{im}A_m = c_{im}p_{mn}B_n = c_{im}p_{mn}c_{on}B'_o = p'_{ik}B'_k.$$

Thus if we put the subscript $o = k$ (since both go through all values 1, 2, 3),

$$p'_{ik}B'_k = c_{im}c_{kn}p_{mn}B'_k,$$

and, for a particular value of k,

$$p'_{ik} = c_{im}c_{kn}p_{mn}.$$

Conversely, if we exchange dashed for undashed letters,

$$p_{ik} = c_{mi}c_{nk}p'_{mn}.$$

Transformation of third-order and higher-order tensors

If two quantities C_{mo} and D_n are related by a third-order tensor q_{nmo} according to the equation

$$C_{mo} = q_{mon}D_n,$$

then the value of q' when the axes are changed is given by

$$q'_{pqr} = c_{pn}c_{qm}c_{ro}q_{nmo}.$$

This may be proved in a similar manner to that employed for second-order tensors. Thus

$$\begin{aligned} C'_{qr} &= q'_{qrp}D'_p \\ &= c_{qm}c_{ro}C_{mo}, \text{ since } C \text{ is a second-order tensor,} \\ &= c_{qm}c_{ro}q_{mon}D_n \\ &= c_{qm}c_{ro}q_{mon}c_{sn}D'_s. \end{aligned}$$

If we put the subscript $s = p$, which we may do since both have all values 1, 2, and 3, we obtain, on equating corresponding coefficients of D'_p,

$$q'_{qrp} = c_{qm}c_{ro}c_{pn}q_{mon}$$

(13, p. 13; 15, Appendix I).

A.4. The formal definition of a group

In the sense used in this book, a 'group' has the following attributes.

(1) Every member of the group when multiplied either by itself or by another member of the group gives rise to a member of the group. The term 'multiplication' as used here always refers to symmetry operations.

(2) The multiplication of three elements A, B, C in a group is associative, i.e.

$$A(BC) = (AB)C.$$

(3) One member of the group is the identity element.

(4) Every member of the group has an inverse which is also a member of the group, so that when multiplied with that element it produces the identity element.

A.5. A theorem on the substitution of characters for classes in equations involving class multiplication coefficients

It follows from the definition of classes of symmetry in a given point group that the product of operations in two classes is equal to the sum of the operations in other classes in the same point group. This may be expressed by the formula

$$\mathscr{C}_i\mathscr{C}_j = \sum_s c_{ijs}\mathscr{C}_s. \tag{1}$$

The replacement of classes by their characters leads to the formula

$$h_i h_j \chi_i \chi_j = \chi_E \sum_s c_{ijs} h_s \chi_s, \tag{2}$$

where h_i, h_j, h_s denote the number of elements of symmetry in classes i, j, s respectively, χ_i, χ_j, χ_s denote corresponding characters, χ_E is the character of the matrix defining the identity operation, and c_{ijs} are the class multiplication coefficients.

Proof of the formula (2)

(a) When $\chi_i = \chi_j = \chi_E = \chi_s = 1$, $\mathscr{C}_i$ and $\mathscr{C}_j$ contain h_i and h_j elements respectively so that their product contains $h_i h_j$ elements. Further, the number of elements corresponding to $c_{ijs}\mathscr{C}_s$ is equal to $c_{ijs}h_s$.

Thus,

$$h_i h_j = \sum_s c_{ijs} h_s$$

and equation (2) is fulfilled.

(b) When $\chi_E = 2$. In this case the representations are 2×2 matrices. For all rotation axes the classes of symmetry (for which $\chi \neq 0$) have 2×2 matrices with equal non-zero terms on the diagonal, and other terms are zero. For two such matrices represented by

$$\begin{bmatrix} a & 0 \\ 0 & a \end{bmatrix} \times \begin{bmatrix} b & 0 \\ 0 & b \end{bmatrix},$$

the product is equal to the matrix

$$\begin{matrix} ab & 0 \\ 0 & ab \end{matrix} .$$

The original pair of matrices are unit matrices with coefficients a and b respectively. The product of their characters is $2a \times 2b = 4ab$, whereas the character of the resulting matrix is $2ab$. The ratio of the two is equal to χ_E.

X

For all classes of symmetry involving planes of symmetry, the characters of the 2×2 matrices are all zero. For example, in point group $4mm$, the class $2\sigma_v$ is composed of the two lines of symmetry perpendicular to the x_1 and x_2 axes respectively. For these lines the matrices are

$$\begin{matrix} -1 & 0 \\ 0 & 1 \end{matrix} \quad \text{and} \quad \begin{matrix} 1 & 0 \\ 0 & -1 \end{matrix}.$$

The resulting diagonal terms for the class composed of both lines of symmetry has two zeros.

Another example may be taken from point group $6mm$ in the class with $3\sigma_v$. The matrices for the three lines of symmetry are

$$\begin{matrix} 1 & 0 \\ 0 & -1 \end{matrix}, \quad \begin{matrix} -\frac{1}{2} & -\frac{\sqrt{3}}{2} \\ -\frac{\sqrt{3}}{2} & \frac{1}{2} \end{matrix}, \quad \text{and} \quad \begin{matrix} -\frac{1}{2} & \frac{\sqrt{3}}{2} \\ \frac{\sqrt{3}}{2} & \frac{1}{2} \end{matrix}.$$

It is clear that for all three taken together the off-diagonal terms cancel one another and the diagonal terms result in zeros also.

Such matrices fall into the same category as those of the classes of rotation axes, in that they are multiples of the unit matrix—in this case with a coefficient equal to zero. Thus whenever $\chi_E = 2$,

$$h_i h_j \chi_i \chi_j = \chi_E \sum_s c_{ijs} h_s \chi_s.$$

To illustrate this we shall consider an example from point group 622, where we have

$$\begin{aligned} \mathscr{C}_2\mathscr{C}_4 &= (C_6 + C_6^5)C_2 \\ &= C_6C_2 + C_6^5C_2 \\ &= C_3^2 + C_3 \\ &= \mathscr{C}_3 \\ &= c_{243}\mathscr{C}_3. \end{aligned}$$

In two dimensions the transformations for C_6 and C_6^5 are

$$\begin{matrix} \frac{1}{2} & \frac{\sqrt{3}}{2} \\ -\frac{\sqrt{3}}{2} & \frac{1}{2} \end{matrix} \quad \text{and} \quad \begin{matrix} \frac{1}{2} & -\frac{\sqrt{3}}{2} \\ \frac{\sqrt{3}}{2} & \frac{1}{2} \end{matrix}.$$

When added together to form class $\mathscr{C}_2$, the off-diagonal terms cancel, so that the character for the class, χ_i, becomes $\frac{1}{2}+\frac{1}{2} = 1$. The transformation for C_2 is

$$\begin{matrix} -1 & 0 \\ 0 & -1 \end{matrix}$$

with a character $\chi_J = -2$, and for C_3 and C_3^2 we have

$$\begin{matrix} -\frac{1}{2} & \frac{\sqrt{3}}{2} \\ -\frac{\sqrt{3}}{2} & -\frac{1}{2} \end{matrix} \quad \text{and} \quad \begin{matrix} -\frac{1}{2} & -\frac{\sqrt{3}}{2} \\ \frac{\sqrt{3}}{2} & -\frac{1}{2} \end{matrix},$$

so that the character for the class $\mathscr{C}_3$ is $-\frac{1}{2}-\frac{1}{2} = -1$.

Thus,

$$h_i h_J \chi_i \chi_J = 2\times 1\times 1\times -2 = -4$$

and

$$\chi_E \sum_s c_{iJs} h_s \chi_s = 2\times 1\times 2\times -1 = -4.$$

(c) When $\chi_E = 3$. The product of two three-dimensional unit matrices may be written as follows.

$$\begin{pmatrix} 1 & 0 & 0 \\ 0 & 1 & 0 \\ 0 & 0 & 1 \end{pmatrix} \times \begin{pmatrix} 1 & 0 & 0 \\ 0 & 1 & 0 \\ 0 & 0 & 1 \end{pmatrix} = \begin{pmatrix} 1 & 0 & 0 \\ 0 & 1 & 0 \\ 0 & 0 & 1 \end{pmatrix}.$$

The characters χ_i, χ_J are both 3 and the character $\chi_s = 3$. Thus χ_s must be multiplied by χ_E to equal $\chi_i \chi_J$.

It is again true that all classes of rotation axes and planes of symmetry in three-dimensional representations have matrices that are multiples of the unit 3×3 matrix. For example, in point group $m3$ the character for class $3C_2$ in representation T_u is -1. The diad axes parallel to X_1, X_2, X_3 have matrices

$$\begin{matrix} 1 & 0 & 0 \\ 0 & -1 & 0 \\ 0 & 0 & -1 \end{matrix}, \quad \begin{matrix} -1 & 0 & 0 \\ 0 & 1 & 0 \\ 0 & 0 & -1 \end{matrix}, \quad \text{and} \quad \begin{matrix} -1 & 0 & 0 \\ 0 & -1 & 0 \\ 0 & 0 & 1 \end{matrix}.$$

The character for each matrix is -1 and when added together they form a multiple of the unit matrix.

Thus the same argument applies here as in the two-dimensional case. As an example we shall take the product $\mathscr{C}_4 \mathscr{C}_5$ in point group $m3$.

$$\mathscr{C}_4\mathscr{C}_5 = (C_{2x}+C_{2y}+C_{2z})I$$
$$= \sigma_x+\sigma_y+\sigma_z = \mathscr{C}_8 = c_{458}\mathscr{C}_8.$$
$$\mathscr{C}_4 = 3C_2 \text{ and has } h_i = 3,\ \chi_i = -1;$$
$$\mathscr{C}_5 = I \quad \text{and has } h_j = 1,\ \chi_i = -3.$$

Thus,

$$h_ih_j\chi_i\chi_j = 3\times1\times-1\times-3 = 9.$$

The matrices for σ_x, σ_y, σ_z are respectively

$$\begin{matrix}-1 & 0 & 0\\ 0 & 1 & 0\\ 0 & 0 & 1\end{matrix}, \qquad \begin{matrix}1 & 0 & 0\\ 0 & -1 & 0\\ 0 & 0 & 1\end{matrix}, \quad \text{and} \quad \begin{matrix}1 & 0 & 0\\ 0 & 1 & 0\\ 0 & 0 & -1\end{matrix},$$

so that the character χ_s is $+1$, $h_s = 3$, and $c_{458} = 1$.
Hence,

$$\chi_E\sum c_{ijs}h_s\chi_s = 3\times1\times3\times1 = 9,$$

and equation (2) is fulfilled (2, p. 226; 8, p. 109).

A.6. Orthogonality rules

There are four rules known as orthogonality rules. The name arises from the general type of formula, which in its simplest form relates to the direction cosines of two mutually perpendicular lines, namely

$$c_{1p}c_{1q}+c_{2p}c_{2q}+c_{3p}c_{3q} = 0.$$

Two of the rules refer to rows and two refer to columns of the character tables given in Appendix 1.

Rule 1

In a given point group, the sum over all the classes of symmetry, i, of the squares of the characters, $\chi_i^{(v)}$, or products of each character and its complex conjugate, in a particular representation, v, multiplied by the number of elements of symmetry in each class, h_i, is equal to the total number of symmetry elements, g, in the group.

Expressed in symbols, this runs

$$\sum_i h_i\chi_i^{(v)2} = g \qquad \text{or} \qquad \sum_i h_i\chi_i^{(v)}\chi_i^{*(v)} = g.$$

Rule 2

In a given point group the sum over all the classes of symmetry, i, of the products of the characters $\chi_i^{(v)}$, $\chi_i^{(w)}$ in the same class but belonging to two representations v, w, multiplied by the number of elements in the class, h_i, is equal to zero.

Or,

$$\sum_i h_i \chi_i^{(v)} \chi_i^{(w)} = 0. \qquad \text{If } v = 1, \sum_i h_i \chi_i^{(v)} = 0.$$

When both representations contain complex quantities, the complex conjugates of the characters in one representation must be used.

Rule 3

(Rules 3 and 4 are similar to Rules 1 and 2 except that they refer to columns instead of rows in the character tables.)

In a given point group the sum over all the representations of the character table, of the squares of all the characters, $\chi_i^{(v)}$ or products of the character and its complex conjugate, belonging to a given class, i, multiplied by the number of elements, h_i, in that class, is equal to the total number of symmetry elements.

This may be written

$$h_i \sum_v \chi_i^{(v)2} = g, \quad \text{or} \quad h_i \sum_v \chi_i^{(v)} \chi_i^{*(v)} = g.$$

The very common application of this rule to the E column results in the expression

$$\sum_v \chi_E^{(v)2} = g,$$

since for this class h_i is necessarily equal to unity.

Rule 4

In a given point group the sum over all the representations of the character table, of the products of the corresponding characters, $\chi_i^{(v)}$, $\chi_j^{(v)}$ in classes i and j is equal to zero.

Or,

$$\sum_v \chi_i^{(v)} \chi_j^{(v)} = 0 \qquad (i \neq j).$$

When both classes contain complex characters the complex conjugates of those in one class must be used (2, p. 230; 8, p. 101; 10, p. 83; 12, p. 31).

A.7. The number of times an irreducible representation is contained in a reducible representation

Reducible representations may be decomposed into an integral number of irreducible representations. This may be expressed symbolically by the equation,

$$\Gamma(R) = \sum_v n_v \Gamma^{(v)},$$

where $\Gamma(R)$ and $\Gamma^{(v)}$ are the reducible and the vth irreducible representations respectively, and n_v is the number of times the latter is contained in the former. As was shown in Appendix 5, we may replace Γ by χ in the above equation, and write

$$\chi_i(R) = \sum_v n_v \chi_i^{(v)}.$$

Multiplying both sides of this equation by $h_i \chi_i^{(v)}$ and summing over all classes, we obtain for a given value of v

$$\sum_i h_i \chi_i(R) \chi_i^{(v)} = n_v \sum_i h_i \chi_i^{(v)2}.$$

From Rule 1, Appendix 5, we have

$$\sum_i h_i \chi_i^{(v)2} = g.$$

Hence

$$n_v = \frac{1}{g} \sum_i h_i \chi_i(R) \chi_i^{(v)}$$

(1, p. 83; 6, p. 34; 8, p. 105; 12, p. 32).

References

THIS is only a brief list of publications which are in book form and are referred to in this book at the ends of certain sections.

1. BHAGAVANTAM, S. *Crystal symmetry and physical properties*. Academic Press, New York (1966).
2. BHAGAVANTAM, S. and VENKATARAYUDU, T. *Theory of groups and its application to physical problems*. Academic Press, New York (1962).
3. BRADLEY, C. J. and CRACKNELL, A. P. *The mathematical theory of symmetry in solids*. Clarendon Press, Oxford (1972).
4. BUERGER, M. J. *Elementary crystallography*. Wiley, New York (1956).
5. COTTON, F. A. *Chemical applications of group theory*. Wiley, New York (1963).
6. CRACKNELL, A. P. *Applied group theory*. Pergamon Press, Oxford (1968).
7. GAY, P. *The crystalline state*. Oliver & Boyd, Edinburgh (1972).
8. HAMERMESH, M. *Group theory and its applications to physical problems*. Addison–Wesley, Reading, Mass. (1960).
9. HENRY, N. F. M. and LONSDALE, K. *International tables for X-ray crystallography*, Vol. 1, 'Symmetry groups'. Kynoch Press, Birmingham (1965).
10. JONES, H. *The theory of Brillouin zones and electronic states in crystals*. North-Holland, Amsterdam (1960).
11. KLEBER, W. *An introduction to crystallography VEB* English edn. Verlag Technik, Berlin (1970).
12. LEECH, J. W. and NEWMAN, D. J. *How to use groups*. Methuen, London (1969).
13. NYE, J. F. *Physical properties of crystals*. Clarendon Press, Oxford (1957).
14. PHILLIPS, F. C. *An introduction of crystallography*. Longmans, London (1963).
15. WOOSTER, W. A. *Crystal physics*. Cambridge University Press (1938).
16. WOOSTER, W. A. and BRETON, A. *Experimental crystal physics*. Clarendon Press, Oxford (1970).

List of Symbols

1,2,3,4,6, rotation axes.
C_1, C_2, C_3, C_4, C_6, rotation axes (anticlockwise rotation).
C_3^2, C_4^3, C_6^5, rotation axes (clockwise rotation).
C_{2a}, C_{2b}, diad axes parallel to the X and Y axes respectively.
C_{2c}, C_{2d} or C_2', C_2'', horizontal diad axes equally inclined to the X and Y axes.
C_{2x}, C_{2y}, C_{2u}, horizontal diad axes inclined at 120° to one another.
$C_{3a}, C_{3b}, C_{3c}, C_{3d}$, triad axes of the cubic point groups.
C_{2x}, C_{2y}, C_{2z}, diad axes parallel to the X, Y, Z, axes respectively.
S_1, S_2, S_3, S_4, S_6, reflection–rotation axes.
$\bar{1}, \bar{2}, \bar{3}, \bar{4}, \bar{6}$, inversion–rotation axes.
$2_1, 3_1, 3_2, 4_1, 4_2, 4_3, 6_1, 6_2, 6_3, 6_4, 6_5$, screw axes.
m, reflection plane.
n, glide plane.
σ_z, σ_h, reflection plane normal to Z axis or horizontal.
σ_x, σ_y reflection planes normal to the X and Y axes respectively.
$\sigma_x, \sigma_y, \sigma_u$, reflection planes in which lie the axes C_{2x}, C_{2y}, C_{2u} respectively.
σ_v, vertical reflection plane.
$\sigma_d, \sigma_{d'}$ vertical reflection planes equally inclined to the X and Y axes.
a, b, c, unit-cell dimensions, glide planes.
α, β, γ, unit-cell angles.
a^*, b^*, c^*, reciprocal unit-cell dimensions.
$\alpha^*, \beta^*, \gamma^*$, reciprocal unit-cell angles.
X_1, X_2, X_3 or X, Y, Z, or x, y, z, coordinate axes.
$\mathbf{h}$, vector of heat flow.
$(\mathrm{d}\theta/\mathrm{d}\mathbf{x})$, vector of temperature gradient.
k_{mn}, coefficients of conductivity; of dielectric polarization.
r_{mn}, coefficients of resistance; components of strain.
α_{mn}, coefficients of thermal expansion; direction cosines in the study of elasticity.
λ, wavelength.
θ, Bragg angle.
(hkl), one face of Miller indices h, k, l.
$\{hkl\}$, a crystal form, i.e. a group of faces linked by symmetry elements.

$[hkl]$, zone axis having indices h,k,l, i.e. a line from the origin passing through a point having coordinates $h\mathbf{a},k\mathbf{b},l\mathbf{c}$.
$\langle hkl \rangle$, a group of zone axes related by symmetry.
t_{mn}, components of stress.
χ_{ik}, coefficients of magnetic susceptibility.
d_{ikl}, piezoelectric modulus.
P_i, components of electric polarization.
E_l, components of electric field strength.
c_{ikmn}, elastic stiffness or elastic constant, also written c_{pq}.
s_{ikmn}, elastic compliance or modulus, also written s_{pq}.
$\delta_{ik(c)}$, a minor (determinant) formed by omission of the ith row and the kth column from the full Δ_c matrix.
Δ_c, full matrix consisting of coefficients of all strain components.
$\delta_{jk(s)},\Delta_s$, corresponding quantities to $\delta_{ik(c)}$ and Δ_c when stress is substituted for strain.
ρ, density; optical rotatory power.
u_p, components of a wave displacement.
$\xi_{(i)p}$, pth component of the amplitude vector of the ith wave.
$\mathbf{K}^*_{(i)}$, wave vector of the ith wave.
ν, frequency of the wave motion.
$\mathbf{n},\mathbf{o},\mathbf{p}$, unit Cartesian vectors in Bravais space.
$\mathbf{n}^*,\mathbf{o}^*,\mathbf{p}^*$, unit Cartesian vectors in reciprocal space.
f_1,f_2,f_3, direction cosines of wave normal.
Φ potential energy due to elastic strain.
A_{pm}, quantities occurring in the Christoffel determinant.
$V_{(i)}$, velocity of the ith wave.
$\mathbf{e}_{(1)},\mathbf{e}_{(2)},\mathbf{e}_{(3)}$, unit vectors along the three vibration directions associated with a given wave normal.
$e_{(i)p}$, the pth component of the unit vector along the vibration direction of the ith wave.
g_{ij}, coefficients of rotatory polarization.
V_H, Hall voltage.
H, magnetic field strength.
I, electric current.
R_H, Hall coefficient.
ρ_{mnp}, Hall coefficient.
a_{ik}, antisymmetric part of the resistivity tensor; polarizability constants.
r_{ik}, symmetric part of the resistivity tensor.
B_{ik}, components of tensor representing refractive index.
B°_{ik}, value of B_{ik} in the unstressed state.
p_{ik}, strain-optical coefficients.
q_{ik}, stress-optical coefficients.
E, identity operator; two-dimensional representation.

I, centre of symmetry.

R_x, R_y, R_z, infinitesimally small rotations about axes X, Y, Z.

T, three-dimensional representation; a translation by one-half of a unit-cell dimension.

ε, $\exp 2\pi i/3$ or $\exp 2\pi i/6$.

ε^*, $\exp -2\pi i/3$ or $\exp -2\pi i/6$.

i, a class of symmetry elements.

$\mathbf{d}_{hkl}$, perpendicular distance between lattice planes of the type hkl.

S, a translation by one unit-cell distance.

χ_i, character of the matrix representing class i in an irreducible representation.

$\chi_i(R)$, character of the matrix representing class i in a reducible representation.

h_i, the number of symmetry elements in class i.

g, the total number of symmetry elements in a given point group.

h, Planck's constant.

Schoenfliess symbols—see Appendix 2.

Index